Urban Sustainability

The Urban Sustainability Book Series is a valuable resource for sustainability and urban-related education and research. It offers an inter-disciplinary platform covering all four areas of practice, policy, education, research, and their nexus. The publications in this series are related to critical areas of sustainability, urban studies, planning, and urban geography.

This book series aims to put together cutting-edge research findings linked to the overarching field of urban sustainability. The scope and nature of the topic are broad and interdisciplinary and bring together various associated disciplines from sustainable development, environmental sciences, urbanism, etc. With many advanced research findings in the field, there is a need to put together various discussions and contributions on specific sustainability fields, covering a good range of topics on sustainable development, sustainable urbanism, and urban sustainability. Despite the broad range of issues, we note the importance of practical and policy-oriented directions, extending the literature and directions and pathways towards achieving urban sustainability.

The series will appeal to urbanists, geographers, planners, engineers, architects, governmental authorities, policymakers, researchers of all levels, and to all of those interested in a wide-ranging overview of urban sustainability and its associated fields. The series includes monographs and edited volumes, covering a range of topics under the urban sustainability topic, which can also be used for teaching materials.

Mohd Alsaleh

Fisheries Sustainability in the EU

Intersections of Competitiveness, Policy, and Food Security

Springer

Mohd Alsaleh
College of Economics and Management
Shanghai Ocean University
Shanghai, Shanghai, China

ISSN 2731-6483 ISSN 2731-6491 (electronic)
Urban Sustainability
ISBN 978-981-92-4495-9 ISBN 978-981-92-4496-6 (eBook)
https://doi.org/10.1007/978-981-92-4496-6

This Springer imprint is published by the registered company Springer Nature Singapore Pte Ltd.
The registered company address is: 152 Beach Road, #21-01/04 Gateway East, Singapore 189721, Singapore

Preface

Discover the future of European fisheries in this definitive analysis of blue growth and marine sustainability—essential for researchers, policymakers, and industry leaders invested in ocean health.

Rooted in 30+ years of EU27 empirical data (1990–2022), the book addresses a critical gap: balancing fisheries productivity with environmental resilience amid global climate pressures and evolving policy landscapes. Its significance lies in bridging fragmented discussions of governance, food security, and competitiveness, offering a unified framework to advance the EU's 2050 blue growth and sustainability goals.

Unlike narrow studies that overlook regional disparities, this work innovates by distinguishing between EU14 developed and EU13 developing states, unpacking how economic structure, fossil fuel reliance, and governance effectiveness shape outcomes uniquely across contexts. It builds on prior scholarship by integrating cutting-edge econometric methods—including method of moments quantile regression (MMQR), FMOLS, and DOLS—to capture heterogeneous impacts across pollution, accident, and food security quantiles, addressing limitations of traditional mean-based analyses.

Guided by clear theoretical frameworks (e.g., Stewardship Theory, growth hypothesis, utilitarian economics), the book tests core hypotheses about fisheries' role in water contamination, climatic occupational risks, and food security. It answers pivotal questions: How do governance and fossil fuel use moderate fisheries' environmental footprint? Can developing EU states leverage aquaculture to enhance food security without sacrificing sustainability? What policy levers align blue growth with marine protection?

With rigorous methodology—including cross-sectional dependence tests, unit root analysis, and panel cointegration—the book measures key variables (fisheries output, governance quality, GDP, fossil fuel consumption) to deliver evidence-based solutions. It proves that sustainable blue economies are not just aspirational, but actionable: through targeted policies (eco-friendly gear, equitable environmental taxes, strengthened governance), EU fisheries can thrive while safeguarding oceans for future generations.

A must-read for anyone seeking to reconcile productive fisheries with planetary health—this book defines the path forward for European marine sustainability.

This book redefines the study of European fisheries with a threefold innovative approach that breaks new ground in sustainability and blue growth research:

1. Heterogeneous Regional and Quantile-Centric Analysis

Unlike conventional studies that treat the EU as a monolith or rely on average-based regression, it pioneers a dual-focused framework:

Explicitly differentiates between EU14 developed and EU13 developing states, uncovering how economic structure, policy capacity, and resource reliance (e.g., fossil fuels) drive divergent outcomes—from food security to marine pollution.

Employs method of moments quantile regression (MMQR) to capture impacts across the full distribution of key outcomes (e.g., marine water contamination, climatic occupational accidents), revealing hidden dynamics (e.g., fisheries' pollution effects are strongest at extreme quantiles) that mean-based models miss.

2. Integrated Theoretical and Empirical Synthesis

It integrates fragmented theories (Stewardship Theory, growth hypothesis, utilitarian economics) into a cohesive framework, testing how governance, economic growth, and industry activity interact to shape marine sustainability. This integration fills a critical literature gap: prior research often siloes "environmental impacts" from "food security" or "governance," while this book demonstrates their interdependence—e.g., how strong governance in EU14 nations reduces pollution and boosts fisheries' contribution to food security.

3. Policy-Relevant, Data-Driven Rigor

Moving beyond descriptive analysis, it combines 30+ years of EU27 panel data (1990–2022) with advanced econometric tools (FMOLS, DOLS, cross-sectional dependence tests) to deliver actionable insights. What's new: it quantifies the "policy levers" that work for distinct EU subgroups—e.g., eco-friendly gear mandates for EU14, equitable environmental taxes for resource-constrained EU13—rather than one-size-fits-all recommendations.

This approach transforms abstract sustainability goals into tangible strategies, making the book indispensable for researchers, policymakers, and industry leaders seeking to balance fisheries productivity with planetary health.

The book offers a comprehensive, empirically grounded exploration of European fisheries' role in blue growth and marine sustainability, with a clear scope organized around cross-cutting themes, regional differentiation, and actionable insights. Its content spans six core pillars, anchored in 30+ years of EU27 data (1990–2022) and structured to bridge theory, evidence, and policy:

1. Foundational Context and Regional Framing

Establishes the EU's fisheries landscape: economic significance (e.g., 6.4 million tons of annual output, 350,000+ jobs), key challenges (overfishing, fossil fuel emissions, marine pollution, climatic occupational risks), and alignment with global sustainability agendas (EU Green Deal, 2050 blue growth targets).

Rigorously differentiates between EU14 developed (e.g., Germany, France) and EU13 developing (e.g., Estonia, Bulgaria) states, highlighting structural disparities in governance capacity, economic reliance on fisheries, and environmental impact.

2. Core Thematic Intersections

Each section unpacks the interdependencies between critical variables, with a focus on EU-specific dynamics:

Fisheries and Environmental Sustainability: Impacts of fisheries operations (e.g., trawling, fossil fuel use) on marine water contamination, ocean acidification, and habitat degradation; heterogeneous effects across EU subgroups.

Food Security and Sectoral Contribution: The role of aquaculture and fisheries in ensuring nutrition access—with EU13 relying more on these sectors than EU14—and how economic growth, governance, and fossil fuel use moderate this relationship.

Governance and Policy Effectiveness: Analysis of EU frameworks (Common Fisheries Policy, Marine Strategy Framework Directive) and global governance factors (voice and accountability, rule of law, regulatory quality) in driving sustainable management, with comparisons between EU14 and EU13 implementation.

Climatic and Occupational Risks: Fisheries' vulnerability to climate change (e.g., extreme weather, ocean warming) and associated occupational hazards (fatal accidents, heat stress), plus mitigation strategies for different EU regions.

3. Methodological and Theoretical Depth

Details of the book's innovative analytical toolkit: Method of moments quantile regression (MMQR) to capture distributional impacts, panel cointegration, unit root tests, and cross-sectional dependence adjustments—explaining how these methods address limitations of traditional fisheries research.

Integrates theoretical frameworks (Stewardship Theory, growth hypothesis, utilitarian economics) to contextualize empirical findings, linking abstract concepts to real-world fisheries dynamics.

4. Policy and Practical Implications

Translates data into targeted recommendations for EU policymakers, industry leaders, and researchers: eco-friendly gear adoption for EU14, equitable environmental taxes for EU13, strengthened governance mechanisms, and fossil fuel transition strategies.

Addresses implementation barriers (e.g., resource constraints in EU13, policy resistance) and identifies opportunities to align blue growth with marine protection (e.g., renewable energy integration, sustainable aquaculture scaling).

5. Future Directions and Limitations

Discusses gaps in current research (e.g., understudied non-EU comparisons, long-term climate impacts) and proposes avenues for extension (e.g., incorporating tourism, globalization, or digital innovation in fisheries).

Overall, the scope balances academic rigor with real-world relevance, covering the full lifecycle of European fisheries sustainability—from foundational context and theoretical framing to empirical analysis and actionable solutions—while centering the EU's unique regional diversity as a core lens for understanding global blue economy challenges.

This book targets a multi-professional, globally engaged audience united by a focus on European fisheries, blue growth, and marine sustainability, including:

Academic Researchers: Fisheries economists, environmental policy scholars, marine sustainability scientists, and EU regional studies experts—seeking rigorous empirical frameworks, innovative methodologies (e.g., MMQR), and cross-thematic analyses (governance, food security, pollution) to inform their work.

EU Policymakers and Regulators: Officials involved in shaping the Common Fisheries Policy (CFP), Marine Strategy Framework Directive, and blue growth agendas—needing data-driven insights to design targeted policies for EU14/EU13 subgroups.

Fisheries Industry Stakeholders: Fisheries managers, aquaculture operators, and industry associations—looking to align operations with sustainability standards and leverage policy opportunities (e.g., renewable energy integration, eco-friendly gear).

International Organizations and NGOs: Teams at the UN, FAO, EU Marine Directorate, and environmental NGOs (e.g., Oceana) focused on ocean conservation and food security—utilizing regional case studies to inform global sustainability initiatives.

Graduate and Advanced Undergraduate Students: Those studying environmental economics, marine policy, EU studies, or sustainable development—gaining access to real-world panel data analysis and applied theoretical frameworks.

The book is positioned at an advanced academic and professional level, balancing scholarly depth with accessibility for non-specialists:

Theoretical Rigor: Integrates specialized frameworks (Stewardship Theory, growth hypothesis, utilitarian economics) and clarifies their application to fisheries dynamics, with sufficient context for readers familiar with basic social science or environmental studies.

Methodological Transparency: Details advanced econometric techniques (MMQR, FMOLS, DOLS) and data processing (cross-sectional dependence tests, unit root analysis) with clear explanations—accessible to researchers and students with intermediate statistical knowledge, while providing technical appendices for deeper dives.

Practical Accessibility: Translates complex empirical findings into actionable policy recommendations and thematic summaries, ensuring non-academic stakeholders (policymakers, industry leaders) can apply key insights without specialized expertise in econometrics.

It avoids overly niche jargon without compromising academic integrity, making it suitable for both classroom use (advanced undergraduate/graduate seminars) and professional decision-making—filling the gap between technical research and real-world implementation.

Chapters 1–3 differentiate between EU14 developed and EU13 developing states while using method of moments quantile regression (MMQR) to capture impacts across the full distribution of outcomes (e.g., marine pollution, food security gaps, occupational risks)—not just average effects. Readers gain hyper-targeted insights (e.g., "EU13 needs equitable environmental taxes; EU14 benefits from eco-friendly gear mandates") rather than one-size-fits-all solutions. For researchers, it demonstrates cutting-edge econometric approaches to address regional disparities; for policymakers/industry, it eliminates guesswork by aligning strategies with a country's economic structure and capacity.

Chapters 4–6 break down silos in existing literature by uniting three critical pillars—fisheries governance, marine environmental sustainability, and food security—into a cohesive framework, showing how they interdepend (e.g., strong governance reduces pollution and boosts food security). Avoid fragmented, single-issue analysis. Academic readers get a holistic theoretical model for future research; practitioners (policymakers, NGOs) can design policies that advance multiple EU goals (2050 blue growth, climate neutrality, zero hunger) simultaneously, maximizing resource efficiency.

Chapters 7–10, anchored in comprehensive panel data (1990–2022) from EU27 nations, combined with advanced tools (FMOLS, DOLS, cross-sectional dependence tests), translate complex statistical findings into clear, implementable recommendations. Ensures credibility (data-driven insights trusted by researchers and policymakers) while solving the "research-to-practice" gap. Industry leaders can adopt evidence-backed operations (e.g., renewable energy integration for vessels); students and early-career researchers gain a model for applying advanced methods to real-world policy challenges. These features position the book as more than a scholarly text—it's a practical roadmap for reconciling fisheries productivity with planetary health, tailored to the EU's unique regional diversity.

Shanghai, China Mohd Alsaleh

Competing Interests The author has no competing interests to declare that are relevant to the content of this manuscript.

Contents

About The Editor

Mohamd Alsaleh serves as Associate Professor of Applied Economics within the College of Economics and Management at Shanghai Ocean University. His current research interests center on natural resource economics and management, environmental economics and governance, and sustainable development policy formulation. His scholarly work is dedicated to developing innovative frameworks and constructing compelling academic narratives surrounding global food security and sustainability issues. Additionally, Dr. Alsaleh holds the position of International Associate Researcher at the Institute of Tropical Agriculture & Food Security (ITAFoS), Malaysia, and serves as Adjunct Associate Professor at several top-ranked academic institutions. Further details regarding his research profile and academic publications are available at the following link: https://jmxy.shou.edu.cn/2023/0301/c851a314242/page.htm

List of Tables

Chapter 1
Fisheries and Environmental Sustainability Toward CO_2 Reduction

Acronyms

EU	European Union
CO_2	Carbon dioxide emissions
MMQR	Method of moments quantile regression
MAP	Multiannual Program (EU Data Collection Framework)
GT	Gross tonnage
FTEs	Full-time equivalents
GVA	Gross value added
DaS	Million days at sea
EUR	Euro
CD	Cross-sectional dependency
FP	Fisheries production
WDI	World Bank Development Indicators
GDP	Gross domestic product
FF	Fossil fuel
PD	Population density
CS	Carbon sink
OLS	Ordinary least square
DOLS	Dynamic ordinary least squares
FMOLS	Fully modified ordinary least squares
FEOLS	Fixed effects ordinary least squares
Europa	European data portal
Eurostat	European Commission's database
LGDP	Natural logarithms of economic growth
LFP	Natural logarithms of fisheries production
LFF	Natural logarithms of fossil fuel
LPD	Natural logarithm of population density
LLC	Lin & Chu Test

M. Alsaleh, *Fisheries Sustainability in the EU*, Urban Sustainability,
https://doi.org/10.1007/978-981-92-4496-6_1

IPPS	Im, Pesaran, and Shin Test
EU27	27 European Union Member States
EU13	13 European Union Member States
BRICS	Brazil, Russia, India, China, South Africa
ASEAN	Association of Southeast Asian Nations
EU14	14 European Union Member States

Every year billions of dollars of tax subsidies are awarded to the fishing fleets of European Union (EU) countries, and the burning of these fuels leads to millions of tons of CO emissions each year—more CO than the total annual emissions produced by the entire fishing fleet in Malta! In this study, the dynamic impact of the fisheries sector on the blue carbon of 27 EU countries during the period 1990–2022 is analyzed in conjunction with other determinants, including fisheries production (FP), fossil fuel (FF) consumption, economic growth, and population density (PD). Applying the advanced method of moments quantile regression (MMQR) technique with fixed effects, we found a strong positive effect of fishery production on carbon sink destruction in most quantiles.

The loss of blue carbon storage becomes increasingly accentuated with fishing output at most quantiles, from the first to the ninth (most pronounced impact at upper quantiles and lowest at the lowest quantiles).

We also showed that fishing output had a more strongly negative influence on carbon sink destruction in EU14 developed countries than in the EU13 developing countries. Economic growth in both EU14 developed and EU13 developing countries tended to alleviate carbon sink deterioration for all quantiles (supporting the growth theory for countries producing fisheries), while fossil fuel usage had the tendency to accelerate the deterioration of carbon sink. The influence of population density in the EU14 developed countries was significantly negative for the first to seventh quantiles, while in the EU13 developing countries it had a significantly positive influence for all quantiles. Policy makers to reduce the damage to blue carbon in EU14 and EU13 could contribute through supporting tidal and wave power for renewable energy use.

1.1 Marine Environment Background

Roughly 30% of the carbon dioxide produced by humans is captured by the oceans and helps temper the effects of global warming. This can happen due to the help of plankton, reefs, fish, seaweed, and green organisms that are small [1]. As a result of the decline in the ability to capture blue carbon due to humans, oceans become warmer and more acidic with a lower oxygen content. The waters have also accumulated more than 90% of excess heat that is being emitted into the climate system. Considered by most as the most important carbon sink, the global oceans have absorbed approximately 40% of all carbon dioxide emissions since the dawn of the industrial era [2]. Over the period from 1994 to 2007, the average for the ocean absorption has reached 314% of global anthropogenic CO_2 emissions and varied from one ocean area to another [2]. The sea accumulates and transfers the majority of the CO_2 (around 93%) on Earth [2]. One issue with natural carbon sinks is that they eventually reach a saturation point, and after reaching that level, it leads to consequences such as ocean acidification [1]. This is caused by the absorption of carbon dioxide that leads to lowering of pH of waters, hurting marine species such as reef habitats, phytoplankton, shellfish, mussels, and mollusks by making them weakened and vulnerable to disease or death [1]. Recent oceanic heat waves at record levels in seas worldwide are producing widespread ocean heat events. The EU fleet had 73,983 vessels with a total gross tonnage (GT) of 1.33 million tons and a total power of 5.3 million kW in 2019 [1]. The MAP program collects data on vessels from Member States: In 2019, there were 57,236 active vessels employing up to 129,540 fishermen, or 92,298 full-time equivalent jobs (FTE), receiving an average annual salary of 23,620 EUR [1]. The estimated Total Gross Value Added (GVA) was EUR 3.4 billion, and total income (subsidies and fishing rights) was EUR 1.25 billion. The gross profit margin was 20%, and GVA in terms of sales was 53% of total sales. 9.0% of income was retained as net gross profit in 2019, totaling nearly EUR 0.6 billion [1]. This is lower than in 2018. In forecasts, EU fleets performance is expected to decrease moderately in 2020 compared with 2019 but improve compared with 2018 performance, including the ones in 2021, but will likely benefit from lower fuel prices to mitigate this loss in 2020 [1].

In addition, EU fleets landed some 4.05 million tons of seafood with an estimated value of € 6.3 billion, using some 6 million days at sea (DaS) and consuming roughly 2.0 billion literes of fuel [2]. These figures can be used to indicate that the active fleet vessels used on average 82 DaS in 2019 and 27,172 literes of fuel on average per vessel in 2019 [2]. Belgian fleets recorded the highest consumption of fuel at 572,485 literes/fleet on average, followed by Lithuanian fleets at 451,621 literes/fleet and Dutch fleets at 307,689 literes/fleet [2]. In regard to DaS, Belgian vessels are first with 207 DaS per vessel, followed by Greek with 162 DaS per vessel, and Spanish with 122 DaS per vessel [2]. Small-scale fleet activity varies by nation; Bulgarian vessels spent just 20 DaS on average in 2019, followed by Malta at 28, Romania at 41 and Croatia at 49 [2]. The fleet activity in terms of days at sea for the entire EU declined by 3.4% from 2018 to 2019 [2]. The total fuel consumption fell by 0.2%, from 2054 million literes to 2010 million literes [2]. Spain led the countries in terms of DaS with 1.9 million estimated DaS, which represents 30.9% of total DaS and 1.01 million more than any other EU country, followed by Italy with 1.2 million days (20.6%) [2]. For fuel consumption, the Spanish fleet consumed approximately 582 million literes (29.0% of the total consumption) [2].

In 2019, EU fishing fleets consumed 2010 million literes of fuel, a figure that is 6% below the average 2009–2018 EU fleet fuel consumption, excluding Croatia and Greece [3]. Total fuel costs amounted to approximately EUR 1003 million, which is 17% below the average 2009–2018 EU fleet fuel costs, excluding Croatia and Greece [3]. The fuel consumption of the EU fishing fleets depends on the number of boats, their target species and equipment [3]. In 2019 there were 73,983 boats at sea with a combined capacity of 1.33 million GT. Between 2008 and 2019, both the number of boats and the total carrying capacity of those boats declined [3].

A study conducted by the Our Fish campaign revealed that EU fishing fleets are granted tax subsidies on their fuel, ranging from 759 million to 1.5 billion EUR annually, but contributing to 7.3 million tons of CO_2 emitted from their fuel usage [4]. These subsidies not only contribute to further pollution, but also pose an immense threat to sea creatures, climate change, and further to small-scale fishers and carbon sinks. The reports "Reports on Climate Impacts and the fishing industry," which outline profits generated through EU tax subsidies on fuel for the fishing fleets at the EU level, including specific examples of fishing industry practices in Spain, France, Italy, the Netherlands, and Portugal, concluded that the largest contributor to fossil fuel subsidy costs for EU fisheries is the fleet that poses the greatest environmental risk, uses most fuel, and is the biggest producer of carbon emissions, thereby negatively impacting the climate, marine habitat, carbon sinks, and the most cost-effective and lowest impact fishers [4]. The core of the issue is that the worst offenders of the EU fisheries contribute to millions of tons of carbon dioxide emissions from the use of fossil fuels each year [4]. Although it emits approximately as much CO_2 as all the countries combined, the industry receives about EUR 759–1.5 billion each year in subsidies. While individual European citizens pay taxes for using gasoline in personal cars, members of the EU fishing fleet do not even have to pay taxes on fossil fuels for ships, which produces around 7.3 million tons of CO_2 per year (almost 6% of the EU's yearly total in 2019). That amount of CO_2 is equivalent to total annual emissions of Malta in 2019 [4]. Generally, global changes to climate and carbon sinks are impacting global marine ecosystem health and distributions of fish species. Even one ton of CO_2 released has a significant impact given the current scope of the global crisis [5]. With the newly implemented EU climate law, EU Member States must radically decrease carbon dioxide emissions to reach net zero in 2050 [5]. The struggle between the needs of commerce and the necessity of environmental preservation has never been greater for ensuring a sustainable future for the world's oceans [5]. In 2016, per capita CO_2 emissions averaged 1.88 tons, an increase from 1.5 tons in 1950 [5]. All the while, total ocean harvests declined from the mid-1990s. Emissions per ton of seafood harvested have increased more in volume for artisanal/subsistence and small-scale fisheries, while remaining the highest in magnitude in the industrial sector, leading to increasing pressure on marine carbon sinks [5]. Low-impact fishing and environmentally destructive practices supported by current subsidies encourage bigger boats that degrade marine habitats, deplete already threatened fish populations, and contribute to carbon sink degradation [6].

Increasing fishing activity across the world is seriously detrimental to ocean health and blue carbon in general. Overfishing commercially attractive fish leads to a disruption of related species and habitats [6]. This overfishing entails large commercial harvests of large quantities of fish and shellfish. However, this practice leads to large unintended losses of non-targeted species—comprising young fish, corals, benthos, sharks, whales, sea turtles, and seabirds [6].

The unintended mortality and losses of non-targeted marine fauna can also pose a strain on the environment and further affect ocean carbon sinks and blue carbon [6]. Fish-catching impacts marine conservation issues that result from overexploitation, depletion of fish stocks, water pollution, degradation of ecosystem structure, and degradation of carbon sinks [6]. In the face of increasing global pressure, some researchers recommend the overall contraction of the global fisheries sector to restore the capacity of a healthy marine environment [6]. The approach contradicts the livelihoods of a wide range of communities dependent on the industry as the main source of income. Nevertheless, evidence suggests that by 2048, no fish may be caught wildly [6] [6]. Stated that a significant component of the carbon sink relies on plankton and a rich assortment of fishes. Therefore, it is significant to address how variations in the carbon source relate to fishing production and harvesting activities. The relationship between fishing practices and carbon sink performance also requires much attention, considering how top-down and bottom-up effects on ecosystem functions and CO_2 regulation vary across ecosystems [6].

The study aims to achieve the following by responding to the following research questions: (1) How does the expansion of the fishing sector impact blue carbon stocks across EU countries between 1990 and 2022? (2) Is the EU a susceptible country to a deterioration in the carbon sink as a result of the expansion of the fishing sector? Is the production of increased fisheries output ecologically sustainable and can it support the EU's sustainable development goals for 2050? The study therefore seeks to achieve the following: (1) Document changes in carbon sinks across the EU countries as fishing production grows from 1990 to 2022. (2) Compare how changes in the fish sector's fishing activities between developed and developing countries affected the storage of blue carbon between 1990 and 2022. (3) How the growth in the fishing sector over the period 1990–2022 has influenced blue carbon levels in both developed and developing economies. The analysis of EU countries suffers from a lack of research, which typically makes the assumption of independent and CD-free disturbance terms across EU nations. The disturbance terms might share many common and/or specific, common or unexpected shocks, or spill-over effects among countries [7–11]. As a result of deeper economic integration in the EU, CD among countries may result in similar effects from a wide range of factors. Factors such as resource crises and fish prices affect both economic growth and fish consumption in EU economies [10], suggesting that CD is properly taken into account for more balanced results regarding the relation between energy, income, and environment. In this paper, we ask if the same growth-mitigating role of natural resources affects the carbon sink characteristics of fish-producing countries. Previous studies used natural resource in a broader, aggregate fashion [12, 13] where no distinction between point-source and diffuse-source resource is made.

Here we used a specific point-source of natural resource, which is the fisheries production, because point-source resource seems to lead to more harmful economic growth consequences than diffuse sources. In this connection, we use the fishery-producing countries, which include developed and developing countries, because strong regulatory frameworks reduce, or even invert, the growth retarding effects of dependence on resources [10, 12]. We also use population density to account for other regulatory controls which help to improve the robustness of inference drawn.

The list of fish-producing EU27 countries, including developed countries (e.g., Germany, Italy, and Sweden) and developing countries (e.g., Bulgaria, Estonia, and Malta), spans the period from 1990 to 2022. We adopted the MMQR method of [14] to investigate the income–carbon sink relationship between fisheries production and the rest of the regressors over different conditional quantiles of the carbon sink with careful regard for the heterogeneous stages of development of these countries. Using fixed effects, this method will give insight into empirical analysis on the variation of the income–carbon sink relations across different conditional quantiles. It helps identify varying relationships across different income-conditional quantiles that conventional average methods are unable to uncover. It is thus a very useful method to rigorously analyze growth theory for fish-producing countries. As far as our knowledge is concerned, this is the first research to analyze the income-induced carbon sink hypothesis in fishery-producing countries using simultaneously the values of seafood output, fossil fuel consumption, population density, and economic development. Analyzing the growth hypothesis over different quantile levels of conditional carbon sinks has importance because; firstly, estimates of conditional quantiles have a more resistant characteristic to be distorted by outliers in the dependent variable than conventional conditional mean estimates; secondly, estimates of conditional mean do not allow us to understand the distributional impact of fisheries production on the carbon sinks and hence quantile regressions provide us more analytically relevant explanation, since they explain the distributional impact of the dependent variable across a range of conditional quantiles and how they vary across different cross-sectional group.

1.2 Fisheries and Marine Pollution: Historical Relationship

In recent years, there has been increased attention to studies on the fishing industry and its relationship with the marine environment and pollution [15]. He was one of the pioneers in providing a fisheries-led development paradigm, and many studies have been undertaken on the issue. Different theories of economic development have been propounded by various economists, but they take into account the contribution of fisheries as a source of both marine and environmental pollution. Building a sustainable fisheries industry could create opportunities for both the economic and social and the environmental development over the long-term; for example, the study of [1–3] examined the relationship between sustainable fisheries production and economic growth in South Asian countries, and it showed that sustainable development is positively affected by the level of fisheries production [4]. Also tested the impacts of partial and complete substitution of fish production on growth performance, feed utilization, histological condition, and economic revenue using a polynomial regression model, and proposed that utilizing underutilized or discarded oceanic resources could be considered a possible avenue for implementing a circular economy and generating high-value products.

[5] Tested the effect of stocking body size of Nile Tilapia (Oreochromis niloticus) on water quality, growth efficiency, and economic return, and it indicated that there was one stocking size that led to the optimal results of the growth performance and financial returns [6]. Carried out relevant research into the role of the skin of Arctic fish in the sustainable development of coastal communities in Southwest Alaska and showed that Indigenous practices using fish skins can play an important role in reducing poverty, achieving gender equality, promoting health and well-being, providing clean water and sanitation, supporting decent work and economic growth, reducing social inequality, advancing progress in the context of climate change, promoting marine conservation, and encouraging responsible consumption and production. Sustainable fishing involves harvesting fish at a rate that does not cause the depletion of the fish stock and the marine environment (including land and water animal life). Sustainable fishing, if actually practiced, protects the marine environment while generating revenue, food, and jobs for coastal communities; for example, [16] explored the linkages between the fishing industry and marine pollution and proposed the usage of processing waste from the fishing industry as an alternative source of biomass fuel in Indonesia [17]. indicated that although fish are the most dominant fauna in the world oceans, we know relatively little about their contribution to carbon export on a regional and global scale, and the estimates of the contribution of fish to the total carbon flux range widely, with the mean value calculated from the model-generated global carbon flux data.

Similarly, [18] investigated the influence of fish stocks on marine carbon cycles in both benthic and pelagic realms, revealing significant effects on net marine CO_2 and CH4 emissions within the benthic zone.

Using a double logit model, [19, 20] tested the influence of marine pollution on the development of marine fisheries in China and found that the marine environmental impact of regulations had a "U-shaped" effect in China's marine fisheries. While there is rapid development in research on the impact of the fishing industry on the marine environment, the impacts of the fishing industry and its surrounding marine environment have not been thoroughly studied or integrated yet [21–23]. in particular explored the plausible underlying mechanisms leading to changes in ecosystem and marine biodiversity. The management of the fishing sector is through legal, social, economic, and political arrangements that control the industry at all three levels (international, national, and local), covering both written laws and social rules.

Illustrative of individual heterogeneity, [24] analyzed, respectively, network resilience and the implications of EU and US fisheries policy for third countries [25]. analyzed the governance impacts of Brexit in the context of Scottish fisheries policy, and found the areas of fisheries management, operation, and research needed more political, social, and relational input [24, 26]. also studied how experiences shape and inform small-scale fisheries governance, referring to the variety of arrangements under which women are involved, including compliance with rules and regulations; participatory structures such as co-management, formal arrangements, and norms such as custom and relations; and relational spaces. Marine seafood supplies 15% of the total protein consumed by humans throughout the world. It takes a heavy toll on the world's oceans to satisfy this growing demand of 15% for 7.5 billion (and expanding) humans each year.

These findings were also found by many previous researchers like [27–29].

Fishing industry alone, according to the statistics, generated 207 million tons of CO_2 during the year 2019, which is approximately equivalent to carbon dioxide generated from 51 coal-fired power plants during the same period [30, 31]. investigated the efficiency of resource usage in warm-water fish farming of different pond sizes under the influence of marine environmental sustainability and energy efficiency [32, 33], discussing fossil fuel consumption and carbon sink in net fisheries, stated that implementing energy audits can enable a structured view of the fuel consumption pattern of the fishing vessels to identify inefficient energy usage. Contrary to the existing studies on fishing industry, this paper has tried to analyze the overall impacts of using aquatic resources on blue carbon and economic development, as well as analyze the impact of the fishing industry on blue carbon emissions and regional economic development in the EU for boosting blue sustainability. Unlike most of the existing studies based on time series regression, this study utilizes static and panel regression models to test how fishing output affects carbon sink and economic development in terms of blue sustainability.

Major contributions in this study are first, it tests the relationship between fishing output, carbon sink dynamics, population size, and economic development within EU countries during the period from 1990 to 2022.

Second, it analyzes the impacts of fishing output on carbon sink and economic development in the EU in a bid to promote blue sustainable development. Two distinct groupings have been established for EU countries, namely industrialized EU countries and developing EU countries, along with their joint total for EU countries region. Based on the analysis of unit root and cointegration to confirm stationarity because of the randomness with drift, method of moments quantile regression (MMQR) has been employed to investigate the relationship among fishing output, depletion in blue carbon storage, and economic development; static and panel regression have been employed to investigate impacts on blue carbon storage and marine pollution respectively; and finally, we forecast that how changes in the level of fishing output lead to blue carbon and marine pollution.

1.3 Method of Moments Quantile Regression (MMQR)

The model used in this study was constructed based on recent work [34–37], and takes the following form:

$$\mathrm{CS2it} = \mathrm{f}\left(\mathrm{FPit,FFit,GDPit,PDit}\right)$$

CS is shorthand for collapse of the blue carbon sink, whereas FP denotes fisheries production. A distinction is also made between fisheries production (FP), calculated as overall fisheries products in tons and decline of the sink, measured as the total decline of millions of tons of dissolved CO_2 in the ocean. The values of FP and CS were retrieved from the World Development Indicators dataset [38]. We also gathered information on Gross Domestic Product (GDP), utilized to measure the overall level of economic development, measured in constant US dollars, from the EU's database [39]. The data for fossil fuel consumption (FF) represents the total amount of fossil fuel consumption expressed in tons of oil equivalent, also obtained from the EU's database [39]. Finally, we retrieved data on population density (PD), measured as population divided by area in km2, from the EU's statistical office [40]. The data collected were used to encompass EU Member States for the period spanning from 1990 to 2022, and all variables were in real values per capita. To fulfill the objectives of this study, annual data were collected for the selected countries from three main sources: the World Bank's World Development Indicators database (WDI), the European Commission's database [39], and the official European data portal [40], covering the years 1990 to 2022. Table 1.1 summarizes the variables, their measurement units, data sources, and transformations applied.

The section then describes our findings. Table 1.2 presents statistical characteristics (maximum, minimum, mean, standard deviation) of the components used. The mean figures are: CS 0.924, FP 4.955, GDP 4.315, FF 2.822, and PD 2.035. The standard deviation of each of the study variables is respectable, amounting to 0.239, 0.833, 0.439, 0.807, and 0.382, respectively. As a consequence of the descriptive characteristics of the factors, the way is now clear to undertake the unit root test.

This study uses the following methods: Dynamic Ordinary Least Squares (DOLS), Fully Modified Ordinary Least Squares (FMOLS), and Fixed Effects

Table 1.1 Data, units, transformation, and source

Data	Measurement unit	Transformation	Source
Carbon sink	Million metric tons	Natural logarithm transformation	WDI
Fishery production	Metric tons	Natural logarithm transformation	WDI
Fossil fuel	Metric tons of oil equivalent	Natural logarithm transformation	Eurostat
Economic growth	Constant US dollars	Natural logarithm transformation	Eurostat
Population density	People per square kilometer	Natural logarithm transformation	Europa

Source: [41]

Table 1.2 Results of descriptive-statistics

Variable	Observations	Mean	Std. Dev.	Min	Max
CS	864	0.924	0.239	0.428	1.871
FP	864	4.955	0.833	2.905	6.310
GDP	864	4.315	0.439	2.182	3.179
FF	864	2.822	0.807	0.133	4.890
PD	864	2.035	0.382	1.214	3.179

Source: [41]

Ordinary Least Squares (FEOLS). FEOLS is estimated using Driscoll and Kraay standard errors that are robust to common forms of cross-sectional dependence and autocorrelation up to lag [42]. As described in [42], the major problems that arise when dealing with estimations of dynamic cointegrated panels include the fact that they suffer from the cross-sectional heterogeneity associated with variations in the cross-section average as well as with the cross-sectional adjustment toward the cointegrating equilibrium. To overcome these problems, in Pedroni's FMOLS estimation method, individual-specific intercepts and the general serialization structure of the errors can be allowed to differ across the panel members. DOLS estimates were found to be more unbiased than both OLS and FMOLS estimators in finite samples according to Monte Carlo simulations, which contributed [43] to the extension of the DOLS estimator to panel data. This method of estimation corrects endogeneity through the inclusion of lead and lag differences. A panel quantile regression methodology was used to test distributional heterogeneity among the effects across quantiles in this research, as earlier methods are subject to these flaws [44]. The panel quantile regression technique was first presented in the seminal paper of [45]. As compared to conventional least squares regression that focuses on estimating the conditional mean of the dependent variable as a function of independent variables, quantile regressions are designed to estimate other quantiles, such as the conditional median. The technique is more robust to the presence of outliers, especially in cases where the conditional means of two variables are not correlated or show only weak correlation [46]. However, in this study, the method of moments quantile regression (MMQR) with fixed effects as developed by [14] is used. Although conventional quantile regression methods are resistant to outliers, they are unable to address individual heterogeneity in panel data models.

The MMQR, on the other hand, overcomes the problem of unobserved individual heterogeneity, as it takes into account that individuals' heterogeneity might affect the entire conditional distribution, rather than just shift means, as is typically seen in [47, 48], and as a result allows us to identify the conditional heterogeneity covariance effect of driving forces on the degradation of CS. The MMQR estimator can be particularly useful in panel data models when they include individual effects, contain endogenous explanatory variables, and when wear interested to obtain non crossing estimates of the regression quantiles [14].

If we have a location-scale model, the conditional quantile estimate QY(τ|X) takes the following form:

$$Y_{it} = \alpha_{i+} X_{it}'\beta' + \left(\delta_i + Z_{it}\gamma' \right) U_{it} \tag{1.1}$$

Where the probability P(i + Zit >0) = 1. The notation (, , ,) refers to the parameters to be estimated. In this equation, (i, i), i = 1,…, n, where "n" represents the fixed effects for a single individual "i," and "Z" is a k-vector comprising the differentiable transformations that form the "X" component, with the "l" element defined as:

$$Z_l = Z_l(X), l = 1,\ldots,k \tag{1.2}$$

The observed element l of the assumed distribution of X (Z) comprises differentiable functions. Let the expression for the fixed effect of individual i be ni, as defined in Eq. (1.1). Xit denote the observed value for individual i at time t. The observations of each individual follow an independent and identically distributed process across t. The error term ui t for individual i at time t is assumed to be independent of ui t over different times and individuals, constant for all individuals over time periods, and uncorrelated to Xit. The normalization of ui t, i, t satisfies the moment condition in [14], which doesn't imply that Xit must be strictly valid over the external world.

$$Q_Y(\tau|X_{it}) = (\alpha_{i+}\ \delta_i q(\tau)) + X_{it}^{'}\beta + Z_{it}^{'}\gamma\, q(\tau) \tag{1.3}$$

Natural logarithms of economic growth (LGDP), fossil fuel production (LFF), and the natural logarithm of population density (LPD). The outcome variable Yit, which we denote by the natural logarithm of CO_2 emissions, exhibits the quantile distribution QY(τ|Xit), according to the placement of regressors Xit. The constant factor Xit − i(τ) i + iq(τ) denotes the qth quantile regression fixed effect under particular values for individuals, and does not equate to a simple linear translation of the intercept as with fixed effects OLS. The individual-specific constant factors vary between individual-specific effects, which can have varying effects according to the distribution of the endogenous variable Y. In order to estimate τ-th, the following problem must be optimized:

$$\min_q \sum_i \sum_t P_\tau \left(R_{it} - \left(\delta_i + Z_{it}^{'}\gamma \right) q \right) \tag{1.4}$$

Where P_τ (A) = (τ − *1*) AI {A ≤ 0} + TAI {A > 0} denotes the check function.

1.4 Current Findings and Outcome

Prior to proceeding to the estimation of unknown coefficients, certain conventional preliminary tests are done in order to detect the time series nature of the variables. We assess whether the panel shows CD (cross-sectional dependency). The panel

Table 1.3 Cross-sectional dependence results

Variables	CO_2	FP	GDP	FF	PD
Pesaran (2004)	(22.37)***	(18.82)***	(83.55)***	(65.35)***	(9.23)***

– Remark: *** refer importance at the 1%, scale
– Source: [41]

cross-sectional data will provide biased estimates of the true parameter coefficients. When neglected, an estimated coefficient obtained through a panel estimation technique, which may implicitly assume that there is an absence of CD, might suffer greatly from cross-sectional dependence of non-observed standard variables that would thus erase much of the appeal from using panel data which offers increased efficiency (Baltagi (2005)). Hence, this particular aspect of our analysis must be addressed to ensure proper estimation of coefficients. For this, we used the CD test by Pesaran (2004a) [49] in order to determine the presence or absence of cross-sectional dependence for the panel. As seen in Table 1.3, the panel shows cross-sectional dependence between most of the countries for all quantitative variables with the exception of the fisheries productivity one. Consequently, we should use methodologies that would allow the use of unit level to mitigate the possibility of size distortions. We need to be careful about the selection of our methods used in the unit root and cointegration testing and the panel estimation methods that are robust to the effects of the CD.

To reliably conduct estimation on the integration property of variables of interest, this study adopts the panel unit root test suggested by LLC and Im (IPS). The aim of the panel unit root test is to boost test power by taking into account both cross-section and time series dimensions. Various panel unit root test frameworks have been developed to test the null of unit root against alternative null hypotheses: [50] test the null of unit root against homogeneous alternative, under which each series in the panel is stationary and is stationary at the same rate. While [49] tests the unit root null against the alternative that at least one series in the panel is stationary. The test presented in this study focuses on panel stationarity, thus inevitably leads one to the LLC framework. The assumption of homogeneity is lifted by [49] test by allowing each member of the panel to have its own autoregressive parameter. Among moderately small panel data, such as the one we examine in this paper, it is demonstrated that [50] LLC panel unit root tests have higher power relative to those of other similar unit root tests. Therefore, we used two tests here; it is also expected to see how cross-sectional dependency impacts panel unit root test results. As we can see from Table 1.4, in all unit root test specifications, all variables are non-stationary in levels, while stationary in first differences, suggesting that every variable used for estimation is an I(1) process.

Table 1.4 Panel unit root test results

	Difference		First difference	
Variable	LLC	IPS	LLC	IPS
CO_2	−10.609	−5.000	−37.388***	−29.637***
FP	−1.770	−21.335	−21.861***	−26.926***
GDP	−210.531	−155.599	−60.128***	−61.753***
FF	−6.340	−20.437	−6.012***	−21.305***
PD	−2.005	−14.672	−8.098***	−20.336***

– Remark: *** refer importance at the 1%, scale
– Levin, Lin & Chu test (LLC), and Im, Pesaran, and Shin W-stat test (IPS)
– Source: [41]

Table 1.5 Panel cointegration test in EU countries

Test	Without trend		With trend	
A). Pedroni residual cointegration test:				
Alternative hypothesis: common AR coefficients (within dimension):				
Panel v-statistic	0.982 (0.162)		0.971 (0.165)	
Panel rho-statistic	−4.538*** (0.000)		−0.665 (0.252)	
Panel PP-statistic	−12.737*** (0.000)		−4.619*** (0.000)	
Panel ADF-statistic	−14.903*** (0.000)		6.060*** (0.000)	
Alternative hypothesis: common AR coefficients (pooled)				
Group rho-statistic	0.286		0.612	
Group PP-statistic	−5.444***		(0.000)	
Group ADF-statistic	−8.237***		(0.000)	
B). Westerlund (2007) bootstrap panel cointegration:				
Statistics	Value	Z-value	p-value	Robust p-value
Gt	−7.884	−4.215	0.000	0.000
Ga	−19.821	−1.503	0.000	0.000
Pt	−56.938	−43.067	0.000	0.000
Pa	−54.503	−24.962	0.000	0.000

– Remark: ***, **, and * refer importance at the 1%, 5%, and 10% scales respectively
– Values in parentheses are *p*-values
– Source: [41]

1.5.2 The use of the method developed in 45 with methodology from 54. The panel cointegration tests are conducted in accordance with 45 method and 54 method, given in the following Table 1.5, to find a long-run relationship between variables if any. The study in 45 proposes an extended panel cointegration methodology that is derived from the Granger's two-step procedure and the Engle method. Initially, the first step is to get rid of the short-run dynamics and the person-specific deterministic pattern while controlling for the heterogeneity, following 45 method. The seven statistics, which are constructed from the estimated residues following this methodology, are divided into two groups: those based on the hypothesis of a single common process, the so-called "pooled" or "within-dimension" tests, and those which assume individual processes, usually designated "grouped" or "between-dimension" tests, as reported in 1.5. There are four tests that are introduced under 54 method, under the null hypothesis that cointegration does not exist. 54 method is more preferred since the structural dynamics (underlying processes) rather than residual dynamics. Since 54 method avoids the commonly used restrictions imposed on individual factors under individual-based analysis for residues, such tests provide strong alternative results if there is any rejection of the null of no cointegration due to common factor restrictions.

The removal of this constraint removes the need for the procedural changes to have same results for the long and short term. The distortionary effects of

Table 1.6 Model 1. Panel estimation for the EU-27 region from 1990 to 2022

Long-run coefficient	DOLS		FMOLS		FEOLS	
	Coefficient	Std. error	Coefficient	Std. error	Coefficient	Std. error
FP	0.049	(0.024)**	0.053	(0.025)**	0.063	(0.019)***
GDP	0.033	(0.017)*	0.112	(0.019)***	0.125	(0.014)***
FF	0.102	(0.028)***	0.071	(0.025)***	0.038	(0.019)**
PD	−0.320	(0.127)**	−0.398	(0.150)***	−0.368	(0.118)***

– Note: ***, **, and * indicate significance at the 1%, 5%, and 10% levels, respectively
Values in parentheses are P-values
– Source: [41]

cross-sectional dependence can be alleviated using [51] Bootstrap method for better estimation of critical values. From Table 1.5, both [42] testing procedure and [51] Bootstrap cointegration testing strongly supported cointegration Table 1.6.

Results of the estimations of the FMOLS, DOLS, and FEOLS estimation techniques. On the whole, the estimates produced by the three estimators on Model 1 in Table 1.6 were roughly the same. On average, the three estimations of the coefficients are broadly similar. By considering the coefficient values and their levels of statistical significance, fish production in the EU region had the most consistent results from the three estimators.

It was found that a 1% increase in fishing output led to 0.06%, 0.05%, and 0.04% deterioration in CS by FEOLS, FMOLS, and DOLS estimators, respectively.

These findings support the conclusions of [30] and other recent literature, such as one recently in science advances, which found fish help store carbon when they die and sink, trapping 12.5% of a fish's carbon in the deep sea after it dies. Furthermore, economic development had a significant impact on the level of reduction in CS in the EU region. A 1% increase in economic development will lead to 0.11% and 0.12% in CS reductions from FEOLS and FMOLS estimators, respectively, supporting the growth hypothesis for the panel of fishery-producing nations, contradicting [5]'s findings on the impact of economic development in the Nile tilapia in a negative and inelastic way but supporting [4]'s conclusions.

Population density in the EU region has had negative coefficients across all estimators, though contradicting [24] while supporting [26]'s Modernization perspective. The growth hypothesis was highly significant statistically on the FEOLS and FMOLS estimation techniques and moderately significant for the DOLS estimations. The EU regions' use of fossil fuels led to a statistically significant positive result for CS deterioration; however, the magnitude of coefficients ranges from 0.03% to 0.07% to 0.10% per 1% rise in production from FEOLS, FMOLS and DOLS estimations respectively.

From Table 1.7, the results of panel quantile regression suggest that the impact of fisheries production has a significant and positive impact on the level of CS deterioration for the first to fourth quantiles and the seventh to ninth quantiles but negligibly on the fifth and sixth quantiles.

Table 1.7 Panel quantile estimation results for EU27 countries

Variables	Method of moments quantile regression with fixed effects										
	Location	Scale	Quantiles								
			0.1	0.2	0.3	0.4	0.5	0.6	0.7	0.8	0.9
FP	(0.063)***	(0.038)***	(0.021)*	(0.024)**	(0.028)***	(0.009)	(0.012)	(0.035)***	(0.056)***	(0.073)***	(0.083)***
GDP	(0.125)***	(0.018)***	(0.088)***	(0.071)***	(0.087)***	(0.092)***	(0.069)**	(0.056)**	(0.013)	(0.017)	(0.173)***
FF	(0.038)***	(0.067)***	(0.089)***	(0.084)***	(0.087)***	(0.054)***	(0.019)**	(0.002)	(0.059)*	(0.104)***	(0.118)***
PD	(−0.368)***	(−0.021)***	(−0.131)***	(−0.105)***	(−0.085)***	(−0.070)***	(−0.076)*	(0.044)	(0.023)	(0.033)	(−0.319)***

Notes: *, **, and *** indicate significance at 10%, 5%, and 1% levels, respectively
Source: [41]

Table 1.8 Model 2. Panel estimation for the EU-14 region from 1990 to 2022

	DOLS		FMOLS		FEOLS	
Long-run coefficient	Coefficient	Std. error	Coefficient	Std. error	Coefficient	Std. error
FP	0.168	0.074**	0.159	0.054***	0.145	0.041***
GDP	0.115	0.059*	0.122	0.040***	0.101	0.030***
FF	0.001	0.028	0.139	0.029***	0.183	0.021***
PD	−0.533	0.281*	−0.028	0.195	−0.070	0.151

– Note: ***, **, and * indicate significance at the 1%, 5%, and 10% levels, respectively
– Values in parentheses are P-values
Source: [41]

The findings show that fisheries regulations targeting output levels work more effectively for more polluting countries. Moreover, the study has shown that the development hypothesis only holds in EU27 and where CS decline is at least median, whereas in sub-median countries, environmental development is prior to economic development. Also, countries below median that are lower in the rank of development—indicated by the level of deterioration—face the greatest obstacle in changing to renewables due to investment. There appear to be no benefits from fisheries production for the countries at the highest CS deterioration levels, possibly due to differences in extraction methods between developed and developing fishery production countries.

Table 1.8 reports the outcomes from FMOLS, DOLS, and FEOLS estimation procedures. Although some of the significance levels are distinguishable statistically, by reviewing Model 2 in Table 1.8, we observe that the estimated coefficients provided by the three estimations are on average comparable. The output from fishing and economic development in EU14 developed countries are most similar between the three estimators, regarding the coefficient values and the statistical significance levels. Both FMOLS, FEOLS, and DOLS estimated a CS degradation of 0.14%, 0.15%, and 0.16%, respectively, when fishing output increases by 1%. The result obtained is similar to the findings of [30] in broad terms; the EU financial concessions made by the EU countries to their fleets damage ecosystems, worsen ocean health and climate change, to the detriment of small and low-impact fishers. Similarly, economic development in EU14 developing countries had a meaningful positive influence on reducing CS degradation. A 1% increase in economic development is estimated to be associated with improvements of 0.10%, 0.12%, and 0.11% from FEOLS, FMOLS, and DOLS, respectively, thus confirming the growth theory for the group of fishery-producing countries. This conclusion is contradictory with the results obtained in [5], who found an inelastic positive relationship between economic development and the increase of CS degradation, but supports the conclusions in [4] on this topic. In contrast, for EU14 industrialized nations, population density appeared to negatively influence CS degradation across all the estimators. This result contradicts the ones in [24] and the growth theory of [26]. While the growth hypothesis is statistically supported by FEOLS and FMOLS at a moderate level of confidence, DOLS finds a strongly supported positive correlation between population density and CS degradation in EU14 developed countries. As is expected,

Table 1.9 Panel quantile estimation results for EU14 countries

Variables	Method of moments quantile regression with fixed effects										
	Location	Scale	Quantiles								
			0.1	0.2	0.3	0.4	0.5	0.6	0.7	0.8	0.9
FP	(0.145)***	(0.096)***	(0.021)*	(0.024)**	(0.028)***	(0.009)	(0.012)	(0.035)***	(0.056)***	(0.073)***	(0.083)***
GDP	(0.101)***	(0.087)***	(0.218)***	(0.255)***	(0.244)***	(0.250)***	(0.232)***	(0.166)***	(0.239)***	(0.250)***	(0.262)***
FF	(0.183)***	(0.099)***	(0.074)**	(0.040)*	(0.014)	(0.015)	(0.048)*	(0.063)***	(0.066)***	(0.072)***	(0.068)***
PD	(−0.070)	(−0.023)***	(−0.060)***	(−0.036)***	(−0.019)**	(0.006)	(0.007)	(−0.026)***	(0.001)	(0.002)	(0.026)

Notes: *, **, and *** indicate significance at 10%, 5%, and 1% levels, respectively
Source: [41]

fossil fuels are related positively and statistically significant to CS degradation in EU14 industrialized countries, although it seems that there is no consistent effect size across different estimation models, from 0.00% (DOLS) to 0.18% (FEOLS) when increasing fossil fuel consumption by 1%.

Table 1.9 reveals the panel quantile estimation. As it seems, fishing output is found to be significant at the first to third and sixth to ninth quantiles, and statistically insignificant between the fourth and fifth quantiles. The first finding supports the validity of fishery regulations about output in countries where pollution is more intensive. The data confirms the growth theory in the developed countries where CS degradation is equal to or greater than the median. It shows a lack of support for the growth theory in nations where CS degradation is below the median. Economic growth is not a factor in lowering levels of marine degradation for below median developing EU14 countries and for developed ones where pollution levels are lower than median, probably due to investments required to change traditional energy to clean energy sources in lower stage of growth nations. Fishery output has no beneficial effect in highly degraded marine environment, likely due to diversity in the technology available in developed marine-producing countries. Policy can infer that from the current evidence this result aligns with those reported by [16, 18, 32], as these authors also supported the growth theory in some industrialized countries highly responsible of both the generation of carbon sinks and degradation of blue carbon, with some of the findings consistent with recent evidence on CO_2 emissions from the industrial sector [19]. We report a large positive association between the increase in the economy in EU14 developed countries and CS degradation in all quantiles. The development theory might be able to apply if policy aims at the regulation of economic activities to preserve marine environments in heavily polluted areas. For fossil fuels, we see a large positive relationship with CS degradation in the first to second and fifth to ninth quantiles. But no statistically significant relationship in the third and fourth quantiles. Just like for the full panel, the positive impact of fossil fuels is bigger in higher quantiles. Population density is found to be negatively associated with CS degradation in all quantiles, but it's statistically significant in nations where CS degradation is higher at quantiles one to four and in quantile six. For lower quantiles and mid-quantiles, especially the third to fifth quantiles, its impact is minimal, whereas from sixth to ninth quantile, population density seems not to matter in determining the CS degradation levels in EU14

Table 1.10 Model 3. Panel estimation for the EU-13 region from 1990 to 2022

	DOLS		FMOLS		FEOLS	
Long-run coefficient	Coefficient	Std. error	Coefficient	Std. error	Coefficient	Std. error
FP	0.087	0.045*	0.066	0.020***	0.069	0.026***
GDP	0.245	0.074***	0.198	0.035***	0.159	0.019***
FF	0.058	0.030*	0.069	0.023***	0.004	0.021
PD	−1.174	0.245***	−1.495	0.169***	−0.263	0.212

– Note: ***, **, and * indicate significance at the 1%, 5%, and 10% levels, respectively
– Values in parentheses are P-values
Source: [41]

developed countries. Those findings could support the idea that density regulations compel governments and organizations to adopt better policies for the marine environment, as the density constraints lead them to adopt environmentally friendly policies [27–29].

Table 1.10: The effect of fishing, economic growth and fossil fuel use on CS deterioration of EU13 developing countries. The effect of fisheries production on CS deterioration is significant and positive across all methods of panel estimation (FEOLS, DOLS, and FMOLS), which give more or less similar times for the coefficients (see Table 1.10). The coefficients indicate that an increase in fishery production by 1% causes the deterioration of CS to increase by 0.06, 0.06, and 0.08 in case of FEOLS, DOLS, and FMOLS, respectively. The positive and statistically significant coefficients of fisheries production in all three approaches reaffirm the presence of the growth hypothesis in the EU13 developing countries.

These findings are in line with [4] for African economies, [30] for global levels, and [6] for Southeast Alaska in terms that marine pollution intensifies as fishery output increases in the early phase of the economy. Table 1.10 also shows that the other driver of CS degradation, GDP (economic growth), is also positively and significantly related to CS deterioration in EU13 developing countries. An increase in GDP by 1% causes CS deterioration to increase by 0.15, 0.24, and 0.19 in FOLS, FMOLS, and DOLS, respectively. The positive and statistically significant coefficients of GDP in all three approaches further confirm the growth hypothesis in EU13 developing countries (and are consistent with the finding of [4]). Fossil fuel use increases CS deterioration within the FMOLS, FEOLS, and DOLS methodologies, whereas 1% increases in fossil fuel use causes CS deterioration to rise by 0.06 and 0.05, respectively. The table below shows the result of the quantile regression of the method of moments (MMQR).

The results clearly confirm that fishing activity has a positive and significant effect on cs deterioration for all the selected quantiles. Moreover, it is shown that fishing output influence varies across quantiles; the coefficient for the effect of fishing is highest for the ninth quantile (0.265) while it is least pronounced at the fourth and the fifth quantiles. Similarly to fishing activity, fossil fuel consumption has a positive effect on cs degradation across all the selected quantiles, while the coefficient of this effect is more pronounced at higher quantiles up to the fifth one and becomes smaller at the higher quantiles up to the ninth.

Unlike previous factors, population density negatively affects CS degradation for all quantiles except the ninth, and it has a statistically significant and negative impact from the first to the ninth quantiles, the magnitude of which increases with the level of quantiles. Lastly, we observe a significant positive impact of economic growth on CS degradation for all the selected quantiles, which increases remarkably as we move toward higher quantiles.

The estimations from the FEOLS panel have been verified by the panel FMOLS and DOLS. As evident from the results, the panel FEOLS coefficients share the same sign and level of significance as the panel FMOLS and DOLS coefficients, confirming that the panel FMOLS and DOLS results are reliable and can be used to draw meaningful conclusions. While the FEOLS coefficients share the same sign as

Table 1.11 Panel quantile estimation results for EU13 countries

Variables	Method of moments quantile regression with fixed effects										
	Location	Scale	Quantiles								
			0.1	0.2	0.3	0.4	0.5	0.6	0.7	0.8	0.9
FP	(0.069)***	(0.056)***	(0.254)*	(0.281)**	(0.272)***	(0.235)	(0.239)	(0.233)***	(0.230)***	(0.229)***	(0.265)***
GDP	(0.159)***	(0.078)***	(0.050)**	(0.044)**	(0.140)***	(0.153)***	(0.090)**	(0.167)**	(0.065)**	(0.133)**	(0.274)**
FF	(0.040)***	(0.010)***	(0.464)***	(0.476)***	(0.495)***	(0.497)***	(0.453)***	(0.433)***	(0.426)***	(0.405)***	(0.326)***
PD	(−0.263)***	(−0.122)***	(−0.222)*	(−0.211)***	(−0.220)***	(0.252)***	(0.359)***	(−0.381)***	(−0.394)***	(−0.359)***	(−0.121)

Notes: *, **, and *** indicate significance at 10%, 5%, and 1% levels, respectively
Source: [41]

the panel FMOLS and DOLS coefficients, they differ slightly in their significance thresholds. Nevertheless, the panel DOLS estimates are dependable and free from issues of endogeneity and serial correlation. To assess the effect of fishing output on CS deterioration based on emission levels, the EU27 nations were divided into two groups—14 developed countries and 13 underdeveloped countries. The projected effect of fishing output on CS deterioration in the EU14 industrialized nations is presented in Tables 1.8 and 1.9, while the corresponding effect for the EU13 developing nations is shown in Tables 1.10 and 1.11. The findings from Tables 1.8 and 1.10 confirm that fishing output significantly accelerates the breakdown of carbon sinks. The results further indicate that EU14 developed countries experience a more pronounced positive effect of fishing output on CS deterioration compared to EU13 developing nations, with precise impact magnitudes of 0.145 and 0.069 for the EU14 and EU13 nations, respectively. This suggests that through the application of marine policy and marine tax instruments targeting the fishing industry, EU14 developed countries are better positioned to achieve more rapid and substantial reductions in CS deterioration than their EU13 developing counterparts.

Similarly, the findings reveal that EU14 developed countries experience a more significant positive effect of fossil fuel use on CS deterioration compared to EU13 developing nations, with precise impact magnitudes of 0.183 and 0.004 for the EU14 and EU13 nations, respectively. This implies that through energy policy measures and taxation on fossil fuel consumption within the fishing industry, EU14 developed countries have greater capacity to substantially reduce CS deterioration than EU13 developing countries. The findings additionally show that EU14 developed countries experience a greater positive impact on CS deterioration than EU13 developing nations, with impact magnitudes of 0.183 and 0.004 for the EU14 and EU13 nations, respectively. This suggests that by adopting sustainable and green energy approaches in fishing-related economic development, EU14 developed countries can achieve considerably greater reductions in CS deterioration compared to EU13 developing countries. On the other hand, the findings also indicate that EU14 developed countries experience a more substantial negative impact of population density on CO_2 emissions than EU13 underdeveloped nations, with precise impact magnitudes of −0.533 and−1.174 for the EU14 and EU13 nations, respectively. This suggests that through legislative and administrative measures targeting the fishing population, EU14 developed countries can achieve more significant reductions in CS degradation relative to EU13 developing countries [53–56].

1.5 Conclusion and Solution for CO_2 Reduction

This research examines the empirical relationship among fisheries output, fossil fuel consumption, economic growth, population density, and CS degradation by taking the growth hypothesis into consideration. This study contributes to the growing body of literature on marine resources and environmental problems related to marine pollution in the fisheries sector and development issues. Based on a 31-year data for 27 fishery-producing countries, this study utilized panel unit root tests, panel cointegration analysis, and panel estimations. Population density (in people/km^2) has been added as the numerical variable on population constraint. Based on panel unit root and cointegration methods, we verified that all the variables are I(1) processes and have a genuine long-run relationship. Although the coefficients' magnitudes across the specifications are not substantially different, using standard panel cointegration estimation methods has revealed differences in the significance of the coefficients. To perform a more comprehensive evaluation over the empirical relationship, we applied the MMQR method, and we explore how external forces have asymmetric impacts across multiple quantiles of the conditional distribution of CS deterioration. Based on the coefficient values of FMOLS, DOLS, and FEOLS, we confirmed that development theories work for fishery-producing countries in the EU27 region.

The impact of fisheries output at various quantiles under MMQR showed a positive and significant relationship from the 1st through the 3rd and 6th through the 9th quantiles of the conditional distribution of CS deterioration, and it decreases at the median quantiles. The positive impact of fossil fuel consumption from the 1st to the 9th quantiles (except the 6th median quantile) was not apparent at that point of the conditional distribution. Perhaps these could show that whereas fisheries output positively affects EU27 countries at the beginning as well as the later stage of transition development, population density only has a numerical contribution to CS degradation between the 1st and 5th quantiles. Also, compared to underdeveloped countries where reliance on seafood imports is heavily needed, fishery-producing countries are much more diversified on fiscals, where their fisheries might proportionally contribute to the generation of higher amount of pollution at high levels, and thus make those EU14 industrialized countries more fiscally diversified.

The MMQR estimates on fishing output and economic growth have the same pattern as shown in Table 1.7. Among the estimations on the relationship between fisheries output and economic development, the highest coefficient was found in the highest quantile, which is stronger from the lowest to the highest quantile. It seems that economic development and fishing output's emission impact more greatly on CS deterioration at higher quantiles, same to higher impact on pollutants. For fishing output and environmental issues on pollutants, its coefficient impact from the lowest to the highest quantiles also presents the same trend, which means that countries with the least fossil fuel dependence of the fishing industry are having greater general consumption of fossil fuel.

Dependence is not equivalent to abundance in this context, thus confirming that there exists a resource curse within deindustrialization processes. For fishing output and population density, their negative and positive coefficients are insignificant at median quantiles, which might be related to the new government's environmental policies that aim to prevent continuous growth of carbon sinks and blue carbon sinks. Since people residing in EU14 industrialized countries generally care more about environmental quality than the residents in EU13 underdeveloped countries where pollution level is lower than EU14, there should be stronger environmental regulations to restrict emission growth that would satisfy people's needs to keep up their support for officials in government. For fisheries output at higher emission quantiles for EU14 developed countries, the insignificance also may be due to diversified concessions applicable to fishing production, and environmental policies for marine pollution control may clash with those for the goal of development. For EU countries at mid-level quantiles (fishing output has a negligibly positive impact on CS degradation), fishing production is no longer causing severe impact on environmental degradation; hence the need to care for such impacts on the environment is reduced due to economic gains obtained from such fishing production, so governments and the public may opt for emission-dependent economic growth more vigorously. In EU13 underdeveloped countries, it seems that population density regulation is overwhelmed by the fishing industry. The significantly high positive impact on marine output for the EU13 developing countries suggests that fishing technology has been upgraded and thus improved working efficiency. Based on [24, 25, 52], improving governance would definitely bring about a decrease in emission pollution. Therefore, we believe improving governance is necessary for seafood-producing countries, especially for EU14 industrialized countries that have the capacity of large-scale carbon sink for marine resources. Economic development has a positive relationship with CS degradation for all the quantiles; the effect of fossil fuel consumption on CS degradation is positive and significant on all quantiles; these are in line with previous growth literature for developed countries. They also can be interpreted differently at different quantiles within fishery-producing EU countries; for low quantiles, deindustrialization may be a factor, that means that when there are many emissions-dependent sectors being displaced by fishing production, the economies have become much more dependent on import-consuming economies whereby a considerable part of economic development is consisted of import of products as well as seafood exports, while transferring their emission-dependent trade production from EU14 industrialized economies to EU13 underdeveloped countries for the goal of acquiring higher returns on profit may improve marine pollution level of the developing economies at higher quantiles. The growth theory works for developed EU14 countries that enforce strict environmental regulations on emission-dependent export-oriented industries of the economy, so the coastal fishing industries may have to limit emissions from developing its work. The difficulty for EU14 developed countries to raise the cost of transitioning toward better energy efficiency and changing fuel is another cause why many of these industries move to EU underdeveloped countries. Since fossil fuel is part of the production processes of almost all productive economic sectors, we can naturally assume that its relationship with CS degradation is positive. While comparing the impacts on CS degradation of fishing output on lower and upper quantiles in EU14, the latter are significantly larger than the former as obtained in Tables 1.9 and 1.11, indicating that at higher quantiles, improving energy efficiency helps the fishing industries save energy, and thus may reduce the cost of environmental control per unit emission of fishing production by better energy systems. Also, the financial cost associated with green technology application, or fuel changing, in EU14 countries can be very high. Also, since there are a lot of investment costs to adopt green technology and change fuel, it is possible for EU14 to obtain much better environmental benefits than EU13 due to EU14's higher ability to bear investment costs compared with developing countries, to develop more energy efficiency and to decrease the reliance of fishing industries on nonrenewable energies and achieve blue sustainability. Promoting electric power and various kinds of green energies, such as wave energy, tidal energy, and ocean energy, is needed for a better fishery production, as for EU13 developing countries, which may not afford more cost on transition toward green energies, to implement the fair tax for fishing industries on emission-dependent fishing, the money may be used for building long-term coastal green energy construction projects.

Finally, the study's span of analysis could also be expanded by adding other important factors like tourism, globalization, and economic growth, which also affect the extent of pollution on the European Union (EU) fisheries sector and the sea environment. The phenomenon could also be analyzed with great depth at a national level (e.g., China) or regional groupings (e.g., ASEAN, BRICS, the Group of Seven and Next 11 countries) with advanced and innovative econometric approaches. Naturally, and similar to all other modeling exercises, this research study had limitations and restrictions; we were unable to bring in the benefits of the blue carbon savings and the capacity of carbon sinks on the analyzed relationships because of data constraints.

Much still needs to be addressed in those subjects. The research could also be broadened to analyze other nation(s) in a comparative manner, at the national level, or other similar scenarios could be addressed. Though we apply some innovative methods to explore the linkage of economic development and marine pollution on the EU marine environment, we have limitations. We rely solely on econometrics and have a restricted range of variables over time from 1990 to 2022. Additional investigation might incorporate the identical dataset with other parameter indices in order to explore the consequences of public as well as private investments in marine industries, as well as ocean power, for fisheries production.

Besides, the methodology is relevant to only the member states of the European Union; this issue could be surmounted through analyzing other countries with elevated levels of emissions. Moreover, it would be worthwhile to analyze possible structural breaks via the research of the complexity of marine pollution economy in the time aspect.

References

1. Rahmizal M (2022) Analysis of Indonesia Marine fisheries with economic growth, population and effort effectiveness. Eur J Formal Sci Eng 5(1):53–61. https://doi.org/10.26417/ejef.v1i1.p17-22
2. Setiyowati H, Nugroho M, Halik A (2022) Developing a blue economy in Depok, West Java, Indonesia: opportunities and challenges of neon tetra fish cultivation. Sustainability 14(20):13028. https://doi.org/10.3390/su142013028

3. Coppola D, Lauritano C, Palma Esposito F, Riccio G, Rizzo C, De Pascale D (2021) Fish waste: from problem to valuable resource. Mar Drugs 19(2):116. https://doi.org/10.3390/md19020116
4. Abo-Taleb HA et al (2021) Effect of a new feed Daphnia magna (Straus, 1820), as a fish meal substitute on growth, feed utilization, histological status, and economic revenue of grey mullet, Mugil cephalus (Linnaeus 1758). Sustainability 13(13):7093. https://doi.org/10.3390/su13137093
5. Shamsuddin M et al (2022) Effects of stocking larger-sized fish on water quality, growth performance, and the economic yield of Nile Tilapia (Oreochromis niloticus L.) in floating cages. Agriculture 12(7):942. https://doi.org/10.3390/agriculture12070942
6. Palomino E, Pardue J (2021) Alutiiq fish skin traditions: connecting communities in the COVID-19 era. Heritage 4(4):4249–4263. https://doi.org/10.3390/heritage4040234
7. Chudik A, Pesaran H (2015) Large panel data models with cross-sectional dependence: a survey. [Online]. https://academic.oup.com/edited-volume/41981/chapter/355335206?searchresult=1&utm_source=TrendMD&utm_medium=cpc&utm_campaign=Oxford_Academic_Books_TrendMD_0. Accessed 20 May 2026
8. Flaten O, Lien G, Koesling M, Løes AK (2010) Norwegian farmers ceasing certified organic production: characteristics and reasons. J Environ Manag 91(12):2717–2726. https://doi.org/10.1016/j.jenvman.2010.07.026
9. Breitung J, Pesaran MH (2008) Unit roots and cointegration in panels. In: Mátyás L, Sevestre P (eds) The econometrics of panel data, Advanced studies in theoretical and applied econometrics, vol 46. Springer Berlin Heidelberg, Berlin, Heidelberg, pp 279–322. https://doi.org/10.1007/978-3-540-75892-1_9
10. Andrews DWK (2005) Cross-section regression with common shocks. Econometrica 73(5):1551–1585. https://doi.org/10.1111/j.1468-0262.2005.00629.x
11. Cerrato M, Sarantis N (2002) The cross-sectional dependence puzzle. London Guildhall University. [Online]. http://fmwww.bc.edu/RePEc/res2002/Cerrato.pdf. Accessed 20 May 2026
12. Balsalobre-Lorente D, Shahbaz M, Roubaud D, Farhani S (2018) How economic growth, renewable electricity and natural resources contribute to CO2 emissions? Energy Policy 113:356–367. https://doi.org/10.1016/j.enpol.2017.10.050
13. Bekun FV, Alola AA, Sarkodie SA (2019) Toward a sustainable environment: nexus between CO2 emissions, resource rent, renewable and nonrenewable energy in 16-EU countries. Sci Total Environ 657:1023–1029. https://doi.org/10.1016/j.scitotenv.2018.12.104
14. Machado JAF, Santos Silva JMC (2019) Quantiles via moments. J Econ 213(1):145–173. https://doi.org/10.1016/j.jeconom.2019.04.009
15. Anderson LG (ed) (2020) Fisheries economics: collected essays, 1st edn. Routledge. https://doi.org/10.4324/9780429288500
16. Reza MS, Azad AK, Abu Bakar MS, Karim MR, Sharifpur M, Taweekun J (2022) Evaluation of thermochemical characteristics and pyrolysis of fish processing waste for renewable energy feedstock. Sustainability 14(3):1203. https://doi.org/10.3390/su14031203
17. Saba GK et al (2021) Toward a better understanding of fish-based contribution to ocean carbon flux. Limnol Oceanogr 66(5):1639–1664. https://doi.org/10.1002/lno.11709
18. Colina M et al (2021) Trophic and non-trophic effects of fish and macroinvertebrates on carbon emissions. Freshw Biol 66(9):1831–1845. https://doi.org/10.1111/fwb.13795
19. McArley TJ, Sandblom E, Herbert NA (2021) Fish and hyperoxia—From cardiorespiratory and biochemical adjustments to aquaculture and ecophysiology implications. Fish Fish 22(2):324–355. https://doi.org/10.1111/faf.12522
20. Li CJ, Kim HH, Yang JW (2021) A comparative study of fishery industry competitiveness in China's coastal provinces. Int J Adv Smart Converg 10(2):158–167. https://doi.org/10.7236/IJASC.2021.10.2.158
21. Chen Y, Lou N, Liu G, Luan Y, Jiang H (2022) Risk analysis of ship detention defects based on association rules. Mar Policy 142:105123. https://doi.org/10.1016/j.marpol.2022.105123

22. Kamau JN et al (2021) Managing emerging fisheries of the North Kenya Banks in the context of environmental change. Ocean Coast Manage 209:105671. https://doi.org/10.1016/j.ocecoaman.2021.105671
23. Chevallier A et al (2021) Diversity of small-scale fisheries in Chile: environmental patterns and biogeography can inform fisheries management. Environ Sci Pol 124:33–44. https://doi.org/10.1016/j.envsci.2021.06.002
24. Drakopulos L (2022) Privatizing the fisheries observer industry: neoliberal science and policy in the U.S. West Coast fisheries. Geoforum 131:116–125. https://doi.org/10.1016/j.geoforum.2022.03.004
25. Connolly J, Zwet AVD, Huggins C, McAngus C (2022) The governance capacities of Brexit from a Scottish perspective: the case of fisheries policy. Publ Policy Admin 37(3):342–362. https://doi.org/10.1177/0952076720936328
26. Galappaththi EK, Susarla VB, Loutet SJT, Ichien ST, Hyman AA, Ford JD (2022) Climate change adaptation in fisheries. Fish Fish 23(1):4–21. https://doi.org/10.1111/faf.12595
27. Mukherjee J, Saha NC, Karan S (2022) Bioaccumulation pattern of heavy metals in fish tissues and associated health hazards in human population. Environ Sci Pollut Res 29(15):21365–21379. https://doi.org/10.1007/s11356-021-17297-6
28. Siddhnath RA, Mohanty BP, Saklani P, Dora KC, Chowdhury S (2022) Dry fish and its contribution towards food and nutritional security. Food Rev Int 38(4):508–536. https://doi.org/10.1080/87559129.2020.1737708
29. Bilal M, Qadir A, Yaqub A, Hassan HU, Irfan M, Aslam M (2023) Microplastics in water, sediments, and fish at Alpine River, originating from the Hindu Kush Mountain, Pakistan: implications for conservation. Environ Sci Pollut Res 30(1):727–738. https://doi.org/10.1007/s11356-022-22212-8
30. Marshall Z, Brockway PE (2020) A net energy analysis of the global agriculture, aquaculture, fishing and forestry system. Biophys Econ Sust 5(2):9. https://doi.org/10.1007/s41247-020-00074-3
31. Firouzi S, Kosari-Moghaddam A, Radgoudarzi M (2022) Resource use efficiency of warm-water fish culture upon different pond sizes. Environ Prog Sustain Energy 41(2):e13778. https://doi.org/10.1002/ep.13778
32. Sala A et al (2022) Energy audit and carbon footprint in trawl fisheries. Sci Data 9(1):428. https://doi.org/10.1038/s41597-022-01478-0
33. Van Neste SL, Couture-Guillet A (2022) The urban politicization of fossil fuel infrastructure: mediatization and resistance in energy landscapes. Environ Plan C Politics Space 40(8):1801–1818. https://doi.org/10.1177/23996544221111657
34. Apergis N, Payne JE (2010) The emissions, energy consumption, and growth nexus: evidence from the Commonwealth of independent states. Energy Policy 38(1):650–655
35. Dogan E, Seker F (2016) Determinants of CO2 emissions in the European Union: the role of renewable and non-renewable energy. Renew Energy 94:429–439. https://doi.org/10.1016/j.renene.2016.03.078
36. Ehigiamusoe KU, Lean HH, Smyth R (2020) The moderating role of energy consumption in the carbon emissions-income nexus in middle-income countries. Appl Energy 261:114215. https://doi.org/10.1016/j.apenergy.2019.114215
37. Anwar A, Siddique M, Dogan E, Sharif A (2021) The moderating role of renewable and non-renewable energy in environment-income nexus for ASEAN countries: evidence from method of moments quantile regression. Renew Energy 164:956–967
38. World Bank Fish to 2030: prospects for fisheries and aquaculture
39. European Commission (2022) Joint Research Centre., Clean Energy Technology Observatory, Ocean energy in the European Union: status report on technology development, trends, value chains and markets : 2022. LU: Publications Office. [Online]. https://data.europa.eu/doi/10.2760/162254. Accessed 30 Dec 2025

40. European Commission The European Green Deal – European Commission. [Online]. https://commission.europa.eu/strategy-and-policy/priorities-2019-2024/european-green-deal_en. Accessed 08 Jun 2025
41. Alsaleh M, Wang X, Nan Z (2024) Toward marine sustainability: Unveiling the effect of the fishery industry on blue carbon sequestration. Sustain Dev 32(1):481–495. https://doi.org/10.1002/sd.2659
42. Pedroni P (2004) Panel cointegration: asymptotic and finite sample properties of pooled time series tests with an application to the PPP hypothesis. Econ Theory 20(03):597–625. https://doi.org/10.1017/S0266466604203073
43. Kao C, Chiang MH (2000) On the estimation and inference of a cointegrated regression in panel data. In: Advances in econometrics, vol 15. Emerald (MCB UP), Bingley, pp 179–222. https://doi.org/10.1016/S0731-9053(00)15007-8
44. Sarkodie SA, Strezov V (2019) Effect of foreign direct investments, economic development and energy consumption on greenhouse gas emissions in developing countries. Sci Total Environ 646:862–871. https://doi.org/10.1016/j.scitotenv.2018.07.365
45. Koenker R, Bassett G (1978) Regression quantiles. Econometrica 46(1):33. https://doi.org/10.2307/1913643
46. Binder M, Coad A (2011) From Average Joe's happiness to Miserable Jane and Cheerful John: using quantile regressions to analyze the full subjective well-being distribution. J Econ Behav Organ 79(3):275–290. https://doi.org/10.1016/j.jebo.2011.02.005
47. Koenker R (2004) Quantile regression for longitudinal data. J Multivar Anal 91(1):74–89. https://doi.org/10.1016/j.jmva.2004.05.006
48. Canay IA (2011) A simple approach to quantile regression for panel data. Econ J 14(3):368–386
49. Pesaran MH (2021) General diagnostic tests for cross-sectional dependence in panels. Empir Econ 60(1):13–50. https://doi.org/10.1007/s00181-020-01875-7
50. Levin A, Lin CF, James Chu CS (2002) Unit root tests in panel data: asymptotic and finite-sample properties. J Econ 108(1):1–24. https://doi.org/10.1016/S0304-4076(01)00098-7
51. Westerlund J (2007) Testing for error correction in panel data*. Oxf Bull Econ Stat 69(6):709–748. https://doi.org/10.1111/j.1468-0084.2007.00477.x
52. Van Der Zwet A, Connolly J, Huggins C, McAngus C (2023) Network resilience and EU fisheries policy engagement in third countries: lessons for post-Brexit governance. Br J Polit Int Rel 25(1):121–138. https://doi.org/10.1177/13691481211067146
53. Isham J (2005) The varieties of resource experience: natural resource export structures and the political economy of economic growth. World Bank Econ Rev 19(2):141–174. https://doi.org/10.1093/wber/lhi010
54. Mehlum H, Moene K, Torvik R (2006) Institutions and the resource curse. Econ J 116(508):1–20. https://doi.org/10.1111/j.1468-0297.2006.01045.x
55. Phillips PCB, Sul D (2003) Dynamic panel estimation and homogeneity testing under cross section dependence. Econ J 6(1):217–259. https://doi.org/10.1111/1368-423X.00108
56. Kremers JJM, Ericsson NR, Dolado JJ (1992) The power of cointegration tests. Oxf Bull Econ Stat 54(3):325–348. https://doi.org/10.1111/j.1468-0084.1992.tb00005.x

Chapter 2
Fisheries Sustainability and Marine Waste Pollution in the EU

Acronyms

EU	European Union
CO_2	Carbon dioxide emissions
MMQR	Method of moments quantile regression
MAP	Multiannual Program (EU Data Collection Framework)
GT	Gross tonnage
FTEs	Full-time equivalents
GVA	Gross value added
DaS	Million days at sea
EUR	Euro
CD	Cross-sectional dependency
FP	Fisheries production
WDI	World Bank Development Indicators
GDP	Gross domestic product
FF	Fossil fuel
PD	Population density
CS	Carbon sink
OLS	Ordinary least square
DOLS	Dynamic ordinary least squares
FMOLS	Fully modified ordinary least squares
FEOLS	Fixed effects ordinary least squares
Europa	European data portal
Eurostat	European Commission's database
LGDP	Natural logarithms of economic growth
LFP	Natural logarithms of fisheries production
LFF	Natural logarithms of fossil fuel
LPD	Natural logarithm of population density
LLC	Lin & Chu Test

M. Alsaleh, *Fisheries Sustainability in the EU*, Urban Sustainability,
https://doi.org/10.1007/978-981-92-4496-6_2

IPPS	Im, Pesaran, and Shin Test
EU27	27 European Union Member States
EU13	13 European Union Member States
BRICS	Brazil, Russia, India, China, South Africa
ASEAN	Association of Southeast Asian Nations
EU14	14 European Union Member States
MW	Marine waste
GR	Governance
UV	Ultraviolet

Marine waste can have a considerable negative economic impact on coastal communities, tourism, shipping, and fisheries. Given its tendency to accumulate and spread, marine waste represents one of the most pressing threats to the sustainability of marine life and the health of EU waters. By examining the roles of fisheries production, fossil fuel use, economic development, and governance over the period from 1990 to 2022, this study seeks to investigate the dynamic influence of the fisheries sector on marine hazardous waste across 27 European nations. Using the method of moments quantile regression (MMQR) with fixed effects, the findings revealed a strong positive association between fisheries production capacity and marine hazardous waste at higher quantiles. From the fifth through the ninth quantiles, fishery output substantially increases marine waste, with the strongest effect observed at the highest quantile and the weakest at the lowest. The results also indicate that EU14 affluent countries experience a more pronounced positive effect on marine waste than EU13 developing nations. Fossil fuel use was found to increase marine waste across all quantiles, while economic development in both EU14 developed and EU13 developing countries was associated with reductions in marine waste, lending support to the growth theory for fishery-producing nations. The influence of governance was significantly negative across all quantiles from the first through the ninth in EU27 nations, with EU14 developed countries found to be more strongly and negatively affected by governance on marine waste. By adopting effective and environmentally friendly technologies, policymakers can work toward reducing marine waste in both EU14 and EU13 nations while advancing sustainable development goals.

2.1 Marine Waste Pollution from Fisheries

The ocean underpins the global systems that make the Earth habitable for humanity [1]. The sea plays an indispensable role in regulating and providing for human needs, encompassing the water we drink, rainfall, atmospheric weather patterns, coastal activities, climatic conditions, much of the food we consume, and even the oxygen we breathe [1]. Ensuring the sustainable future of humanity therefore requires the professional and responsible management of the ocean, which is undoubtedly one of the most fundamental global resources. It has been empirically established that the normal functioning of the natural environment and biodiversity is being increasingly undermined by rising ocean acidification and the ongoing degradation of coastal waters resulting from marine pollution. Fisheries at various scales are equally affected by these harmful pollutant activities [1]. The health of the Earth and the well-being of its inhabitants depend significantly on marine biodiversity—the variety of plant and animal life in aquatic habitats. Accordingly, combating overfishing, marine pollution, and high levels of ocean acidity requires the establishment of well-resourced and properly enforced marine protection zones [1]. In 2019, the EU's 73,983 fishing vessels had a combined gross tonnage of 1.33 million tonnes and an engine capacity of 5.3 million kilowatts [2]. According to data provided by member states under the EU MAP, 57,236 operational vessels in 2019 directly employed 129,540 fishers, equivalent to 92,298 FTEs, with an average annual salary of EUR 23,620. The EU fleet landed 4.05 million tonnes of seafood valued at an estimated EUR 6.3 billion after spending 6.0 million days at sea and consuming 2.0 billion liters of fuel [2]. Gross value added (GVA) was calculated at EUR 3.4 billion and gross profit at EUR 1.25 billion, with a gross profit margin of 20% and GVA as a percentage of sales at 53% [2]. In 2019, 9.0% of total sales were retained as profit, yielding a net profit exceeding EUR 0.6 billion. These figures reflect a decline compared to 2018, and nowcast projections indicate that the EU fisheries fleet's performance will decline moderately relative to 2019 and more significantly compared to 2018 in both 2020 and 2021, though the drop in fuel prices in 2020 will partially offset this decline [2]. In 2022, the European Commission and several other nations formally joined the Clean Seas Campaign as a demonstration of their commitment to ending marine waste pollution [3]. Through this commitment, the European Commission acknowledged the urgent need to reduce marine litter and plastic pollution entering lakes, rivers, and the ocean, thereby increasing its engagement with the world's largest campaign aimed at reversing the tide of waste and plastic pollution [3]. The European Commission has also set a collection target of 90% for single-use plastic beverage bottles by 2029, along with a 25% recycling rate for PET beverage bottles from 2025 and a 30% recycling rate for all plastic beverage bottles from 2030 [3].

Marine pollution is fundamentally a global issue that threatens the world's oceans, with millions of tons of hazardous waste discharged into the ocean each year, generating environmental, economic, public health, and aesthetic consequences [4]. It is estimated that more than 150 million tons of plastic waste have already accumulated in the world's seas, with a further 4.6 to 12.7 million tons added annually [4]. Recent projections suggest that plastic waste entering the ocean each year could quadruple to 29 million metric tons by 2040, equivalent to 50 kg of plastic debris for every meter of shoreline globally [4]. While the majority of marine pollution is believed to originate from land-based sources—with notable regional variation, such as in the Northeast Atlantic where fisheries and fish shipping are significant contributors—sea-based sources are attracting increasing attention due to the substantial volumes of lost or abandoned fishing gear and the resulting harm to marine life sustainability and the fishing economy [5]. Marine pollution can have a severe negative economic impact on coastal communities, tourism, shipping, and fisheries, and its buildup and spread make it one of the most pressing threats to ocean health globally. For the tourism and fisheries sectors alone, the costs of marine waste are estimated to range from 259 million to 695 million euros [6]. Plastic waste from the fisheries industry is particularly pervasive, accounting for 27% of all marine waste and posing a serious threat to marine biodiversity and sustainability [6]. Birds, sea turtles, and marine mammals frequently mistake plastics for food or become entangled in abandoned fishing gear, and over 11,000 tons of abandoned, lost, or discarded fishing gear are generated each year, representing nearly a third of all marine waste collected in European waters [6]. Such gear makes up more than half of the so-called "great waste patch" in the Pacific Ocean, and an estimated 20% of EU fishing gear is lost at sea, giving rise to the phenomenon of "ghost fishing gear" [7]. The ingestion of fisheries waste particles can impair marine animals' ability to digest normal food and expose them to hazardous chemical contaminants [7]. Lost fishing gear also continues to trap and kill marine species, making ghost gear particularly damaging to ocean health [7]. Beyond its ecological harm, ghost gear carries significant economic consequences, including losses to coastal tourism, reduced catch for target fisheries, navigational safety hazards, and shipping delays, as well as the ongoing capture of vulnerable and endangered species and ecosystem destruction [8]. Scientific research has shown that plastic increases the buoyancy of phytoplankton excrement, causing it to float higher in the water column and sink more slowly, thereby extending the time available for carbon to escape into the atmosphere rather than being sequestered [8]. The ocean surface is also increasingly covered with micro- and nanoplastics, and the potential role of whales in carbon sequestration and phytoplankton growth is an area of ongoing scientific inquiry [8]. A typical great whale is estimated to accumulate approximately 30 tons of CO_2 over its lifetime, and upon death it sinks to the ocean floor, locking that carbon away for hundreds of years [9]. Plastic pollution endangers whales, and premature whale deaths reduce the amount of carbon that can be captured. A 2018 study also found that plastic waste releases carbon when exposed to sunlight, and microplastics not only disrupt the proper sinking of carbon but also float on the water's surface, further releasing carbon upon exposure to sunlight [9]. Marine waste pollution and climate change together pose a compounding threat to the ocean, as both plastic manufacturing and the generation of fisheries waste diminish the ocean's capacity to absorb CO_2 emissions, with fisheries waste itself being derived from fossil fuels [9]. Storms and extreme weather events also disperse existing fisheries waste to wider areas of the ocean [9], contributing to the formation of floating plastic islands and the blanketing of beaches and shorelines with waste [10]. This visible accumulation, however, represents only the most apparent symptom of the ocean's ongoing fisheries waste crisis. Research shows that 94% of fisheries plastic waste entering the ocean sinks to depths where low temperatures and the absence of light prevent decomposition, allowing it to persist for generations [10].

The primary research questions guiding this study are: (1) Between 1990 and 2022, how does the fisheries sector influence the sustainability of marine life in European waters? (2) Are EU member states more vulnerable to the effects of hazardous waste generated by the fisheries industry? (3) Can the fisheries sector meet EU marine sustainability goals by 2050 while remaining ecologically responsible?

The main objectives of this study are to: (1) investigate the extent of marine pollution caused by the fisheries industry in EU waters between 1990 and 2022; (2) provide a comprehensive analysis of the fisheries sector's effects on marine life sustainability in both EU13 developing and EU14 developed nations between 1990 and 2022; and (3) examine the scope of the impact of hazardous waste generated by the fisheries industry on both developed and developing nations between 1990 and 2022.

A fundamental weakness in prior research on EU states is the assumption that disturbance terms are independent and free of cross-sectional dependency (CD) among variables across countries. Such dependencies may arise from common or unexpected shocks, spillover effects, or a combination of these factors [11–15]. Accounting for CD is essential in this analysis, as a cross-sectional group of EU economies is likely to be similarly shaped by deepening economic integration. EU countries are also exposed to shared shocks—such as resource crises and fish price fluctuations—that affect both economic development and fish consumption. According to [14], properly incorporating CD into research leads to more balanced and reliable findings regarding the relationships between fossil fuels, economic development, and the environment. The present study aims to determine whether the growth-mitigating influence of natural resources similarly shapes the marine waste profiles of fishery-producing nations. Earlier studies used natural resources as a broad aggregate variable [16, 17], without distinguishing between point-source and diffuse-source resources. In contrast, this study uses a more specific, point-source natural resource variable—fishery production and fossil fuel use. Since point-source resources generally carry more adverse consequences than diffuse ones, the study focuses on fishery-producing EU economies [18, 19]. By incorporating governance as a regulatory control variable, the sample spans both developed and developing fishery-producing economies, enabling more robust conclusions, given that sound policies have been shown to reduce or even reverse the growth-dampening effects of resource dependence [20]. The EU27 fishery-producing nations in the sample include developed economies such as Germany, Italy, and Sweden, alongside emerging economies such as Estonia, Bulgaria, and Malta, covering the period from 1990 to 2022. Given that the selected countries are at varying stages of economic development, the study adopts the cutting-edge method of moments quantile regression developed by Machado and Silva to examine the validity of the marine waste–economic growth nexus within the fisheries industry and across other control parameters at diverse conditional quantiles of marine waste distribution. By incorporating fixed effects, the approach offers empirical insight into the distributional heterogeneity of these relationships, capturing patterns that conventional mean regression methods would fail to detect. As a result, the growth theory is rigorously examined across fishery-producing nations. This article is the first to investigate marine waste driven by economic expansion in the EU27 nations while simultaneously accounting for fisheries production levels, fossil fuel use, economic growth, and governance. Evaluating the growth hypothesis across diverse quantiles of marine waste conditional distributions is necessary for several reasons. First, conditional quantile estimates demonstrate greater resistance to outliers in the outcome variable compared to conditional mean estimates, which are more susceptible to their distorting effects [21]. Conditional mean estimates also fail to capture the full distributional effects of the fisheries industry on hazardous waste. Quantile regression, by contrast, divides the distributional influence of independent variables over the dependent variable into multiple quantile ranges, offering greater intuitive appeal—particularly in panel regressions—and making it easier to categorize the varied effects across different cross-sectional groups. As a result, conditional quantile estimates yield information that conditional mean estimations cannot provide. The remainder of the paper is structured as follows: Sect. 2.2 briefly reviews the growth concept within the associated literature; Sect. 2.3 outlines the data and techniques; Sect. 2.4 presents and discusses the estimation outcomes; and Sect. 2.5 provides a concluding assessment and examines potential policy implications based on the empirically established findings.

2.2 Marine Waste Pollution: Historical Background

Recent academic and economic attention has increasingly turned to studies on the fisheries industry and its relationship to marine pollution. [4] Conducted empirical research on the impact of antibiotics and aquaculture waste feeds on marine benthic ecosystems in the Mediterranean Sea, finding that feed waste altered habitat conditions, reduced macro-invertebrate diversity, and significantly influenced the organization of microbial communities [5, 6]. examined the effects of fisheries waste on marine life sustainability, highlighting the key role of ghost fishing gear in marine pollution and the decline of marine ecosystem health [7]., in a study conducted in Bangladesh on the sustainable utilization of fishery waste to achieve marine conservation goals, argued that fish waste—also known as fishery by-products—should not be regarded as inferior to the fish itself, but rather as a valuable and commercially viable resource capable of generating health, social, economic, and marine benefits [8]., who studied the sustainable intensification of aquaculture through nutrient recycling and circular economies, concluded that improved nutrient recycling practices—achieved through technological advances in wastewater harvesting, by-product processing, and the optimization of feeding composition—can enhance sustainable productive output while reducing the harmful impact of toxic waste on marine life. Similarly, [9] investigated the repurposing of fish processing waste for the production of prodigiosin through microbial technology, describing fish waste as a promising material for cost-effective bio-production and a potential contributor to sustainable bioeconomic growth and blue development with reduced marine pollution. [10], by contrast, examined the challenges associated with low-value fish and fish waste, noting that while fish waste can contribute significantly to marine degradation, it can also be harnessed for bioenergy production, blue growth, and environmental preservation.

Building a sustainable fisheries industry opens up greater opportunities for long-term growth across economic, social, and environmental dimensions. For instance,

[22–24] examined the connection between fisheries output and economic growth in South Asian countries, finding a direct positive effect on long-term expansion [25]. investigated the effects of partial and complete substitution of fish production on growth performance, feed utilization, histological condition, and economic revenue, using a polynomial regression model to argue that underutilized or discarded marine material may offer a viable pathway for achieving a circular economy and generating high-value commodities [26]. examined how initial stocking body sizes of Nile Tilapia (Oreochromis niloticus) affect water quality, growth efficiency, and economic output, concluding that a specific stocking fish size yielded the best outcomes in terms of growth performance and financial return [27]. investigated how Arctic fish skin crafts contributed to the sustainable development of coastal communities in Southwest Alaska, identifying Indigenous fish skin practices as having a considerable positive impact on a range of challenges, including poverty reduction, health and well-being, gender equality, clean water and sanitation, decent employment and economic growth, social inequality, climate change, and marine sustainability.

Fisheries governance encompasses the legal, social, economic, and political frameworks that regulate the fishing sector, operating across international, national, and local levels and incorporating both statutory laws and social conventions. As illustrations of individual heterogeneity, [28, 29] examined network resilience and the involvement of EU and US fisheries policies in third countries, respectively [30, 31]. assessed the governance implications of Brexit from the perspective of Scottish fisheries policy, identifying fisheries management, operations, and research as areas requiring greater investment at the political, communicative, and relational levels [28, 32]. also examined how women's experiences shape small-scale fisheries governance, referencing the institutional contexts in which women participate—spanning a broad range of arrangements including regulatory interactions, co-management structures, informal norms, customary practices, and relational spaces.

Research indicates that in 2019 alone, marine fisheries vessels released 207 million tons of CO_2 into the atmosphere, an amount roughly equivalent to the emissions of 51 coal-fired power plants over the same period [33, 34]. investigated resource use efficiency in warm-water fish farming across different pond sizes in the context of environmental advancement and sustainable energy [35, 36]., in their studies of fossil fuel inputs and carbon footprints in trawl fisheries, noted that systematic energy auditing of fishing vessel fuel profiles is an effective method for identifying inefficiencies. While earlier research has primarily concentrated on the fisheries industry itself, this study explores the varied effects of fisheries sector expansion on marine life sustainability with the aim of achieving blue growth. To advance sustainable development, it also examines the impact of fisheries output on hazardous waste and economic expansion in EU waters. Unlike most of the existing literature, this study employs both static and panel regressions to assess how fisheries production affects marine waste and economic development. The key contributions of this paper are as follows: first, it focuses on the relationships between fisheries production, marine waste, governance, and economic development in EU waters from 1990 to 2022; second, and more critically, it investigates the varied effects of the fisheries sector on marine waste and economic development across

EU27 waters with the goal of achieving sustainable blue growth. Two distinct groups are used for the EU27 region: EU14 developed nations and EU13 developing countries. Given that the data spans 1990 to 2022 and exhibits random walk behavior with drift, unit root and cointegration analyses are employed to assess stationarity. The method of moments quantile regression (MMQR) is then used to examine the interconnections between fish production, marine waste, governance, and economic development. Static and panel regression analyses are also conducted on the factors influencing marine pollution and waste, and the study concludes with projections of how varying levels of fisheries production will affect future marine waste and pollution.

2.3 Marine Waste Pollution Modeling

The model used in this study was built upon recent investigations by [37–40], and takes the following form:

$$MW_{it} = f\left(FP_{it}, FF_{it}, GDP_{it}, GR_{it}\right),$$

The variables MW and FP represent marine waste and fisheries production, respectively. More specifically, MW refers to marine hazardous waste measured in tons, while FP denotes total fisheries output measured in metric tons. Data for these variables were sourced from the World Development Indicators [41]. Gross domestic product (GDP), used as a measure of economic growth and expressed in constant US dollars, was obtained from the European Commission's database [42]. Fossil fuel consumption (FF) is proxied by the volume of fuel used, measured in metric tons of oil equivalent, also drawn from the European Commission's database [42]. Governance (GR), which reflects the effectiveness of government, was sourced from the official European statistics gateway [43]. The data used in this analysis covers EU economies over the period from 1990 to 2022, with all variables expressed in real per capita terms.

To fulfill the objectives of this study, annual data were collected for the selected countries from three main sources: the World Bank's World Development Indicators database (WDI), the European Commission's database (Eurostat), and the official European data portal (Europa), covering the years 1990 to 2022. Table 2.1 summarizes the variables, their descriptive statistics, measurement units, data sources, and transformations applied.

We go over the results in this section. First, the statistical properties of the selected components are shown in Table 2.2 (e.g., mean, standard deviation, maximum, and minimum). MW, FP, GDP, FF, and GR have mean values of 6.088, 4.955, 4.315, 2.822, and 1857, respectively. Each of the study's components had an exceptional standard deviation of 1.357, 0.807, 0.439, 0.833, and 0.107, respectively. We can move on to the unit root test because of the factors' descriptive qualities.

Table 2.1 Data, units, transformation, and source

Data	Measurement unit	Transformation	Source
Marine waste	Tons	Natural logarithm transformation	WDI
Fishery production	Metric tons	Natural logarithm transformation	WDI
Fossil fuel	Metric tons of oil equivalent	Natural logarithm transformation	Eurostat
Economic growth	Constant US dollars	Natural logarithm transformation	Eurostat
Governance	% of confidence for governance	Natural logarithm transformation	Europa

Source: [44]

Table 2.2 Cross-sectional dependence results

Variables	MW	FP	GDP	FF	GR
Pesaran (2004)	(2.62)***	(18.82)***	(83.55)***	(65.35)***	(31.30)***

– Remark: *** refer importance at the 1%, scale
– Source: [44]

This study employs dynamic ordinary least squares (DOLS), fully modified ordinary least squares (FMOLS), and fixed effects ordinary least squares (FEOLS) for comparative purposes. The FEOLS approach is complemented by Driscoll and Kraay standard errors, which demonstrate strong resilience to common forms of cross-sectional dependency and autocorrelation up to a specified lag. As noted by [45], the primary sources of concern in estimating dynamic cointegrated panels are heterogeneity issues arising from variations in cross-sectional averages and cross-sectional adjustments toward the cointegrating equilibrium. To address these concerns, Pedroni's FMOLS model incorporates individually specific intercepts and accommodates diverse serial correlation structures of the error processes across panel members. Based on Monte Carlo simulation findings, DOLS estimations are determined to be more impartial compared to both FMOLS and OLS estimators in finite samples, which [46] extends the DOLS estimator to panel data settings. The DOLS estimator also controls for endogenous feedback by incorporating lead and lag differences, thereby correcting for endogeneity.

Given the limitations of earlier estimation methods, a panel quantile regression strategy was adopted to assess distributional and heterogeneous influences across quantiles [47]. The panel quantile regression technique was originally introduced in the foundational work of [48]. Unlike conventional least-squares regressions, which estimate the conditional mean of the endogenous variable given specific values of the exogenous variables, quantile regressions are used to estimate either the conditional median or a range of different quantiles of the response variables. Quantile regressions are more resistant to the occurrence of outliers in estimation and are particularly well-suited when there is little or no relationship between the conditional means of two variables [49]. However, the current study adopts the method of moments quantile regression (MMQR) of [50] with fixed effects. While standard quantile regression is robust to outliers, it does not account for potential unobserved heterogeneity among individuals in a panel. The MMQR technique allows individual-level effects to influence the entire distribution rather than merely adjusting means, as in the approaches of [21] and [51], thereby making it feasible to identify the conditional heterogeneous covariance effects of the drivers of marine waste. The MMQR estimation approach is particularly appropriate in panel data models that incorporate individual effects and contain endogenous explanatory variables. It also has the practical advantage of producing non-crossing estimates of the regression quantiles, making the results straightforward to interpret. For a location-scale variation model, the estimate of the conditional quantiles QY(τ|X) takes the following form:

$$Y_{it} = \alpha_{i+} X_{it}' \beta' + \left(\delta_i + Z_{it}' \gamma' \right) U_{it} \tag{2.1}$$

In which the probability, $P(\delta_i + Z_{it}\gamma' > 0) = 1$. $(\alpha, \beta', \delta, \gamma')$ denotes parameters which are to be estimated. $(\alpha_i, \delta_i), i = 1, \ldots, n$, in this equation, "n" stands for the fixed effects of a single "i," and "Z" is a k-vector that includes the differentiable transformations that make up the "X" component, with the "l" element stated.

$$Z_l = Z_l(X), l = 1, \ldots, k \tag{2.2}$$

The recognized parameters of X (Z) contain element l and consist of differentiable transformations, while the fixed effects (n) for individual i are mathematically expressed in Eq. (2.1). The observed variable for individual i at time t is denoted as Xit, which follows an independent and identically distributed process that does not vary with time (t). The unobserved error term for individual i at time t is represented by Uit, which is independent, constant across all individuals (i) and time periods (t), and orthogonal to Xit. The normalization of Uit also satisfies the moment conditioning requirement in [50], which does not necessarily demand strict external validity.

$$Q_Y(\tau|X_{it}) = (\alpha_{i+} \delta_i q(\tau)) + X_{it} \beta' + Z_{it} \gamma' q(\tau) \tag{2.3}$$

In this analysis, the vector of explanatory variabless Xit/ in Eq. (2.3) includes the natural logarithm of FP (LFP), the natural logarithms of economic growth (LGDP) and fossil fuel production (LFF), and the natural logarithm of governance (LGR). The dependent variable Yit, representing the natural logarithm of marine waste, has a quantile distribution expressed as QY(τ|Xit), conditional on the position of the independent variables Xit. The scalar coefficient Xit—i(τ) i + iq(τ) captures the quantile-τ effect under fixed conditions for individual i, which does not imply a simple intercept shift as in the case of fixed effects least squares. These time-invariant components produce varying effects that may differ across quantiles of the conditional distribution of the endogenous variable Y. To estimate the τ-th sample quantile q(τ), the following optimization problem must be solved:

$$\min_q \sum_i \sum_t P_\tau \left(R_{it} - \left(\delta_i + Z_{it} \gamma' \right) q \right) \tag{2.4}$$

Where $P_\tau(A) = (\tau - 1) AI \{A \leq 0\} + TAI \{A > 0\}$ denotes the check function.

2.4 Current Condition of Marine Waste Pollution

Before estimating unknown parameters, a series of standard preliminary tests is conducted to determine the time-series characteristics of the variables. The first step involves assessing whether cross-sectional dependency (CD) is present within the panels. The original coefficient estimates of the variables may be subject to distortions caused by cross-sectional dependency. If left unaddressed, cross-sectional dependency—which may arise from unobserved common variables—can significantly diminish the efficiency advantages of panel data [52]. It is therefore essential to account for this concern to obtain reliable coefficient estimates. To evaluate cross-sectional dependency within the panel, this study adopts [53] CD analysis. The quantitative parameters presented in Table 2.2 indicate the presence of cross-sectional dependency among nations across the variables, except fisheries production. Consequently, to reduce the risk of size distortions, the cointegration analysis, unit root testing, and panel estimation techniques employed in this study will need to adopt methodologies that are sufficiently robust to the effects of cross-sectional dependency.

To reliably assess the integration properties of the variables under investigation, this study employs the panel unit root tests proposed [54] (LLC) and [55] (IPS). Panel unit root tests are designed to enhance testing efficiency by combining both cross-sectional and time-series data. Several panel unit root frameworks have been developed to test the unit root null hypothesis against various distinct competing alternatives [54]. Test the unit root null hypothesis against a homogeneous alternative, under which each series in the panel is stationary and shares the same rate of mean reversion [55]. On the other hand, test the unit root null hypothesis against the alternative that at least one series in the panel is stationary. The test introduced in this study places emphasis on panel stationarity, which naturally draws attention to the LLC framework. This restrictive assumption is relaxed in the [55] test, which permits each panel member to have its own autoregressive parameter. For moderately small panel datasets such as the one used in the present study, the [54] (LLC) panel unit root test has been shown to outperform comparable unit root tests in terms of statistical power. Both tests are applied here to assess how cross-sectional dependency influences the panel unit root test results. As presented in Table 2.3, all

Table 2.3 Panel unit root test results

Variable	Difference		First difference	
	LLC	IPS	LLC	IPS
MW	−17.039	−14.155	−34.852***	−35.810***
FP	−3.076	−3.700	−22.737***	−26.926***
GDP	−1.571	−0.210	−18.74***	−68.362***
FF	−9.255	−4.970	−20.610***	−21.409***
GR	−1.090	−0.254	−23.471***	−23.479***

– Remark: *** refer importance at the 1%, scale
– Levin, Lin & Chu test (LLC), and Im, Pesaran, and Shin W-stat test (IPS)
Source: [44]

Table 2.4 Panel cointegration test for EU countries

Test	Without trend		With trend	
A). Pedroni residual cointegration test:				
Alternative hypothesis: common AR coefficients (within dimension):				
Panel v-statistic	0.797 (0.787)		2.225 (0.987)	
Panel rho-statistic	−0.253 (1.000)		−0.132 (0.447)	
Panel PP-statistic	−3.854*** (0.000)		−4.319*** (0.000)	
Panel ADF-statistic	−4.194*** (0.000)		5.514*** (0.000)	
Alternative hypothesis: common AR coefficients (between dimensions)				
Group rho-statistic	1.602		0.945	
Group PP-statistic	−4.174***		(0.000)	
Group ADF-statistic	−5.364***		(0.000)	
B). Westerlund (2007) Bootstrap panel cointegration:				
Statistics	Value	Z-value	p-value	Robust p-value
Gt	−5.284	−14.118	0.000	0.000***
Ga	−19.146	−1.090	0.138	0.000
Pt	−24.852	−10.957	0.000	0.000***
Pa	−18.063	−2.565	0.005	0.000***

– Remark: ***, ** and * refer importance at the 1%, 5%, and 10% scales respectively
– Values in parentheses are *p*-values
Source: [44]

variable parameters across all unit root test specifications are non-stationary at levels but become stationary at first differences, confirming that each variable in the estimation is integrated of order one, I(1).

To establish the presence of a genuine long-run relationship among the variable parameters, this study employs the panel cointegration testing framework of [45] and the bootstrapped panel cointegration testing approach of [56], as presented in Table 2.4 [45]. provides a comprehensive panel cointegration testing framework grounded in the principles of the two-step technique of Engle and Granger. In the first stage of the process, Pedroni's method controls for heterogeneity by removing short-run factors and person-specific deterministic patterns. From the estimated residuals, Pedroni derives seven alternative test statistics, which fall into two categories: those that assume a single common process, most commonly referred to as "within-dimension" or "pooled" tests, and those that account for individual processes, most commonly referred to as "between-dimension" or "grouped" tests, as presented in Table 2.4. The Westerlund method introduces four novel and distinctive approaches that are not commonly found in the existing literature, adopted under the null hypothesis that no cointegration exists. Since the underlying dynamics are structural rather than residual in nature, the testing framework removes the common factor restrictions typically imposed on residual-based dynamic tests. The failure to satisfy common factor restrictions can substantially undermine the effectiveness of residual-based cointegration analyses, making the use of structural dynamics necessary [57].

With this restriction lifted, it is no longer necessary for the short-run and long-run adjustment processes to remain the same. By applying Westerlund's bootstrap methodology, it becomes possible to reduce the distortionary effects of cross-sectional dependency and obtain more reliable critical values. The findings presented in Table 2.4 confirm that cointegration is strongly supported by both [45] and the bootstrapped cointegration testing analysis of [56].

Table 2.5 presents the outcomes of the FMOLS, DOLS, and FEOLS estimation procedures. Although all three differ with respect to their levels of statistical significance, the coefficient estimates produced by the three estimators for Model 1 in Table 2.5 are broadly comparable on average. In terms of both statistical significance and coefficient magnitude, fish production in the EU region demonstrates the greatest consistency across all three estimators. According to the FEOLS, FMOLS, and DOLS estimators, a one percentage point increase in fisheries output is associated with reductions in marine waste of 0.06%, 0.12%, and 0.84%, respectively. This finding is consistent with the key conclusions of [4–6]. Over 11,000 tons of abandoned, misplaced, or discarded fisheries gear are generated each year in European waterways, accounting for nearly a third of all marine waste collected. Such gear makes up more than half of the so-called "great waste patch" in the Pacific Ocean, and marine waste—which can harm or kill animals through ingestion

Table 2.5 Model 1. Panel estimation for the EU-27 region from 1990–2022

Long-run coefficient	DOLS		FMOLS		FEOLS	
	Coefficient	Std. error	Coefficient	Std. error	Coefficient	Std. error
FP	0.843	(0.197)***	0.124	(0.019)***	0.066	(0.020)***
GDP	0.576	(0.203)***	0.337	(0.014)***	0.300	(0.030)***
FF	0.034	(0.093)	0.212	(0.029)***	0.309	(0.070)***
GR	−1.268	(0.723)*	−0.195	(0.017)***	−0.243	(0.040)***

– Note: ***, **, and * indicate significance at the 1%, 5%, and 10% levels respectively
– Values in parentheses are P-values
Source: [44]

or entanglement and can destabilize the ecosystems they depend on—has severely affected hundreds of marine species. As anticipated, fossil fuels in the EU region show a statistically significant positive relationship with marine waste, though the strength of this relationship varies across DOLS specifications. For the FMOLS and FEOLS estimators, a percentage increase in fossil fuel production is associated with increases in marine waste ranging from 0.21% to 0.30%. These findings are consistent with those of [33]. Active marine fisheries are energy-intensive food production operations, and fossil fuel use has a considerable influence on their economic viability.

Economic development also plays a significant role in reducing marine waste across the EU region. A one percentage point increase in economic growth is associated with reductions in marine waste of 0.30%, 0.33%, and 0.57% for the FEOLS, FMOLS, and DOLS estimators, respectively, lending support to the growth theory for the panel of fishery-producing nations. These findings build on the main conclusions of [22–24], who identified an inelastic and positive influence of economic expansion on fisheries waste. Governance in the EU region, on the other hand, is found to have a considerable negative impact across all estimation parameters. This result diverges from the findings of [28] but aligns with the modernization perspective advanced by [31, 32, 29]. The growth hypothesis receives strong statistical support in the FMOLS and FEOLS specifications, while the DOLS specification provides only moderate statistical support. A Green Deal, as a voluntary agreement governed by private law between businesses, non-profits, and local and regional governments, aims to reduce ship-generated pollution, promote clean-sea fisheries, and maintain pristine coastlines across European nations. As shown in Table 2.6, the panel quantile regression estimates indicate that the effect of fisheries output on marine waste is statistically significant at the fifth, sixth, and ninth quantiles, while the impact is negligible at the earlier quantiles—specifically from the first through the fourth. This finding supports the effectiveness of marine regulations targeting the fisheries sector in nations with higher levels of marine pollution. The results further demonstrate that the growth hypothesis holds only in the EU27, where marine waste levels are at or above the median, suggesting that in below-median nations, economic growth takes precedence over marine environmental quality. Furthermore, nations below the median that are at lower stages of development and with lower levels of hazardous waste may encounter greater difficulty transitioning the fisheries sector from non-recycled to recycled inputs due to the associated investment costs. In nations with the most severe levels of hazardous waste, fisheries have only a modest beneficial impact on marine waste, which may be attributable to differences in extraction techniques between the established EU14 countries and the growing EU13 economies. These findings carry significant implications for policy strategy, as they align with the conclusions of [4–6], who tested the growth hypothesis for several nations producing very high levels of toxic and hazardous marine waste. The growth hypothesis was also validated for another leading contributor to harmful marine waste in recent empirical research by [8], consistent with the present findings. The results further agree with those of [9, 10], who demonstrated the validity of the growth hypothesis for the primary sources of toxic and

Table 2.6 Panel quantile estimation results for EU27 countries

Variables	Method of moments quantile regression with fixed effects										
	Location	Scale	Quantiles								
			0.1	0.2	0.3	0.4	0.5	0.6	0.7	0.8	0.9
FP	(0.118)***	(0.020)***	(0.056)	(0.001)	(0.084)	(0.014)	(0.254)***	(0.264)***	(0.054)	(0.023)	(0.239)***
GDP	(0.270)	(0.030)***	(0.022)	(0.325)*	(0.275)*	(0.270)	(0.333)*	(0.450)***	(0.634)***	(0.625)***	(1.289)***
FF	(0.227)***	(0.070)***	(0.963)***	(0.978)***	(0.756)***	(0.601)***	(0.635)***	(0.487)***	(0.573)***	(0.675)***	(1.140)***
GR	(−0.242)*	(0.040)***	(−3.239)***	(−4.512)***	(−4.018)***	(−3.858)***	(−3.347)***	(−3.289)***	(−2.607)***	(−2.221)***	(−6.677)***

Notes: *, **, and *** indicate significance at 10%, 5%, and 1% levels, respectively
Source: [44]

hazardous marine waste. Across all quantiles, a substantial and positive relationship is observed between fossil fuel use and marine waste, though the magnitude of this effect diminishes at higher quantiles. The influence of economic expansion on marine waste, as measured by GDP, is evident from the second through the ninth quantiles but is relatively limited at the first and fourth quantiles, suggesting that marine ecological regulations may support economic development in more heavily polluted nations. Regarding governance, the findings indicate that its impact on marine waste is negative across all quantiles and statistically significant in nations with high levels of fisheries waste from the first through the ninth quantiles. This negative association between governance and marine waste—a key element of the modernization hypothesis—may account for this outcome, as the effectiveness of marine policies, according to [30, 29], compels governments and their administrators to adopt ecologically sound fisheries strategies.

Table 2.7 presents the findings from the FMOLS, DOLS, and FEOLS estimation procedures. Although there are statistically distinguishable differences in significance levels, the coefficient estimates produced by the three specifications used in Model 2 in Table 2.7 are broadly comparable on average. In terms of both statistical significance and coefficient magnitude, fish output and economic development in the EU14 developed nations demonstrate the greatest consistency across all three estimators. According to the FEOLS, FMOLS, and DOLS estimators, a one percentage point increase in fisheries production is associated with reductions in marine waste of 1.27%, 1.52%, and 2.56%, respectively. This finding is consistent with the main conclusions of [7, 8]. The tax relief extended by EU governments to the fisheries industry encourages overfishing and endangers ocean health through the accumulation of plastic debris. Under the influence of solar UV radiation, wind, ocean currents, and other natural processes, plastic breaks down into tiny particles known as microplastics—particles smaller than 5 mm—or nanoplastics, which are particles smaller than 100 nm. Their minute size makes them easy to inadvertently ingest, with consequences for the sustainability of marine life. As anticipated, fossil fuels in the EU14 developed countries show a statistically significant and positive relationship with marine waste, though the coefficient varies between the FMOLS, FEOLS, and DOLS specifications. For the DOLS and FEOLS estimators, a percentage increase in fossil fuel consumption is associated with increases in marine waste ranging from 0.03% to 0.71%. These results are consistent with earlier findings by

Table 2.7 Model 2. Panel estimation for the EU-14 region from 1990 to 2022

Long-run coefficient	DOLS		FMOLS		FEOLS	
	Coefficient	Std. error	Coefficient	Std. error	Coefficient	Std. error
FP	2.565	(1.105)**	1.523	(0.036)***	1.274	(0.053)***
GDP	1.852	(0.633)***	0.525	(0.023)***	0.744	(0.119)***
FF	1.228	(0.714)*	0.025	(0.028)	0.101	(0.034)***
GR	−6.494	(2.753)**	−0.285	(0.036)***	−1.009	(0.194)***

– Note: ***, **, and * indicate significance at the 1%, 5%, and 10% levels, respectively
– Values in parentheses are P-values
Source: [44]

[35, 36]. Beyond altering the atmosphere, the burning of fossil fuels has also substantially modified the chemistry of the world's oceans. Economic expansion also significantly reduces the volume of marine waste in the EU14 developed nations, with a one percentage point increase in economic growth associated with reductions of 0.74%, 0.52%, and 1.85% for the FEOLS, FMOLS, and DOLS estimators, respectively, lending support to the growth theory for the panel of EU14 developed nations. This finding contrasts with those of [26, 27], who identified a positive and inelastic influence of economic expansion on marine pollution, but supports the main conclusions of [25]. Marine pollution affects a wide range of economic activities, including shipping, fisheries, aquaculture, and leisure—for instance, the cost of repairing vessels and equipment damaged by abandoned fishing gear represents a considerable annual expense for the fisheries sector. Governance in the EU14 developed nations is found to have a significant negative impact across all estimated parameters, a result that diverges from [28] but supports the modernization perspective of [32]. According to the DOLS estimate, a unit increase in governance is associated with a 6.49% reduction in marine waste. The FMOLS and FEOLS specifications provide moderately significant statistical support for the growth hypothesis, while the DOLS specification offers strong statistical support. As shown in Table 2.7, the panel quantile regression estimates indicate that the effects of the fisheries sector on marine waste are statistically significant from the first through the ninth quantiles, supporting the effectiveness of marine policies in promoting fisheries production in the EU14 developed nations. The findings confirm that the growth hypothesis holds across all quantile levels of marine waste in the EU14 developed nations, suggesting that marine policy takes precedence over fisheries production in these countries. Furthermore, owing to the financial implications involved, transitioning fisheries gear from conventional to recycled sources may be more challenging for EU14 developed nations at lower quantiles where hazardous waste levels are relatively low. In EU14 nations with the highest levels of marine waste, the fisheries industry was found to have a less favorable impact on hazardous waste, which may be attributable to the varying extraction techniques available within EU14 developed fisheries economies. These findings carry significant policy implications, as they support the growth hypothesis validated by [7] for several industrialized nations with high levels of toxic and hazardous waste production. The growth hypothesis was also confirmed for another major source of hazardous waste in recent empirical research by [10], consistent with the present findings. The results further support the growth hypothesis for the largest producers of hazardous and toxic waste, as demonstrated by [8, 9]. The effects of economic expansion on marine waste are substantial only at the first and second quantiles, becoming inconsequential from the third through the ninth quantiles, suggesting that environmental regulations may support economic development in more heavily polluted nations and that economic growth takes precedence over marine environmental quality in the EU14 developed nations. Regarding fossil fuel use, marine waste results show a strong and positive association from the first to the sixth quantiles and from the seventh to the ninth quantiles, while the influence of fossil fuel use is minimal at the middle fifth quantile. The magnitude of the fossil fuel effect also diminishes at higher quantiles. Governance,

on the other hand, is found to have a detrimental and statistically significant impact on marine waste in nations with high levels of marine waste at the first, third, fifth, and sixth quantiles, while its impact is negligible at the remaining quantiles—specifically the second, fourth, seventh, eighth, and ninth. These results are consistent with the perspectives [31], who argued that effective governance encourages executive boards and governments to foster and strengthen ecologically friendly policies within the fisheries sector of the economy (Table 2.8).

Table 2.9 presents the estimation results of diverse panel econometric techniques, namely FMOLS, DOLS, and FEOLS, under Model 3. These estimation methods yield relatively similar coefficient outcomes. The empirical evidence reveals that fisheries production exerts a statistically significant positive effect on marine waste across the 13 developing EU member countries. Specifically, a 1% growth in fisheries production corresponds to a 0.37%, 0.22%, and 0.83% rise in marine waste under FEOLS, DOLS, and FMOLS estimation separately. The significantly positive coefficients of fisheries production derived from the three empirical approaches validate the presence of the growth hypothesis within these 13 developing EU economies. This empirical outcome aligns with prior research conducted by [9, 10], whose studies demonstrated that marine pollution tends to escalate alongside fisheries production during the initial economic growth stage. In accordance with FEOLS, FMOLS, and DOLS estimation outcomes, fossil fuel consumption also exhibits a strong positive correlation with marine waste generation. A 1% expansion in fossil fuel usage contributes to a 0.18%, 0.17%, and 1.03% increase in marine waste for FEOLS, FMOLS, and DOLS, respectively, thereby suggesting that fossil fuel consumption aggravates marine pollution in developing economies, which is consistent with existing scholarly conclusions. Moreover, the statistical results indicate that GDP also serves as a positive and significant determinant of marine waste in the EU13 developing nations. To be specific, an increase of 1% in GDP brings about a 0.26%, 0.12%, and 0.30% growth in marine waste in the FEOLS, FMOLS, and DOLS models, respectively. The positively significant GDP coefficients across all three estimation methods further verify the growth hypothesis for these EU developing countries. Such findings are consistent with the empirical evidence documented by [22–24]. In addition, the empirical analysis establishes a negative association between governance quality and marine waste in the 13 developing EU states. An improvement of 1% in governance level can reduce marine waste by 0.67% and 0.44% under DOLS and FMOLS estimation, respectively.

Table 2.10 reports the quantile regression estimation results. Firstly, fisheries production is identified as a positive driving factor of marine waste. The statistical data indicates that the significant influence of fisheries production on marine waste declines from 1.533 to 1.146 between the 1st and 3rd quantiles. Nevertheless, its impact becomes pronounced from the 6th to the 9th quantile, with the coefficient value rising again from 9.909 to 1.039. Furthermore, fossil fuel consumption shows a statistically robust positive correlation with marine pollution across all quantiles. The influence magnitude of fossil fuels is higher at lower quantiles (1st quantile) and continues to grow at higher quantiles (9th quantile). Similarly, governance presents a negative influence on marine waste at higher quantiles, and this negative

Table 2.8 Panel quantile estimation results for EU14 countries

Variables	Method of moments quantile regression with fixed effects										
	Location	Scale	Quantiles								
			0.1	0.2	0.3	0.4	0.5	0.6	0.7	0.8	0.9
FP	(0.049)***	(0.053)***	(0.402)***	(0.379)***	(0.369)**	(0.427)**	(0.537)***	(0.484)***	(0.315)***	(0.244)***	(0.089)**
GDP	(0.953)	(0.119)***	(1.283)*	(1.670)**	(0.677)	(0.112)	(0.693)	(1.268)	(0.040)	(0.259)	(0.464)
FF	(0.112***)	(0.034)***	(1.065)***	(0.992)***	(0.778)***	(0.592)***	(0.528)***	(0.325)***	(0.223)	(0.380)***	(0.456)***
GR	(−1.179)**	(0.194)***	(−3.844)**	(−1.385)	(−3.140)*	(1.694)	(−3.836)***	(−2.890)**	(1.205)	(1.212)	(0.124)

Notes: *, **, and *** indicate significance at 10%, 5%, and 1% levels, respectively
Source: [44]

Table 2.9 Model 3. Panel estimation for the EU-13 region from 1990 to 2022

Long-run coefficient	DOLS		FMOLS		FEOLS	
	Coefficient	Std. error	Coefficient	Std. error	Coefficient	Std. error
FP	0.839	(0.110)***	0.220	(0.032)***	0.371	(0.108)***
GDP	0.304	(0.133)**	0.129	(0.014)***	0.262	(0.096)*
FF	1.038	(0.077)**	0.174	(0.039)***	0.181	(0.094)***
GR	−0.677	(0.330)**	−0.446	(0.031)***	−0.518	(1.059)

– Note: ***, **, and * indicate significance at the 1%, 5%, and 10% levels, respectively
– Values in parentheses are P-values
Source: [44]

correlation is statistically significant from the 6th to the 9th quantile, with the governance impact magnitude intensifying at higher quantile levels. Additionally, economic growth is confirmed to positively affect marine waste. As illustrated in Table 2.9, the pollution effect of economic growth remains statistically significant, with the coefficient value increasing from 0.349 to 0.458 throughout the 6th to 9th quantiles.

Panel FMOLS and DOLS confirmed the estimations obtained from the FEOLS panel. As observed, the coefficients of panel FEOLS exhibit the same sign and level of significance as those of panel FMOLS and DOLS. This indicates that the panel FMOLS and DOLS results are dependable and appropriate for concluding. Although the FEOLS coefficients maintain the same sign as the panel FMOLS and DOLS coefficients, there is a slight variation in the significance threshold. Nonetheless, the panel DOLS estimations are considered reliable and free from issues related to endogeneity and serial correlation. To investigate the effects of the fisheries sector on hazardous waste according to the degree of marine pollution in each country, the EU27 nations were divided into two categories—14 developed countries and 13 developing countries. The anticipated impact of the fisheries industry on hazardous waste in the developed EU14 countries is presented in Tables 2.7 and 2.8.

Simultaneously, Tables 2.9 and 2.10 present the estimated impacts of the fisheries sector on hazardous waste in the developing EU13 countries. The results reported in Tables 2.7 and 2.9 indicate that the fisheries sector has a significant effect on hazardous waste. Moreover, the findings reveal that developed EU14 countries exert a significantly greater positive influence on marine waste compared to the developing EU13 nations. The exact effect magnitudes for the EU14 and EU13 countries are 1.274 and 0.371, respectively. This suggests that developed EU14 nations are more capable of reducing hazardous waste than emerging EU13 nations through the implementation of fisheries regulations and marine taxation policies within the fisheries sector. However, the results further indicate that fossil fuel consumption exerts a greater positive influence on marine hazardous waste in the developing EU13 countries than in the developed EU14 countries. The precise impact magnitudes for the EU13 and EU14 nations are 0.18 and 0.101, respectively. This implies that EU13 developing countries could substantially decrease marine hazardous waste relative to EU14 developed nations by adopting energy policies and imposing energy taxes on fossil fuel usage within the fisheries sector.

Table 2.10 Panel quantile estimation results for EU13 countries

Variables	Method of moments quantile regression with fixed effects										
	Location	Scale	Quantiles								
			0.1	0.2	0.3	0.4	0.5	0.6	0.7	0.8	0.9
FP	(0.187)***	(0.108)***	(1.533)*	(1.219)**	(1.146)***	(1.174)	(1.174)	(0.909)***	(0.788)***	(1.078)***	(1.039)***
GDP	(0.188)**	(0.096)*	(0.446)	(0.504)	(0.341)	(0.268)	(0.265)	(0.349)**	(0.392)***	(0.444)***	(0.458)***
FF	(0.312)***	(0.094)***	(0.466)***	(1.177)***	(1.293)***	(1.266)***	(1.260)***	(1.239)***	(1.176)***	(0.860)***	(0.837)***
GR	(−0.484)*	(1.059)	(−0.872)	(−0.524)	(−0.542)	(1.085)	(0.925)	(−2.190)*	(−3.745)***	(−4.151)***	(−4.498)***

Notes: *, **, and *** indicate significance at 10%, 5%, and 1% levels, respectively
Source: [44]

In addition, the findings reveal that developed EU14 countries demonstrate a stronger and more significant positive impact on marine hazardous waste than developing EU13 nations. The impact magnitudes for the EU13 and EU14 countries are 0.262 and 0.744, respectively. This indicates that increased investment in green and environmentally sustainable fisheries in developed EU14 nations may considerably reduce marine hazardous waste compared to underdeveloped EU13 countries. Nevertheless, the results also highlight a significant negative effect of governance on marine hazardous waste in developed EU14 nations, whereas no such impact is evident in underdeveloped EU13 countries. The exact magnitude of this effect for the developed EU14 nations is −1.009. This demonstrates that EU14 developed countries are more effective in reducing marine hazardous waste than EU13 developing nations through governance-based fisheries policies. One possible explanation is that, under the current amendments to tax regulations, the fisheries sector in EU13 developing countries remains exempt from taxation.

2.5 Conclusion and Solution for Marine Waste Pollution

This research empirically explores the long-run nexus between fisheries output, fossil fuel consumption, economic expansion, governance quality, and marine hazardous waste, with a particular focus on validating the economic growth hypothesis. It contributes to the existing literature on fisheries economics and environmental sustainability. Based on a 31-year panel dataset covering 27 EU member states, this study adopts panel unit root tests, panel cointegration procedures, and advanced panel estimation techniques. Governance, quantified by public confidence indicators, was incorporated as a control variable to measure sustainable development performance and regulate marine waste fluctuations. Preliminary econometric tests confirm that all research variables follow an I(1) integration order and possess stable long-term cointegration relationships. Despite minor numerical differences in coefficient values across estimation models, conventional panel cointegration methods display varied significance levels for regression coefficients. To capture the heterogeneous influence of explanatory variables across the conditional distribution of marine waste, the method of moments quantile regression (MMQR) was employed for in-depth empirical analysis. The coefficient results of FEOLS, FMOLS, and DOLS collectively validate the growth hypothesis for fishery-dependent EU27 economies. The MMQR estimation outcomes reveal that fisheries production yields significantly positive impacts on marine waste from the 5th to the 9th quantiles, with pronounced effects observed in upper quantiles. Fossil fuel consumption maintains a statistically significant positive correlation with marine waste across all quantiles (1st to 9th), indicating its persistent pollution contribution throughout different waste concentration levels. This phenomenon reflects the dual impact of fisheries production during both early and advanced economic transition stages within EU economies. Conversely, governance demonstrates a consistent waste reduction effect across all quantiles. Economically diversified fishery-producing nations generate more marine waste than import-dependent emerging economies, which explains the substantial contribution of fisheries sectors to waste accumulation in high-income regions. Consistent variation trends between fisheries production and economic growth are documented in Table 2.6. The highest quantile witnesses the maximum coefficient value of fisheries output, with the correlation intensity between fishery production and economic growth gradually strengthening from lower to upper quantiles. This indicates that regions with severe marine waste typically exhibit stronger interactive effects between fishery activities and economic expansion. Similarly, the pollution magnification effect of fossil fuel consumption peaks at the highest waste quantiles. This counterintuitive outcome suggests that nations with low fossil fuel reliance in fisheries may have higher overall fossil fuel consumption, supporting the deindustrialization-induced resource curse theory. The insignificant governance inhibition effect in lower quantiles can be attributed to the late implementation of environmental regulatory policies targeting hazardous waste. Residents in high-income EU14 countries prioritize environmental quality, compelling local governments to enforce stringent waste reduction measures to gain public support. Additionally, high-waste EU14 economies impose stricter constraints on fishery industries, explaining the prominent influence of fishery production in upper quantiles. Marine pollution abatement policies may hinder economic growth to a certain extent. For quantiles near the median, the visible deterioration of marine environments raises public environmental concerns, prompting both citizens and policymakers to prioritize fishery-driven economic development over environmental governance. The resource curse prevalent in fishery-related industries also weakens governance effectiveness in lower quantiles for both developed and developing EU economies. The significant positive coefficients of fisheries production in upper quantiles reflect the technological optimization of fishery industries within EU14 nations. Consistent with the findings of [28, 29, 31] (2022), improved governance plays a critical role in marine waste mitigation. Therefore, this study recommends governance enhancement for fishery-producing countries, especially high-pollution developed EU14 economies. Economic growth maintains a positive correlation with marine waste across most quantiles, while fossil fuel consumption shows universal significant positive impacts except for the 1st and 4th quantiles, further verifying the growth hypothesis proposed in prior studies. Heterogeneous quantile-level mechanisms are identified among EU fishery economies. In lower quantiles, deindustrialization leads fisheries sectors to replace other pollution-intensive tradable industries, increasing national dependence on fish import and export trades. In upper quantiles, the cross-regional transfer of labor-intensive and export-oriented fishery industries from low-income EU13 economies to high-income EU14 nations effectively reduces waste accumulation in developing regions. Stringent environmental regulations in developed nations also constrain onshore fishery pollution emissions and validate the growth hypothesis. In contrast, high costs of energy upgrading and fuel switching drive pollution-intensive fishery enterprises in EU13 countries to relocate to regions with lenient taxation and environmental constraints. As a universal production input, fossil fuels inevitably aggravate marine pollution across all quantiles. The impact intensity of fisheries production peaks at the highest quantile and bottoms out at the lowest quantile. The rising pollution intensity per unit of energy consumption across quantiles stems from the slow progress in the energy transition in developed EU14 nations, which retain traditional fossil fuel energy systems. Given the persistent positive correlation between fishery output and marine waste, all EU member states need to accelerate the energy transformation of fisheries sectors by adopting renewable energy sources such as marine biomass, wave energy, tidal energy, and ocean thermal energy. For developing EU13 countries that face barriers to renewable energy adoption, targeted environmental taxation for high-pollution fishery enterprises is recommended. Tax revenues should be reinvested into sustainable marine energy infrastructure construction to achieve long-term pollution control. This research acknowledges several limitations and provides directions for future improvement. Firstly, constrained by data availability, this study fails to quantify the potential benefits of marine waste reduction. Secondly, the research scope is limited to EU economies with a narrow selection of research variables covering the period from 1990 to 2022. Future studies can incorporate indicators such as tourism development, globalization, and financial expansion, and explore the influence of public-private investment on fishery production and ocean energy utilization. Thirdly, subsequent research can expand the research scope to economies with high marine pollution emissions, including ASEAN, BRICS, and G7 countries. Moreover, advanced econometric models can be adopted to account for data structural breaks and conduct comparative analysis on marine pollution complexity across different economic cycles.

References

1. Arora NK, Mishra I (2019) United Nations Sustainable Development Goals 2030 and environmental sustainability: race against time. Environ Sustain 2(4):339–342. https://doi.org/10.1007/s42398-019-00092-y
2. Verlé K (2020) Scientific, Technical and Economic Committee for Fisheries (STECF): the 2020 annual economic report on the eu fishing fleet (STECF 20-06). In: JRC scientific and policy reports. Publications Office of the European Union. [Online]. https://pureportal.ilvo.be/nl/publications/scientific-technical-and-economic-committee-for-fisheries-stecf-t-2/. Accessed 20 May 2026
3. Domínguez Razo AN, Segura Campos MR (2022) Ibero-American grains as a source of biomaterials for the manufacture of films and coatings: green alternative of the XXI century for sustainable development. Food Process Preserv 46(7):e16709. https://doi.org/10.1111/jfpp.16709
4. González-Gaya B, García-Bueno N, Buelow E, Marin A, Rico A (2022) Effects of aquaculture waste feeds and antibiotics on marine benthic ecosystems in the Mediterranean Sea. Sci Total Environ 806:151190. https://doi.org/10.1016/j.scitotenv.2021.151190
5. Kim S, Kim YT, Oh LS, Kim HJ, Lee J (2022) Marine waste upcycling—recovery of nylon monomers from fishing net waste using seashell waste-derived catalysts in a CO_2 -mediated thermocatalytic process. J Mater Chem A 10(37):20024–20034. https://doi.org/10.1039/D2TA02060B
6. Charter M, Carruthers R (2022) Products from waste fishing nets: accessories, clothing, footwear, home ware and recreation. [Online]. https://research.uca.ac.uk/5734/1/Products-from-waste-fishing-gear-March-2022.pdf. Accessed 20 May 2026

7. Mozumder M et al (2022) Sustainable utilization of fishery waste in Bangladesh—a qualitative study for a circular bioeconomy initiative. Fishes 7(2):84. https://doi.org/10.3390/fishes7020084
8. Campanati C, Willer D, Schubert J, Aldridge DC (2022) Sustainable intensification of aquaculture through nutrient recycling and circular economies: more fish, less waste, blue growth. Rev Fish Sci Aquacult 30(2):143–169. https://doi.org/10.1080/23308249.2021.1897520
9. Nguyen VB, Wang SL, Nguyen AD, Phan TQ, Techato K, Pradit S (2021) Bioproduction of prodigiosin from fishery processing waste shrimp heads and evaluation of its potential bioactivities. Fishes 6(3):30. https://doi.org/10.3390/fishes6030030
10. Racioppo A, Speranza B, Campaniello D, Sinigaglia M, Corbo MR, Bevilacqua A (2021) Fish loss/waste and low-value fish challenges: state of art, advances, and perspectives. Foods 10(11):2725. https://doi.org/10.3390/foods10112725
11. Chudik A, Pesaran H (2015) Large panel data models with cross-sectional dependence: a survey. [Online]. https://academic.oup.com/edited-volume/41981/chapter/355335206?searchresult=1&utm_source=TrendMD&utm_medium=cpc&utm_campaign=Oxford_Academic_Books_TrendMD_0. Accessed 20 May 2026
12. Flaten O, Lien G, Koesling M, Løes AK (2010) Norwegian farmers ceasing certified organic production: characteristics and reasons. J Environ Manag 91(12):2717–2726. https://doi.org/10.1016/j.jenvman.2010.07.026
13. Breitung J, Pesaran MH (2008) Unit roots and cointegration in panels. In: Mátyás L, Sevestre P (eds) The econometrics of panel data, Advanced Studies in Theoretical and Applied Econometrics, vol 46. Springer Berlin Heidelberg, Berlin, Heidelberg, pp 279–322. https://doi.org/10.1007/978-3-540-75892-1_9
14. Andrews DWK (2005) Cross-section regression with common shocks. Econometrica 73(5):1551–1585. https://doi.org/10.1111/j.1468-0262.2005.00629.x
15. Cerrato M, Sarantis N (2002) The cross sectional dependence puzzle. London Guildhall University. [Online]. http://fmwww.bc.edu/RePEc/res2002/Cerrato.pdf. Accessed 20 May 2026
16. Balsalobre-Lorente D, Shahbaz M, Roubaud D, Farhani S (2018) How economic growth, renewable electricity and natural resources contribute to CO2 emissions? Energy Policy 113:356–367. https://doi.org/10.1016/j.enpol.2017.10.050
17. Bekun FV, Alola AA, Sarkodie SA (2019) Toward a sustainable environment: nexus between CO2 emissions, resource rent, renewable and nonrenewable energy in 16-EU countries. Sci Total Environ 657:1023–1029. https://doi.org/10.1016/j.scitotenv.2018.12.104
18. Isham J (2005) The varieties of resource experience: natural resource export structures and the political economy of economic growth. World Bank Econ Rev 19(2):141–174. https://doi.org/10.1093/wber/lhi010
19. Mehlum H, Moene K, Torvik R (2006) Institutions and the resource curse. Econ J 116(508):1–20. https://doi.org/10.1111/j.1468-0297.2006.01045.x
20. Leite MC, Weidmann J (1999) Does mother nature corrupt? Natural resources, corruption, and economic growth. International Monetary Fund. [Online]. https://books.google.com/books?hl=en&lr=&id=UFYYEAAAQBAJ&oi=fnd&pg=PA3&dq=Leite,C.,+Weidmann,+J.+(1999).+Does+Mother+Nature+Corrupt:+Natural+Resources,+Corruption,+and+Economic+Growth+IMF+Working+Papers.+International+Monetary+Fund.&ots=o29DeZhGdV&sig=-yH1IhlzkIN4hEXDP5n8cCGrHWo. Accessed 20 May 2026
21. Koenker R (2004) Quantile regression for longitudinal data. J Multivar Anal 91(1):74–89. https://doi.org/10.1016/j.jmva.2004.05.006
22. Rahmizal M (2022) Analysis of Indonesia marine fisheries with economic growth, population and effort effectiveness. Eur J Formal Sci Eng 5(1):53–61. https://doi.org/10.26417/ejef.v1i1.p17-22
23. Setiyowati H, Nugroho M, Halik A (2022) Developing a blue economy in Depok West Java, Indonesia: opportunities and challenges of neon tetra fish cultivation. Sustainability 14(20):13028. https://doi.org/10.3390/su142013028

24. Coppola D, Lauritano C, Palma Esposito F, Riccio G, Rizzo C, De Pascale D (2021) Fish waste: from problem to valuable resource. Mar Drugs 19(2):116. https://doi.org/10.3390/md19020116
25. Abo-Taleb HA et al (2021) Effect of a new feed Daphnia magna (Straus, 1820), as a fish meal substitute on growth, feed utilization, histological status, and economic revenue of grey mullet, Mugil cephalus (Linnaeus 1758). Sustainability 13(13):7093. https://doi.org/10.3390/su13137093
26. Shamsuddin M et al (2022) Effects of stocking larger-sized fish on water quality, growth performance, and the economic yield of Nile Tilapia (Oreochromis niloticus L.) in floating cages. Agriculture 12(7):942. https://doi.org/10.3390/agriculture12070942
27. Palomino E, Pardue J (2021) Alutiiq fish skin traditions: connecting communities in the COVID-19 era. Heritage 4(4):4249–4263. https://doi.org/10.3390/heritage4040234
28. Drakopulos L (2022) Privatizing the fisheries observer industry: neoliberal science and policy in the U.S. West Coast fisheries. Geoforum 131:116–125. https://doi.org/10.1016/j.geoforum.2022.03.004
29. Van Der Zwet A, Connolly J, Huggins C, McAngus C (2023) Network resilience and EU fisheries policy engagement in third countries: lessons for post-Brexit governance. Br J Polit Int Rel 25(1):121–138. https://doi.org/10.1177/13691481211067146
30. Sorensen J, Kincl L, Weil R, Dzugan J, Christel D (2022) Fisheries governance and associated health implications: current perspectives from US commercial fishermen. Mar Policy 141:105119. https://doi.org/10.1016/j.marpol.2022.105119
31. Connolly J, Zwet AVD, Huggins C, McAngus C (2022) The governance capacities of Brexit from a Scottish perspective: the case of fisheries policy. Public Policy Adm 37(3):342–362. https://doi.org/10.1177/0952076720936328
32. Galappaththi M, Armitage D, Collins AM (2022) Women's experiences in influencing and shaping small-scale fisheries governance. Fish Fish 23(5):1099–1120. https://doi.org/10.1111/faf.12672
33. Marshall Z, Brockway PE (2020) A net energy analysis of the global agriculture, aquaculture, fishing and forestry system. Biophys Econ Sust 5(2):9. https://doi.org/10.1007/s41247-020-00074-3
34. Firouzi S, Kosari-Moghaddam A, Radgoudarzi M (2022) Resource use efficiency of warm-water fish culture upon different pond sizes. Environ Prog Sustain Energy 41(2):e13778. https://doi.org/10.1002/ep.13778
35. Sala A et al (2022) Energy audit and carbon footprint in trawl fisheries. Sci Data 9(1):428. https://doi.org/10.1038/s41597-022-01478-0
36. Van Neste SL, Couture-Guillet A (2022) The urban politicization of fossil fuel infrastructure: mediatization and resistance in energy landscapes. Environ Plan C Politics Space 40(8):1801–1818. https://doi.org/10.1177/23996544221111657
37. Apergis N, Payne JE (2010) The emissions, energy consumption, and growth nexus: evidence from the commonwealth of independent states. Energy Policy 38(1):650–655. https://doi.org/10.1016/j.enpol.2009.08.029
38. Dogan E, Seker F (2016) Determinants of CO2 emissions in the European Union: the role of renewable and non-renewable energy. Renew Energy 94:429–439. https://doi.org/10.1016/j.renene.2016.03.078
39. Ehigiamusoe KU, Lean HH, Smyth R (2020) The moderating role of energy consumption in the carbon emissions-income nexus in middle-income countries. Appl Energy 261:114215. https://doi.org/10.1016/j.apenergy.2019.114215
40. Anwar A, Siddique M, Dogan E, Sharif A (2021) The moderating role of renewable and non-renewable energy in environment-income nexus for ASEAN countries: evidence from method of moments quantile regression. Renew Energy 164:956–967
41. World Bank Fish to 2030: prospects for Fisheries and Aquaculture
42. European Commission, The European Green Deal – European Commission. [Online]. https://commission.europa.eu/strategy-and-policy/priorities-2019-2024/european-green-deal_en. Accessed 08 Jun 2025

43. European Commission Joint Research Centre (2022) Clean Energy Technology Observatory, Ocean energy in the European Union: status report on technology development, trends, value chains and markets: 2022. LU: Publications Office. [Online]. https://data.europa.eu/doi/10.2760/162254. Accessed 30 Dec 2025
44. Alsaleh M, Wang X, Nan Z (2025) Estimating the impact of fishery industry on marine pollution: new insights from method of moments quantile regression. Energy Environ 36(4):1797–1823. https://doi.org/10.1177/0958305X231200572
45. Pedroni P (2004) Panel cointegration: asymptotic and finite sample properties of pooled time series tests with an application to the PPP hypothesis. Econ Theory 20(03):597–625. https://doi.org/10.1017/S0266466604203073
46. Kao C, Chiang MH (2000) On the estimation and inference of a cointegrated regression in panel data. In: Advances in econometrics, vol 15. Emerald (MCB UP), Bingley, pp 179–222. https://doi.org/10.1016/S0731-9053(00)15007-8
47. Sarkodie SA, Strezov V (2019) Effect of foreign direct investments, economic development and energy consumption on greenhouse gas emissions in developing countries. Sci Total Environ 646:862–871. https://doi.org/10.1016/j.scitotenv.2018.07.365
48. Koenker R, Bassett G (1978) Regression quantiles. Econometrica 46(1):33. https://doi.org/10.2307/1913643
49. Binder M, Coad A (2011) From Average Joe's happiness to Miserable Jane and Cheerful John: using quantile regressions to analyze the full subjective well-being distribution. J Econ Behav Organ 79(3):275–290. https://doi.org/10.1016/j.jebo.2011.02.005
50. Machado JAF, Santos Silva JMC (2019) Quantiles via moments. J Econ 213(1):145–173. https://doi.org/10.1016/j.jeconom.2019.04.009
51. Broadley A, Stewart-Koster B, Burford MA, Brown CJ (2022) A global review of the critical link between river flows and productivity in marine fisheries. Rev Fish Biol Fish 32(3):805–825. https://doi.org/10.1007/s11160-022-09711-0
52. Phillips PCB, Sul D (2003) Dynamic panel estimation and homogeneity testing under cross section dependence. Econ J 6(1):217–259. https://doi.org/10.1111/1368-423X.00108
53. Pesaran MH (2021) General diagnostic tests for cross-sectional dependence in panels. Empir Econ 60(1):13–50. https://doi.org/10.1007/s00181-020-01875-7
54. Levin A, Lin CF, James Chu CS (2002) Unit root tests in panel data: asymptotic and finite-sample properties. J Econ 108(1):1–24. https://doi.org/10.1016/S0304-4076(01)00098-7
55. Im KS, Pesaran MH, Shin Y (2003) Testing for unit roots in heterogeneous panels. J Econ 115(1):53–74
56. Westerlund J (2007) Testing for error correction in panel data*. Oxf Bull Econ Stat 69(6):709–748. https://doi.org/10.1111/j.1468-0084.2007.00477.x
57. Kremers JJM, Ericsson NR, Dolado JJ (1992) The power of cointegration tests. Oxf Bull Econ Stat 54(3):325–348. https://doi.org/10.1111/j.1468-0084.1992.tb00005.x

Chapter 3
Fisheries Activities and Environmental Work Risk Analysis

Acronyms

EU	European Union
CO_2	Carbon dioxide emissions
MMQR	Method of moments quantile regression
FTEs	Full-time equivalents
GVA	Gross value added
DaS	Million days at sea
EUR	Euro
CD	Cross-sectional dependency
FP	Fisheries production
WDI	World Bank Development Indicators
GDP	Gross domestic product
PF	Fossil fuel
PD	Population density
CS	Carbon sink
DOLS	Dynamic ordinary least squares
FMOLS	Fully modified ordinary least squares
FEOLS	Fixed effects ordinary least squares
Europa	European data portal
Eurostat	European Commission's database
LGDP	Natural logarithms of economic growth
LFP	Natural logarithms of fisheries production
LFF	Natural logarithms of fossil fuel
LPD	Natural logarithm of population density
LLC	Lin & Chu Test
IPPS	Im, Pesaran, and Shin Test
EU27	27 European Union Member States
EU13	13 European Union Member States
BRICS	Brazil, Russia, India, China, South Africa
ASEAN	Association of Southeast Asian Nations

M. Alsaleh, *Fisheries Sustainability in the EU*, Urban Sustainability,
https://doi.org/10.1007/978-981-92-4496-6_3

EU14	14 European Union Member States
MW	Marine waste
GR	Governance
UV	Ultraviolet
ILO	International Labor Office
HSE	Health and Safety Executive
ESI	The European Structural and Investment
EMFAF	European Marine, Fisheries, and Aquaculture Fund
IMO	International Maritime Organization
OM	Occupational mortality
IUU	Illegal, unreported, and unregulated

Fishing is considered one of the most hazardous occupations worldwide, with elevated rates of injuries and fatalities. This is attributed to several contributing factors. Therefore, this study investigates the effects of fisheries production, fossil fuel consumption, economic growth, and governance on occupational mortality from 1990 to 2022. By applying the advanced method of moments quantile regression (MMQR) approach with fixed-effect components, the results revealed a significant association between the expansion of the fisheries sector and occupational mortality at higher quantiles. The findings further indicated that developed EU14 member states perceive the expansion of the fisheries industry as having a greater impact on occupational mortality compared to the developing EU13 countries. Concerning the influence of fossil fuel consumption on maritime occupational mortality, the results demonstrated that most quantiles of this variable are positively associated with the occurrence of risks and threats. However, economic growth in both the developed and emerging EU14 and EU13 countries was found to reduce occupational mortality, thereby supporting the growth hypothesis in fisheries-producing nations. Moreover, the association between governance and occupational mortality was identified as significantly negative from the first to the ninth quantiles across the EU27 countries. Occupational mortality across European nations has risen considerably, most likely due to drowning incidents, maritime accidents, and severe weather conditions. This study emphasizes occupational mortality as a crucial component of sustainable fisheries management, aiming to promote a future in which fishing activities are safer and contribute more effectively to social welfare.

3.1 The Dangers of Fisheries Work Environment

According to a 2022 official report released by the International Labor Office (ILO), approximately 24,000 fishermen and fisheries practitioners lose their lives annually, ranking fisheries and aquaculture among the most high-risk industries globally [1]. Statistical records from fisheries administrative departments and fishery associations indicate that mortality rates continue to rise within artisanal fisheries sectors in developing countries, posing severe public safety concerns [1]. The increasing occupational mortality can be attributed to multiple industry changes, including the overexploitation of coastal fishery resources, technological upgrades of fishing vessels and fishing gear, insufficient vocational training, inadequate working experience and professional skills, intense commercial operational pressure, and the iterative adjustment of fisheries management regimes [1]. The fisheries industry possesses an extremely high fatal accident incidence rate of 23.82, which far exceeds the average industrial accident rate of 1.65 [2]. In particular, the European Union's fisheries sector is recognized as one of the riskiest industries across the region, with its accident rate four times higher than that of the construction industry—a universally acknowledged high-risk sector. While most industrial sectors witnessed a continuous decline in accident frequency from 2010 to 2020, the fisheries and aquaculture industry stood as an exception with sustained safety risks [2]. Nevertheless, the limited statistical samples of the fisheries industry may lead to biased data interpretation and hinder the identification of universal accident patterns [3]. National occupational safety statistics present consistent outcomes: during 2021–2022, the construction industry accounted for 25% of total work-related fatalities, followed by the manufacturing sector with 18%, and agriculture, forestry, and fisheries jointly contributing 25% of occupational deaths [3]. Statistical data demonstrate that nearly half of fisheries workplace accidents occur in coastal farming, breeding, and fishing operational areas. Approximately 5.74% of fatal incidents take place on waterborne fishing vessels, and fisheries practitioners generally face higher exposure to occupational hazards than workers in other industries [3]. The extreme risk of the fisheries sector is not proportional to its industrial scale. In the United Kingdom, the fatality rate of the fisheries subsector is 15 times the national industrial average [4]. Between 2013 and 2018, the average crew fatality rate of UK-registered fishing vessels reached 55 deaths per 100,000 crew members, with the figure dropping slightly to 43 deaths per 100,000 crew members in 2017–2018. Despite the declining trend, the mortality rate remains prominently higher than in other industries. For reference, the Health and Safety Executive (HSE) recorded an average fatality rate of merely 0.45 deaths per 100,000 employees across all British industrial sectors in 2017 [4]. Similar risk disparities are observed in other European countries: the fisheries fatality rate in Sweden is 22 times the national average, while Spain records an eight-fold difference. In terms of non-fatal accidents, the incident rate of the UK fisheries and aquaculture industry is more than twice the national average, rising from 1556.86 to 3514.05 within a single year. Unlike other industries with improving safety conditions, the fisheries sector fails to present a sustained downward trend in accident frequency [5]. Fisheries workers suffering from severe injuries require longer sick leave compared with employees in other industries. Most non-fatal injuries in the fisheries sector result in work absences of 7 days or longer, while minor injuries with shorter recovery periods account for a relatively low proportion [5]. Such frequent and severe accidents trigger additional socioeconomic burdens, including increased social security expenditures and reduced industrial production efficiency. The European Structural and Investment (ESI) Funds are designed to facilitate employment recovery and economic development across Europe. Specifically, the European Marine, Fisheries, and Aquaculture Fund (EMFAF) provides financial support for coastal employment expansion and living standard improvement, guiding fishermen to adopt sustainable fishery and aquaculture management modes. Approved by the European Parliament and the EU Council, a total budget of €6.108 billion has been allocated to EMFAF for the period 2021–2027. Although the growth rate of fishery accident cases has gradually slowed down, problematic policy loopholes remain [5]. While the EU27 commits to fishery industry development and crisis relief, some financial policies expand fleet capacity—including subsidies for vessel renovation and stringent entry restrictions for new vessel owners. These measures impede the implementation of sustainable fisheries management and hinder the progress of EU blue growth strategies. In terms of resource exploitation, overfishing has become increasingly severe across the EU, with some fishing fleets operating at three times the sustainable capacity. Additionally, public financial funds are used to purchase new engines and fishing vessels (up to 24 m in length) to maintain fishing production. Such policies not only damage marine biodiversity but also violate the EU Green Deal and conflict with established marine environmental protection and blue growth targets. In terms of air pollution, ship emissions from fishing vessels in the North Sea and Baltic Sea caused €22 billion in annual European health damages in 2000 [2]. Thanks to the International Maritime Organization (IMO)'s 2008 sulfur emission restriction policies for marine fuels, the health-related economic losses are projected to decrease to €14.1 billion by 2020. However, data from the Danish Centre of Energy indicates that the overall health cost of international marine transportation in Europe increased from €58.4 billion in 2000 to €61.4 billion in 2020 [2]. EU maritime incident statistics recorded 18 fatalities involving EU-registered vessels in 2021, up from 15 deaths in 2020, with crew members accounting for 95% of all marine fatal casualties [2]. This study proposes three core research questions and three corresponding research objectives. The research questions are as follows: (1) whether a measurable correlation exists between fisheries industry expansion and occupational mortality across European countries from 1990 to 2022; (2) whether EU member states can optimize sustainable fisheries management to adapt to the expanding fishery economy; (3) whether standardized fishery activities can assist the EU in achieving its 2050 blue growth targets. The research objectives include: (1) quantifying the occupational mortality effects induced by fisheries industry growth in EU marine areas between 1990 and 2022; (2) systematically analyzing the impact of occupational mortality on sustainable fisheries management in developed EU14 and developing EU13 nations; (3) comparing the magnitude of mortality risks triggered by fishery expansion across developed and emerging EU economies. This research presents substantial innovations and contributions to existing literature on EU sustainable fisheries management. Firstly, this study adopts analytical tools including ecosystem models and advanced regression techniques to evaluate the occupational mortality impacts of fishery activities under sustainable management frameworks. Secondly, it focuses on the socioeconomic influences of fisheries management policies, covering coastal community welfare and resident livelihoods. Thirdly, this research integrates multidisciplinary perspectives from ecology, economics, sociology, and political science to comprehensively analyze complex industrial issues and propose targeted optimization solutions. Fourthly, different from traditional regression models that only estimate average variable correlations, the MMQR method enables comprehensive analysis across the entire conditional distribution of dependent variables, capturing dynamic correlation changes at different quantiles. Fifthly, the MMQR model accommodates heteroskedasticity, allowing for variable volatility caused by fishery heterogeneity such as species differences, geographical locations, and time variations. This paper is structured in a clear sequential framework. Section 3.2 reviews relevant theoretical literature concerning the growth hypothesis. Section 3.3 elaborates on research methodologies and raw data. Section 3.4 presents empirical results and conducts an in-depth statistical analysis. Section 3.5 summarizes research conclusions and proposes targeted policy implications based on empirical findings.

3.2 Fisheries Work Environment Background

Marine fishing is universally recognized as the world's most dangerous occupation. Official statistical data from countries with complete safety recording systems demonstrate that fisheries occupational mortality rates are significantly higher than national industrial averages. According to Canadian Transportation Safety Board data compiled by [6], marine accidents occurred 236 times annually on average between 2010 and 2020, among which fishing vessels accounted for 33% of all recorded incidents [6]. The study also evaluated the safety supervision intensity of fisheries occupational health standards and identified persistent hazards and challenges within the industry. A Norwegian-based study [1] explored fisheries management's impacts on fishermen's occupational health and safety. The research revealed that despite high injury rates and widespread health hazards, the regulatory constraints for fisheries occupational safety are less stringent than those of other industrial sectors [2] conducted field research on fishery workplace hazards and physical injuries at Tanji Fishing Port in The Gambia. The statistical results categorized on-site hazards into three primary types: ergonomic risks (25%), environmental risks (23%), and physical climate-related risks (21%). In terms of injury types, muscular strains and falls ranked the highest, accounting for 19% and 17% of total accidents, respectively [3]. Investigated occupational injury risks in England's commercial Dungeness crab fishery, concluding that repetitive operational movements and unnatural shoulder and back postures substantially increase the probability of severe musculoskeletal injuries [4]. Analyzed the driving factors of safety measure adoption in the fisheries industry and identified five core favorable determinants: professional safety knowledge, organizational management commitment, industrial cooperative behavior, supportive policies, and strict legal supervision. Adopting secondary data analysis, [5] explored the correlation between socioeconomic structural factors and fishermen's physical illnesses in Mexico. The study confirmed that deregulation-induced policy adjustments reduce fishermen's economic income and generate long-term adverse impacts on their physical health and living quality. Fossil fuel combustion in fishery production also triggers severe air pollution, inducing multiple chronic diseases such as asthma, cancer, and cardiovascular disorders, and even premature death [7] systematically reviewed the interactive relationships between fossil fuel consumption, atmospheric pollution, climate change, and public health risks, verifying the severe health hazards of fossil fuel usage [8]. Explored the causal linkage between energy consumption, environmental pollution, economic growth, and public health in Pakistan. The empirical results indicated a one-way causal relationship from energy consumption and pollution to health deterioration, while two-way causality exists between economic growth and health outcomes. Similarly, [9] studied the health risks of petrochemical exposure in the United States, pointing out that industrial petrochemical pollutants invade the human body through food, water, air, and skin contact, resulting in irreversible physiological damage [10] applied the PMG-ARDL model to examine the impacts of urbanization and carbon emissions on public welfare in South Asia from 2000 to 2020, proving that these pollutants significantly raise under-five child mortality rates [11] adopted the ARDL model and bound cointegration test to explore the health implications of industrial carbon emissions in Nigeria (1980–2016), with a particular focus on fishery energy consumption. Combined with a sensitivity analysis, the research confirmed that total carbon emissions exert a prominent adverse effect on neonatal and under-five mortality [12]. Sound governance systems are the fundamental guarantee for high-quality medical and healthcare services. Governance encompasses a series of institutional norms, policies, and supervision mechanisms that constrain industrial operational behaviors, including those of the fisheries sector [13, 14] analyzed the correlation between urban governance capabilities and climate-related public health, emphasizing the critical role of public health supervision in promoting industrial safety standardization [15, 16] integrated international relations theories to explore the influence of institutional authority on global health decision-making and negotiation processes, confirming that international governance mechanisms can optimize global public health safety standards [17, 18] studied the mitigating effects of cross-institutional cooperation on health inequality in developed countries. The collective legitimacy and credibility of multi-party alliances facilitate intersectional collaboration, thereby improving public health equity. Economic development is also closely correlated with public health conditions. Economic research demonstrates that sustained economic growth contributes to increased life expectancy [19] analyzed the interactive relationship between income level, medical service quality, and public health in low- and middle-income countries, noting that developed economies possess superior medical resources while still facing inherent disease risks and food insecurity problems [20] verified the significant influence of income levels on public health outcomes and evaluated the practical effects of household pollution reduction measures in underdeveloped regions [21] pointed out that high-income countries with sound medical insurance systems generally achieve better public health performance [22] further confirmed that income stability positively correlates with self-reported health conditions [23, 24] explored the connection between digital infrastructure construction and public health, highlighting the significant optimization effect of digitalization on medical services in high-income economies. Distinguished from prior literature that merely focuses on industrial fishery production, this study comprehensively analyzes the ecological and bioeconomic impacts of aquatic resource exploitation. It further investigates the correlation between fishery output growth, individual economic development, and occupational mortality in EU marine areas through static and panel regression methods. The core research scope covers the interactive linkage between fisheries production, occupational mortality, governance quality, and economic growth across the EU27 from 1990 to 2022. This research presents multiple innovative improvements. Firstly, it categorizes the 27 EU member states into two subgroups: 14 developed economies and 13 developing economies for comparative analysis. Secondly, considering the stochastic volatility and trending characteristics of the 32-year research dataset, this study adopts unit root tests and cointegration tests for data stationarity verification. Thirdly, the MMQR method is applied to identify the quantile heterogeneity of research variables. Fourthly, static and panel regression models are supplemented to systematically analyze marine risk and mortality determinants. Lastly, this paper predicts the dynamic impacts of fishery output changes on marine safety risks, occupational mortality, and industrial sustainability.

This research empirically explores the long-run nexus between fisheries output, fossil fuel consumption, economic expansion, governance quality, and marine hazardous waste, with a particular focus on validating the economic growth hypothesis. It contributes to the existing literature on fisheries economics and environmental sustainability. Based on a 31-year panel dataset covering 27 EU member states, this study adopts panel unit root tests, panel cointegration procedures, and advanced panel estimation techniques. Governance, quantified by public confidence indicators, was incorporated as a control variable to measure sustainable development performance and regulate marine waste fluctuations. Preliminary econometric tests confirm that all research variables follow an I(1) integration order and possess stable long-term cointegration relationships. Despite minor numerical differences in coefficient values across estimation models, conventional panel cointegration methods display varied significance levels for regression coefficients. To capture the heterogeneous influence of explanatory variables across the conditional distribution of marine waste, the method of moments quantile regression (MMQR) was employed for in-depth empirical analysis. The coefficient results of FEOLS, FMOLS, and DOLS collectively validate the growth hypothesis for fishery-dependent EU27 economies. The MMQR estimation outcomes reveal that fisheries production yields significantly positive impacts on marine waste from the 5th to the 9th quantiles, with pronounced effects observed in upper quantiles. Fossil fuel consumption maintains a statistically significant positive correlation with marine waste across all quantiles (1st to 9th), indicating its persistent pollution contribution throughout different waste concentration levels. This phenomenon reflects the dual impact of fisheries production during both early and advanced economic transition stages within EU economies. Conversely, governance demonstrates a consistent waste reduction effect across all quantiles. Economically diversified fishery-producing nations generate more marine waste than import-dependent emerging economies, which explains the substantial contribution of fisheries sectors to waste accumulation in high-income regions. Consistent variation trends between fisheries production and economic growth are documented in Table 2.7. The highest quantile witnesses the maximum coefficient value of fisheries output, with the correlation intensity between fishery production and economic growth gradually strengthening from lower to upper quantiles. This indicates that regions with severe marine waste typically exhibit stronger interactive effects between fishery activities and economic expansion. Similarly, the pollution magnification effect of fossil fuel consumption peaks at the highest waste quantiles. This counterintuitive outcome suggests that nations with low fossil fuel reliance in fisheries may have higher overall fossil fuel consumption, supporting the deindustrialization-induced resource curse theory. The insignificant governance inhibition effect in lower quantiles can be attributed to the late implementation of environmental regulatory policies targeting hazardous waste. Residents in high-income EU14 countries prioritize environmental quality, compelling local governments to enforce stringent waste reduction measures to gain public support. Additionally, high-waste EU14 economies impose stricter constraints on fishery industries, explaining the prominent influence of fishery production in upper quantiles. Marine pollution abatement policies may hinder economic growth to a certain extent. For quantiles near the median, the visible deterioration of marine environments raises public environmental concerns, prompting both citizens and policymakers to prioritize fishery-driven economic development over environmental governance. The resource curse prevalent in fishery-related industries also weakens governance effectiveness in lower quantiles for both developed and developing EU economies. The significant positive coefficients of fisheries production in upper quantiles reflect the technological optimization of fishery industries within EU14 nations. Consistent with the findings of [25–27], improved governance plays a critical role in marine waste mitigation. Therefore, this study recommends governance enhancement for fishery-producing countries, especially high-pollution developed EU14 economies. Economic growth maintains a positive correlation with marine waste across most quantiles, while fossil fuel consumption shows universally significant positive impacts except for the 1st and 4th quantiles, further verifying the growth hypothesis proposed in prior studies. Heterogeneous quantile-level mechanisms are identified among EU fishery economies. In lower quantiles, deindustrialization leads fisheries sectors to replace other pollution-intensive tradable industries, increasing national dependence on fish import and export trades. In upper quantiles, the cross-regional transfer of labor-intensive and export-oriented fishery industries from low-income EU13 economies to high-income EU14 nations effectively reduces waste accumulation in developing regions. Stringent environmental regulations in developed nations also constrain onshore fishery pollution emissions and validate the growth hypothesis. In contrast, high costs of energy upgrading and fuel switching drive pollution-intensive fishery enterprises in EU13 countries to relocate to regions with lenient taxation and environmental constraints. As a universal production input, fossil fuels inevitably aggravate marine pollution across all quantiles. The impact intensity of fisheries production peaks at the highest quantile and bottoms out at the lowest quantile. The rising pollution intensity per unit of energy consumption across quantiles stems from the slow progress in the energy transition in developed EU14 nations, which retain traditional fossil fuel energy systems. Given the persistent positive correlation between fishery output and marine waste, all EU member states need to accelerate the energy transformation of fisheries sectors by adopting renewable energy sources such as marine biomass, wave energy, tidal energy, and ocean thermal energy. For developing EU13 countries that face barriers to renewable energy adoption, targeted environmental taxation for high-pollution fishery enterprises is recommended. Tax revenues should be reinvested into sustainable marine energy infrastructure construction to achieve long-term pollution control. This research acknowledges several limitations and provides directions for future improvement. Firstly, constrained by data availability, this study fails to quantify the potential benefits of marine waste reduction. Secondly, the research scope is limited to EU economies with a narrow selection of research variables covering the period from 1990 to 2022. Future studies can incorporate indicators such as tourism development, globalization, and financial expansion, and explore the influence of public-private investment on fishery production and ocean energy utilization. Thirdly, subsequent research can expand the research scope to economies with high marine pollution emissions, including ASEAN, BRICS, and G7 countries. Moreover, advanced econometric models can be adopted to account for data structural breaks and conduct comparative analysis on marine pollution complexity across different economic cycles.

3.3 Utility Functions Model for Fisheries Occupations

Derived from the seminal utilitarian work, utilitarianism constitutes a philosophical doctrine prioritizing the maximization of collective happiness and societal well-being, and it has profoundly shaped contemporary economic thinking. This ethical framework established the fundamental concept of utility, which refers to the subjective satisfaction and gratification obtained from the consumption of commodities and services. In modern economic research, utility functions are widely adopted to quantify individual consumption preferences and behavioral decision-making. Within the economic discipline, utilitarian thought has guided multiple theoretical frameworks and policy formulations, with a core emphasis on optimizing overall social welfare. Accordingly, this study proposes two utilitarian-based economic hypotheses:

H1: Occupational mortality in the fisheries industry does not maximize societal welfare.

H2: Occupational mortality in fisheries activities does not maximize well-being.

Utilitarianism belongs to consequentialist ethics, focusing on the maximization of universal pleasure and benefit. The core principle indicates that morally acceptable behaviors are those capable of generating the greatest advantages for the majority of the public. This research adopts the empirical econometric model constructed in prior scholarly studies by [28–31], and the functional formula is specified as follows:

$$OM_{it} = f(FP_{it}, FF_{it}, GDP_{it}, GR_{it})$$

In this equation, OM represents occupational mortality, while FP denotes fisheries production. Specifically, fisheries production (FP) is quantified as the total marine catch measured in metric tons, and occupational mortality (OM) records the frequency of maritime collisions, navigation hazards, fatalities, and physical injuries. Raw data for these two indicators were extracted from the World Development Indicators [32]. Gross domestic product (GDP), which reflects economic growth at constant US dollar prices, was sourced from Eurostat [33]. Fossil fuel consumption (FF), proxied by petroleum consumption measured in metric tons of oil equivalent, was also obtained from the Eurostat database [33]. Moreover, governance effectiveness (GR), serving as a comprehensive governance indicator, was collected from the official European statistical database [33]. The research sample covers 27 EU economies, with the research period spanning from 1990 to 2022. All compiled datasets are validated for cross-sectional accuracy and individual-country applicability. This study acquires balanced annual panel data from three authoritative databases: the World Bank WDI database, Eurostat, and the official Europa statistical platform. Detailed variable definitions, measurement units, data sources, and relevant modifications are systematically summarized in Table 3.1.

We go over the findings in this section. To begin, Table 3.2 displays the statistical characteristics of the selected parts (i.e., the maximum, minimum, mean, and standard deviation). FI, FP, GDP, FF, and GR have relative mean values of

Table 3.1 Data, units, transformation, and source

Data	Measurement unit	Variable	Source
Occupational mortality	Fatalities rate	Death and injury incidents from agriculture, fishery, and forestry	WDI
Fisheries production	Metric tons	Fisheries and aquaculture production	WDI
Fossil fuel	Metric tons of oil equivalent	Energy consumption in vessel activities	Eurostat
Economic growth	Constant US dollars	GDP of (agriculture, fishery, forestry)	Eurostat
Governance	% of confidence in governance	Confidence for governance in agriculture, fishery, and forestry	Europa

Source: [34]

Table 3.2 Results of descriptive statistics

Variable	Observations	Mean	Std. Dev.	Min	Max
FI	864	2.125	1.189	1.792	2.642
FP	864	4.926	0.825	3.211	6.310
GDP	864	4.378	0.395	3.129	5.253
FF	864	2.752	0.761	0.145	4.600
GR	864	1.868	0.070	1.482	1.979

Source: [34]

2.125, 4.926, 4.378, 2.752, and 1868 respectively. For every component of the research, the best standard deviation is 1.189, 0.825, 0.395, 0.761, and 0.070, in that order. The precise properties of the factors allow us to go on to the unit root test.

This study adopts three mainstream panel estimation techniques for comparative analysis, namely fixed effects ordinary least squares (FEOLS), dynamic ordinary least squares (DOLS), and fully modified ordinary least squares (FMOLS). The FEOLS model is integrated with Driscoll and Kraay standard errors, which can effectively eliminate common statistical biases, including cross-sectional dependence and serial correlation. As highlighted by [35, 36], heterogeneous cross-sectional characteristics and varied equilibrium adjustment speeds frequently trigger estimation biases in dynamic cointegrated panel data. To mitigate such drawbacks, Pedroni's FMOLS framework incorporates individual intercepts and allows serial correlation calibration for heterogeneous error terms across cross-sections. Based on Monte Carlo simulation evidence from [37], the DOLS estimator exhibits superior statistical performance compared with FMOLS and FEOLS, particularly for small panel datasets [38] further optimized the DOLS approach to adapt to panel data applications. Since this method considers potential endogeneity issues, it eliminates endogenous estimation bias through the addition of lead and lag difference terms. Given the inherent limitations of conventional mean-based regression models, this research further employs panel quantile regression to capture the heterogeneous distributional influences of explanatory variables [39]. Originally proposed by [40], quantile regression differs from traditional least-squares regression, which merely calculates the conditional mean of dependent variables. Instead, this technique evaluates the conditional median and multiple quantiles of explained variables. Quantile regression possesses stronger robustness against dependent-variable errors and performs optimally when the conditional means of variables exhibit weak correlation [41]. Nevertheless, this paper adopts the fixed-effects method of moments quantile regression (MMQR) developed by [42]. Although conventional quantile regression is outlier-resistant, it fails to capture unobserved individual heterogeneity. In contrast, the MMQR model enables individual-specific effects to affect the entire conditional distribution rather than merely shifting mean values, which facilitates the identification of heterogeneous covariance effects of marine waste determinants [43]. This estimation method is highly applicable for panel models containing individual fixed effects and endogenous explanatory variables. Additionally, MMQR generates non-crossing quantile coefficients, thereby simplifying the empirical interpretation of quantile regression results. The conditional quantile function QY(τ|X) under the location-scale MMQR model is defined in the formula below:

$$Y_{it} = \alpha_{i+} X_{it}' \beta' + \left(\delta_i + Z_{it} \gamma' \right) U_{it} \tag{3.1}$$

Where $P(\delta_i + Z_{it}\gamma' > 0) = 1.(\alpha, \beta', \delta, \gamma')$ are the values that need to be approximated; Z is a k-vector of identified components of X that are differentiable changes with element *l* provided by: (α_i, δ_i), $i = 1,\ldots, n$ denotes the individual i fixed effects.

$$Z_l = Z_l(X), l = 1,\ldots,k \tag{3.2}$$

For any given i, Xit is independently distributed, identifiable, and independent across time (t). It is normalized to meet the [42] moment criteria, which do not need strict homogeneity among other things. Additionally, it is distributed independently and identically across people (i) and throughout time (t), and Uit is orthogonal to Xit. Equation (3.1) suggests the following;

$$QY(\tau|Xit) = (i + iq(\tau)) + Xit + Zit\,q(\ ) \tag{3.3}$$

In this empirical framework, the vector of explanatory variables X_{it} consists of the natural logarithmic forms of core research indicators, including logarithmic fisheries production (LFI), logarithmic economic growth (LGDP), logarithmic fossil fuel consumption (LFF), and logarithmic governance quality (LGR). Within the MMQR specification, the quantile fixed effect for each cross-section i is captured by the scalar coefficient X_{it} − i(\tau) i + iq(\tau). The conditional quantile of the dependent variable Y_{it}—namely the natural logarithm of marine waste (LMW)—is expressed as Q_{Y}(\tau|X_{it}), which indicates that the distributional characteristics of marine waste are determined by the variations in the aforementioned independent variables. Distinct from conventional least-squares fixed-effect models, the individual-specific intercept remains constant throughout estimation. The heterogeneous empirical outcomes across quantiles reflect the fluctuating influences of explanatory variables along the conditional distribution of the endogenous variable Y. Notably, these individual heterogeneous components are assumed to be time-invariant. Moreover, Y.q(\tau) denotes the τ-th sample quantile, which can be calculated by solving the optimized objective function presented below:

$$\min_q \sum_i \sum_t P_\tau \left(R_{it} - \left(\delta_i + Z_{it}\gamma'\right) q\right) \tag{3.4}$$

Where $P_\tau(A) = (\tau - 1)\,AI\{A \le 0\} + TAI\{A > 0\}$ denotes the check function.

3.4 Fisheries Work Environment Risk Analysis

Before the empirical estimation of unknown parametric coefficients, a series of preliminary statistical tests was performed to examine the intrinsic characteristics of the panel dataset. Firstly, this research conducted a cross-sectional dependence

Table 3.3 Cross-sectional dependence results

Variables	FI	FP	GDP	FF	GR
Pesaran (2005)	(97.71)***	(18.82)***	(83.55)***	(65.35)***	(31.30)***

– Remark: *** refer importance at the 1%, scale
– Source: [34]

(CD) test to detect potential cross-sectional correlation across the analyzed variables. According to [44], unaddressed cross-sectional dependence can generate biased coefficient estimations and undermine the efficiency of panel data analysis, as unobserved common factors may interfere with true parametric values. Therefore, rigorous preliminary testing is essential to ensure estimation accuracy. This study adopts the cross-sectional dependence (CD) test proposed by [45] to evaluate cross-sectional correlation among the selected variables. As presented in Table 3.3, the empirical results verify the presence of significant cross-sectional dependence across most variables, with the exception of fisheries production. To effectively mitigate statistical errors caused by cross-sectional correlation, this research employs robust panel unit root tests and cointegration procedures that are capable of accommodating cross-sectional dependence issues [46, 47].

To accurately identify the statistical properties and integration characteristics of core research variables, two widely adopted panel unit root tests were utilized in this study: the [48] (LLC) test developed in 2002 and the [49] test proposed in 2003. Panel unit root testing integrates cross-sectional and time-series information to improve the overall statistical efficiency of data examination. These testing methods are constructed to verify the null hypothesis of unit root existence against corresponding alternative hypotheses. Specifically, LLC proposed a homogeneous alternative hypothesis, which assumes that all panel series are stationary with identical convergence rates, thereby statistically rejecting the null hypothesis of unit root presence. In comparison, the IPS test relaxes the homogeneous restriction, allowing each cross-sectional unit to possess independent autoregressive parameters and hypothesizing that not all panel series contain non-stationary traits. The LLC test is incorporated in this research due to its reliable statistical performance for small sample datasets, which matches the data characteristics of the current study. Combined with the earlier CD test outcomes, the two unit root tests were selected to effectively account for cross-sectional dependence interference during stationarity examination. As summarized in Table 3.4, the empirical results demonstrate that all researched variables exhibit non-stationary characteristics at the level form but become stationary after first-order differentiation. Collectively, every variable incorporated in the regression model follows an I(1) integration order.

Table 3.5 presents the empirical outputs of the [35] panel cointegration test and the [50] bootstrapped panel cointegration test, which are jointly employed to verify the presence of long-term equilibrium correlations among the analyzed variables. Developed based on the foundational two-step cointegration framework proposed by Engle and Granger, [35] constructed an extended panel cointegration methodology. This empirical approach accommodates cross-sectional

Table 3.4 Panel unit root test results

Variable	Difference		First difference	
	LLC	IPS	LLC	IPS
IF	2.99	10.559	−8.170***	−13.160***
FP	−3.076	−3.700	−24.136***	−22.737***
GDP	−1.571	−0.210	−18.74***	−68.362***
FF	−9.255	−4.970	−20.610***	−21.409***
GR	−1.090	−0.254	−23.471***	−23.479***

– Remark: *** refer importance at the 1%, scale
– Levin, Lin, & Chu test (LLC), and Im, Pesaran, and Shin W-stat test (IPS)
Source: [34]

Table 3.5 Panel cointegration test for EU countries

Test	Without trend	With trend
A). Pedroni residual cointegration test:		
Alternative hypothesis: common AR coefficients (within dimension):		
Panel v-statistic	19.354 (0.000)***	25.170 (0.000)***
Panel rho-statistic	−0.227 (0.589)	−1.078 (0.859)
Panel PP-statistic	−7.456*** (0.000)	−6.284*** (0.000)
Panel ADF-statistic	−7.784*** (0.000)	−6.637*** (0.000)
Alternative hypothesis: common AR coefficients (between dimensions)		
Group rho-statistic	2.840	0.997
Group PP-statistic	−8.585***	(0.000)
Group ADF-statistic	−8.287***	(0.000)
B). Westerlund (2007) Bootstrap panel cointegration:		

– Remark: ***, **, and * refer importance at the 1%, 5%, and 10% scales respectively
– Values in parentheses are *p*-values
– Source: [34]

heterogeneity by eliminating short-term variable fluctuations and capturing individual-specific evolutionary features within the initial estimation stage. Relying on estimated residual sequences, [35] established seven distinct statistical test indicators. As displayed in Table 3.5, these statistics are categorized into two major groups: pooled within-dimension tests and grouped between-dimension tests. Furthermore, the [50] cointegration method conducts four statistical tests under the null hypothesis of no cointegration. Different from traditional residual-based cointegration techniques, this approach removes conventional constraint limitations on common factors, considering both fundamental and residual structural changes within panel data. According to [51], invalid common factor constraints severely weaken the robustness of residual-based cointegration examinations; hence, incorporating flexible structural dynamic adjustments is essential for reliable cointegration verification.

Table 3.6 Model 1. Panel estimation for the EU-27 region from 1990 to 2022

Long-run coefficient	DOLS		FMOLS		FEOLS	
	Coefficient	Std. error	Coefficient	Std. error	Coefficient	Std. error
FP	0.074	(0.045)	0.041	(0.010)***	0.048	(0.012)***
GDP	−1.081	(0.240)***	−0.189	(0.085)***	−0.160	(0.010)***
FF	0.290	(0.025)***	0.116	(0.071)***	0.118	(0.090)***
GR	−0.842	(0.220)***	−0.085	(0.046)*	−0.072	(0.060)

– Note: ***, **, and * indicate significance at the 1%, 5%, and 10% levels respectively
– Values in parentheses are P-values
Source: [34]

Table 3.6 summarizes the estimation outcomes of the three panel regression techniques, namely FEOLS, FMOLS, and DOLS, with observable disparities in statistical significance across estimated coefficients. The three estimators yield relatively similar coefficient magnitudes for the baseline model, despite minor differences in significance levels. Consistent with theoretical expectations, fisheries production exhibits a statistically positive association with occupational mortality across EU member states. Nevertheless, the magnitude of this relationship varies across estimation specifications; specifically, a 1% increase in fisheries production raises occupational mortality rates by 0.041% (FEOLS) and 0.048% (FMOLS). These empirical outcomes align with the prior research findings documented by [1, 2, 5, 6]. Unregulated fishing behaviors, including illegal, unreported, and unregulated (IUU) fishing, overfishing, unauthorized coastal fishing, and the utilization of prohibited fishing gear, breach both legal regulations and ethical standards. Such unsustainable fishing practices degrade marine ecosystems and undermine the long-term stability of the fisheries industry. Among all explanatory variables, fossil fuel consumption demonstrates the most stable coefficient magnitude and consistent statistical significance throughout the EU sample. The empirical results indicate that a 1% rise in vessel fossil fuel consumption increases fisheries occupational mortality by 0.118%, 0.116%, and 0.290% under FEOLS, FMOLS, and DOLS estimations, respectively. These findings are consistent with the conclusions of [7–9, 52, 53]. Fluctuating fuel prices impose substantial economic pressure on fishing enterprises, elevating operational expenditures related to labor compensation, vessel maintenance, and loan repayment obligations. To sustain revenue margins under high fuel costs, fishermen may resort to hazardous and unsustainable fishing behaviors, which further elevate on-site occupational risks. Conversely, economic growth exerts a significant negative influence on fisheries occupational mortality across the EU region. The estimation results reveal that a 1% increase in economic growth reduces mortality rates by 0.16%, 0.18%, and 0.08% for FEOLS, FMOLS, and DOLS models, respectively. This statistical evidence validates the growth hypothesis for fishery-dependent economies. The current study replicates the core conclusions of [19–21], emphasizing that standardized ethical fishing practices enhance long-term economic sustainability and operational viability within the fisheries sector.

Moreover, the governance indicator presents a statistically negative impact on occupational mortality. Empirically, a 1% improvement in governance quality reduces fisheries fatality rates by 0.22% (FEOLS) and 0.04% (FMOLS). Although FMOLS and FEOLS provide solid statistical support for the growth hypothesis, the DOLS estimation fails to generate statistically significant outcomes for governance. These results are consistent with prior scholarly works by [16, 18, 54, 55]. The literature highlights that the fisheries sector—spanning small-scale artisanal operations to large commercial fisheries—possesses intricate institutional and ethical characteristics that are heavily governed by regulatory frameworks. Table 3.7 presents the panel quantile regression outputs, illustrating the heterogeneous nexus between fisheries expansion and occupational mortality. Fisheries production significantly increases mortality rates at the 3rd, 4th, 5th, 8th, and 9th quantiles, while insignificant correlations are observed at the 1st and 2nd quantiles. This heterogeneous distribution indicates that economies located within lower quantiles prioritize fishery output expansion over occupational safety protocols. Additionally, countries below the median quantile require external financial and technical assistance to upgrade industrial safety standards, owing to the high capital investment required for safety infrastructure optimization. Fossil fuel consumption maintains a strong positive correlation with occupational mortality across most quantiles, with the exception of the 6th, 7th, and 8th quantiles. Furthermore, the marginal impact of fossil fuel usage gradually weakens at higher quantile levels [10]. Similarly, economic growth yields significant negative effects on mortality rates for the majority of quantiles, with non-significant coefficients recorded at the 6th, 7th, and 9th quantiles. This empirical heterogeneity implies that marine environmental regulations can facilitate economic progress for countries with intensive marine pollutant emissions. Lastly, governance exhibits a consistent negative and statistically significant influence on marine occupational hazards across all quantiles (1st to 9th), particularly within high-risk fishery regions. This phenomenon can be interpreted through the modern institutional perspective, which explains the complex interactive relationship between governance efficiency and fishery safety regulation. The empirical evidence reinforces the arguments proposed by [13, 15], stating that effective governmental supervision compels the fisheries industry to balance blue economic growth with standardized occupational safety and quality management protocols.

Table 3.8 presents the estimation outputs of FEOLS, FMOLS, and DOLS for Model 2, focusing on the empirical analysis of developed EU14 economies. Although slight discrepancies in statistical significance exist across estimators, the three models generate generally comparable coefficient magnitudes. Within the EU14 developed sample, fisheries production and economic growth display the most stable statistical performance in terms of coefficient magnitude and significance level. The regression outcomes indicate that a 1% increase in fisheries output reduces occupational mortality by 0.15%, 0.14%, and 0.13% under FEOLS, FMOLS, and DOLS estimation techniques, respectively. These empirical findings are consistent with the scholarly conclusions documented by [3, 5]. IUU fishing activities, including overfishing, unauthorized exploitation of marine protected areas, and the deployment of prohibited fishing equipment, violate legal constraints and ethical norms. Such destructive fishing behaviors deteriorate marine ecological environments and hinder the sustainable development of commercial fisheries.

Table 3.7 Panel quantile estimation results for EU27 countries

Variables	Method of moments quantile regression with fixed effects										
	Location	Scale	Quantiles								
			0.1	0.2	0.3	0.4	0.5	0.6	0.7	0.8	0.9
FP	(0.012)***	(0.013)***	(0.016)	(0.010)	(0.020)**	(0.020)***	(0.013)*	(0.008)	(0.009)	(0.031)***	(0.040)***
GDP	(−0.010)***	(0.018)*	(−0.240)***	(−0.211)***	(−0.193)***	(−0.180)***	(−0.178)***	(−0.097)	(−0.079)	(−0.073)***	(−0.019)
FF	(0.090)***	(0.066)***	(0.096)***	(0.086)***	(0.084)***	(0.057)***	(0.035)***	(0.018)	(0.003)	(0.012)	(0.032)**
GR	(−0.060)	(0.349)***	(−0.380)***	(−0.575)***	(−0.587)***	(−0.612)***	(0.565)***	(−0.997)**	(−2.143)***	(−2.703)***	(−2.904)***

Notes: *, **, and *** indicate significance at 10%, 5%, and 1% levels, respectively
Source: [34]

Table 3.8 Model 2. Panel estimation for the EU-14 region from 1990 to 2022

Long-run coefficient	DOLS		FMOLS		FEOLS	
	Coefficient	Std. error	Coefficient	Std. error	Coefficient	Std. error
FP	0.138	(0.083)*	0.147	(0.051)**	0.155	(0.034)***
GDP	−0. 267	(0.028)***	−0.293	(0.038)***	−0.106	(0.026)***
FF	0.128	(0.052)***	0.106	(0.018)***	0.283	(0.016)***
GR	−0.468	(0.290)	−0.388	(0.180)**	−0.230	(0.123)*

– Note: ***, **, and * indicate significance at the 1%, 5%, and 10% levels respectively
– Values in parentheses are P-values
Source: [34]

Fossil fuel consumption retains a positive and statistically significant correlation with occupational mortality within the EU14 developed economies. The empirical coefficients demonstrate that a 1% increase in vessel fossil fuel consumption elevates fatality rates by 0.28%, 0.10%, and 0.12% according to FEOLS, FMOLS, and DOLS estimations. These results align with prior research conducted by [10, 12]. Government fuel subsidies tied to vessel tonnage or fossil fuel consumption may incentivize excessive fishing intensity and environmentally unfriendly fishing practices. Such subsidy policies potentially lower industrial safety standards, thereby elevating occupational injury and fatality risks for onboard fishermen.

Furthermore, economic growth exerts a detectable negative influence on fisheries occupational mortality across the EU14 developed nations. The statistical results show that a 1% economic growth rate corresponds to mortality fluctuations of 0.02%, 0.03%, and 0.26% in FEOLS, FMOLS, and DOLS models, respectively, thereby validating the growth hypothesis for developed EU fishery economies. This empirical evidence supports the research outcomes of [23], who confirmed the robust negative association between economic expansion and fishery mortality. The findings are also consistent with the conclusions of [22]. Notably, economic growth is jointly determined by global market fluctuations, technological innovation, and governmental regulatory policies; hence, isolating the standalone influence of fishery mortality on macroeconomic development remains empirically challenging. Similarly, governance quality yields a significant negative impact on occupational mortality within the EU14 sample. This empirical outcome contradicts the findings of [14] but supports the theoretical utility hypothesis proposed by [15]. Specifically, FEOLS and FMOLS results indicate that a 1% improvement in governance reduces fishery mortality by 0.12% for both estimation methods. While FEOLS and FMOLS provide strong statistical validation for the utility hypothesis, the DOLS estimator only generates marginally significant evidence. Despite prevalent industrial challenges including overfishing, marine habitat degradation, climate pressure, and illegal harvesting, the fisheries sector possesses substantial transformative potential. Strict ethical constraints and optimized governance frameworks can effectively promote equitable, environmentally friendly, and sustainable development within the fishing industry. Table 3.9 reports the panel quantile regression results for the EU14 developed economies. Fisheries production exhibits statistically significant effects on occupational mortality from the 1st to the 5th quantiles. This heterogeneous

Table 3.9 Panel quantile estimation results for EU14 countries

Variables	Method of moments quantile regression with fixed effects										
	Location	Scale	Quantiles								
			0.1	0.2	0.3	0.4	0.5	0.6	0.7	0.8	0.9
FP	(0.034)***	(0.012)**	(0.024)**	(0.009)	(0.014)	(0.022)***	(0.011)*	(0.007)	(0.005)	(0.010)*	(0.010)
GDP	(−0.026)***	(0.014)**	(−0.333)***	(−0.322)***	(−0.356)***	(−0.371)***	(−0.386)***	(−0.359)***	(−0.343)***	(−0.295)***	(−0.246)***
FF	(0.016)***	(0.049)**	(0.057)***	(0.026)	(0.013)	(0.020)**	(0.004)	(0.005)	(0.018)**	(0.008)	(0.022)
GR	(−0.123)*	(0.253)*	(−0.155)	(−0.211)	(−0.376)**	(−0.445)***	(−0.598)***	(−0.594)***	(−0.574)***	(−0.531)**	(−0.269)

Notes: *, **, and *** indicate significance at 10%, 5%, and 1% levels, respectively
Source: [34]

empirical outcome reflects the successful implementation of marine strategic policies within developed European fisheries sectors. Significant mortality fluctuations recorded in low and median quantiles validate the applicability of the utility hypothesis for the EU14 economies. The statistical heterogeneity implies that fishery output expansion in developed EU14 nations is not entirely governed by industrial fishery policies. Economies situated in lower quantiles at preliminary development stages require additional institutional and technical support to upgrade fishery safety supervision systems. Moreover, the marginal influence of fisheries production weakens substantially in the highest quantile groups, which can be attributed to the diversified fishing technologies and operational mechanisms adopted across EU14 member states. Economic growth, proxied by GDP, presents statistically significant negative effects on occupational mortality across all quantiles (1st to 9th). This evidence highlights the necessity of sustainable fishery management frameworks to facilitate blue economic growth for countries suffering from severe marine environmental degradation. It further indicates that economic development prioritization generally outweighs occupational safety optimization within the industrial operational framework of developed EU14 nations. In terms of fossil fuel consumption, significant positive correlations between hydrocarbon usage and mortality rates are observed at the 1st, 4th, and 7th quantiles, whereas insignificant coefficients are recorded for the remaining quantile levels. Consistent with empirical characteristics identified in the full EU sample, the marginal mortality contribution of fossil fuels gradually declines at higher quantiles. Finally, governance presents significant negative impacts on fishery mortality from the 3rd to the 8th quantiles within the EU14 dataset. In contrast, governance coefficients are statistically insignificant at the 1st, 2nd, and 9th quantiles, reflecting heterogeneous governance effectiveness across different marine risk levels in developed European economies.

Table 3.10 presents the empirical outputs of FEOLS, FMOLS, and DOLS for Model 3, focusing specifically on the empirical examination of developing EU13 economies. The three panel estimators yield highly consistent coefficient magnitudes with minor numerical discrepancies. The regression results reveal that fisheries production exerts a statistically positive effect on occupational mortality within the EU13 developing sample. Specifically, a 1% expansion in fisheries output elevates fatality rates by 0.04%, 0.15%, and 0.17% under FEOLS, DOLS, and FMOLS estimations, respectively. The robust positive coefficients across all three estimation

Table 3.10 Model 3. Panel estimation for the EU-13 region from 1990–2022

Long-run coefficient	DOLS		FMOLS		FEOLS	
	Coefficient	Std. error	Coefficient	Std. error	Coefficient	Std. error
FP	0.177	(0.044)***	0.157	(0.028)***	0.045	(0.011)***
GDP	−0.011	(0.074)	−0.078	(0.048)***	−0.214	(0.010)***
FF	0.110	(0.029)**	0.016	(0.022)	0.084	(0.009)***
GR	−0.656	(0.1710)**	−0.697	(0.119)***	−0.095	(0.052)

– Note: ***, **, and * indicate significance at the 1%, 5%, and 10% levels, respectively
– Values in parentheses are P-values
Source: [34]

techniques validate the presence of the utility hypothesis among developing EU economies. This empirical consensus aligns with the research conducted by [1, 56], who emphasized that fisheries expansion inevitably increases occupational mortality risks during the preliminary industrial development stage. Likewise, fossil fuel consumption maintains a significant positive correlation with fishery mortality in FEOLS and DOLS specifications. A 1% increase in vessel fossil fuel usage raises occupational mortality by 0.08% (FEOLS) and 0.11% (DOLS). Consistent with [2], such positive linkage can be attributed to vessel heterogeneity and fuel-dependent operational hazards. Developing economies commonly utilize aging and low-efficiency fishing vessels with inadequate safety infrastructure. Additionally, continuous exposure to toxic combustion pollutants imposes chronic health threats to onboard fishermen, further exacerbating occupational vulnerability. Economic growth presents a statistically negative influence on occupational mortality across the EU13 developing nations. Empirically, a 1% increase in GDP corresponds to mortality reductions of 0.21% and 0.07% for FEOLS and FMOLS models. The negative and significant GDP coefficients consistently observed across all estimators further confirm the applicability of the utility hypothesis within developing European fisheries. These findings corroborate the conclusions of [20, 21]. Although standardized ethical fishing practices generate long-term economic sustainability, fishermen in less developed regions are often incentivized to adopt hazardous, unsustainable harvesting methods for immediate short-term economic gains. Moreover, governance quality demonstrates a distinct negative association with occupational mortality throughout the EU13 sample. The estimation outcomes indicate that a 1% improvement in governance efficiency reduces fishery fatalities and accident rates by 0.69% (FMOLS) and 0.65% (DOLS). This evidence supports the perspective proposed by [57], highlighting that sustainable fisheries governance frameworks should prioritize social equity. Effective institutional supervision ensures equitable benefit distribution among small-scale fishermen, coastal residential communities, and seafood consumers, thereby mitigating man-made operational risks. Table 3.11 summarizes the heterogeneous estimation outcomes of the MMQR technique for the EU13 developing economies. Firstly, fisheries production maintains statistically significant positive effects on occupational mortality across all quantiles. The marginal mortality impact of fishery output exhibits an upward quantile trend, with coefficient values rising steadily from 0.068 (1st quantile) to 0.206 (9th quantile). Secondly, fossil fuel consumption retains a robust positive correlation with mortality rates throughout every quantile, indicating that the detrimental influence of hydrocarbon usage intensifies gradually at higher mortality quantiles. Thirdly, governance exerts consistent negative inhibitory effects on occupational mortality across the full conditional distribution, with significant risk-reduction performance observable at upper quantile levels within the EU13 developing panel.

The empirical findings derived from the FEOLS estimation are strongly validated by the complementary outputs of FMOLS and DOLS estimators. All three techniques yield consistent coefficient signs and statistical significance levels, confirming the statistical reliability and inferential validity of the FMOLS and DOLS empirical results. Although minor disparities in significance thresholds are detected across models, the directional consistency of coefficients remains intact. Notably, the DOLS

Table 3.11 Panel quantile estimation results for EU13 countries

Variables	Method of moments quantile regression with fixed effects										
	Location	Scale	Quantiles								
			0.1	0.2	0.3	0.4	0.5	0.6	0.7	0.8	0.9
FP	(0.011)***	(0.108)***	(0.068)*	(0.145)***	(0.096)***	(0.128)***	(0.191)***	(0.190)***	(0.190)***	(0.205)***	(0.206)***
GDP	(−0.010)***	(−0.096)	(−0.152)	(−0.020)	(−0.181)***	(−0.109)	(−0.035)	(−0.035)	(−0.021)	(−0.037)	(−0.010)
FF	(0.009)***	(0.094)***	(0.094)***	(0.066)***	(0.115)***	(0.119)***	(0.122)***	(0.123)***	(0.125)***	(0.092)**	(0.103)**
GR	(−0.052)	(−1.059)**	(−1.143)***	(−0.639)	(−0.362)***	(−0.485)***	(−0.714)***	(−0.728)***	(−0.708)***	(−0.783)***	(−0.720)***

Notes: *, **, and *** indicate significance at 10%, 5%, and 1% levels, respectively
Source: [34]

estimator exhibits superior robustness by eliminating endogenous bias and serial correlation interference. To conduct a comparative empirical analysis based on economic development heterogeneity, the full EU27 sample was categorized into 14 developed economies and 13 developing economies. Tables 3.8 and 3.9 present the empirical nexus between fisheries expansion and occupational mortality for the EU14 developed group, while Tables 3.10 and 3.11 illustrate the corresponding relationships within the EU13 developing group. The comparative statistical evidence extracted from Table 3.8 and Table 3.10 confirms that fisheries production expansion exerts a significant impact on occupational mortality across both sub-samples. Developed EU14 economies gain greater safety benefits from fishery expansion relative to developing EU13 nations, reflected by lower mortality risks. Specifically, the average impact magnitude stands at −0.155 for EU14 countries and 0.045 for EU13 economies. This evident divergence indicates that developed European member states effectively mitigate unsustainable fishing behaviors through rigorous industrial safety and quality control protocols, whereas developing economies lack such regulatory advantages. Furthermore, fossil fuel consumption generates stronger mortality-induced effects within the developed EU14 sample compared with the EU13 developing group. The average coefficient magnitude of fossil fuel usage is recorded at 0.283 and 0.084 for EU14 and EU13 nations, respectively. This empirical distinction demonstrates that developed EU economies enforce stricter regulatory constraints on vessel fossil fuel combustion, establishing more sophisticated sustainable fishery management frameworks than their developing counterparts. Economic growth presents a stronger negative mitigating effect on occupational mortality within the EU14 developed panel. The average magnitude of economic growth is −0.214 for EU14 economies and −0.106 for EU13 economies. Such comparative outcomes emphasize that sustainable fishery management mechanisms in high-income European countries deliver more substantial mortality-reduction effects than those implemented in developing member states. Similarly, governance quality yields consistent negative inhibitory impacts on occupational mortality across both sub-samples, with a more pronounced mitigating effect observed in developed economies. The average governance coefficient is calculated at −0.230 for EU14 nations and −0.095 for EU13 nations. The empirical disparity implies that institutional optimization generates more prominent safety improvements in developed countries. This comparative advantage originates from streamlined fiscal policies in advanced European economies, including rationalized taxation systems and targeted governmental subsidies, which are specifically designed to promote sustainable fishery governance.

3.5 Conclusion and Solutions for Healthier Fisheries Work Environment

This empirical research primarily explores the causal relationships between fisheries production, fossil fuel consumption, economic growth, governance quality, and occupational mortality, while systematically verifying the applicability of the utility hypothesis within the European fisheries context. The study contributes to the existing literature on sustainable fishery management, with a particular emphasis on marine resource conservation and marine environmental sustainability under economic development frameworks. Based on a 30-year balanced panel dataset covering 27 European Union economies, this research implemented multiple econometric procedures, including cross-sectional dependence tests, panel unit root examinations, and cointegration tests. Governance indicators were incorporated as institutional proxy variables to control for marine operational risks. All researched variables conform to an I(1) integration order, and cointegration validation confirms the existence of stable long-run equilibrium relationships among parameters. Conventional panel regression techniques reveal heterogeneous coefficient significance across models; therefore, the MMQR estimation approach was adopted to ensure empirical robustness. The MMQR method effectively captures the distributed heterogeneous influences of explanatory variables across different conditional quantiles of occupational mortality, marine fatality rates, and marine ecological unsustainability. Collectively, the FEOLS, FMOLS, and DOLS estimation outcomes statistically validate the utility hypothesis for the full EU27 sample. The MMQR empirical results demonstrate that fisheries expansion exerts significant positive effects on occupational mortality across selected quantiles within the EU27 panel. Significant positive correlations are identified from the 3rd to 5th quantiles and from the 8th to 9th quantiles of the mortality conditional distribution. Fossil fuel consumption maintains robust positive associations with fatal fishing accidents predominantly within lower-to-medium quantiles (1st to 5th quantiles), reflecting higher fuel-induced operational risks during preliminary industrial development stages. In contrast, governance quality consistently reduces occupational mortality throughout the entire quantile distribution (1st to 9th quantiles). Economically advanced EU nations exhibit higher economic integration and diversified industrial structures, while developing European economies remain heavily reliant on fishery imports accompanied by elevated marine operational hazards. Accordingly, fisheries production constitutes a fundamental component of marine risk accumulation within developed economies.

Visual interpretation of MMQR outputs in Table 3.7 indicates analogous distributed dynamics between economic growth and occupational mortality. The strongest mitigating coefficient of economic expansion is observed at the 8th quantile. The marginal influence of fisheries production intensifies gradually from lower to upper quantiles, signifying that marine risk exposure reaches its peak when fishery output and economic expansion exert the most substantial mortality impacts. Conversely, the detrimental effects of fossil fuel consumption are weakest at the lowest risk quantiles. This empirical heterogeneity implies that nations with low fossil fuel dependence in fisheries generally exhibit higher overall fossil fuel consumption intensity. The universal negative governance effect across all quantiles can be attributed to standardized industrial safety regulations, which constrain uncontrolled fisheries expansion and mitigate occupational hazards. Developed EU14 economies prioritize industrial safety and quality standards to a greater extent than high-risk EU13 developing nations. To maintain public credibility and political legitimacy, governments in advanced economies actively implement targeted policies to reduce fishery fatalities. Nevertheless, the highest mortality quantiles within developed EU14 fisheries still face persistent operational challenges, necessitating continuous optimization of risk supervision mechanisms. Near-median quantiles exhibit significant positive correlations between fishery output and mortality rates, as environmental hazards generate immediate tangible impacts on public well-being. Consequently, both citizens and policymakers prioritize risk-adaptive economic development within fishery-dependent industries. Owing to inefficient resource management in low-quantile risk groups, the mortality-reduction effects of governance remain insignificant for lower-quantile observations across both EU14 and EU13 sub-samples. Significant negative governance impacts observed at upper quantiles confirm the maturity and optimization of fishery regulatory systems within developed European economies. Consistent with [17, 18], institutional governance improvements effectively alleviate marine environmental risks; hence, intensified governance intervention is strongly recommended for high-mortality fishery sectors, particularly within the EU14 region.

Economic growth yields significant negative mortality-reduction effects from the 1st to 5th quantiles, whereas fossil fuel consumption generates positive mortality impacts within the same quantile interval. These empirical outcomes align with the established progress hypothesis in existing marine economic literature. The quantile-based heterogeneity can be explained through industrial evolution mechanisms. At lower quantiles, rapid fishery output expansion drives export-oriented economic growth, increasing industrial dependence on fishery commodities. At higher quantiles, labor-intensive and export-driven fishery industries gradually relocate from high-regulatory EU14 economies to less-restrictive EU13 nations, thereby reducing mortality pressure on developed economies and shifting economic dependence toward importation. Advanced safety and quality standards in technologically mature EU14 nations enforce industrial upgrading; however, stringent regulatory compliance and technological renovation costs incentivize industrial relocation toward regions with lower taxation and weaker environmental constraints. Fossil fuel consumption universally influences all industrial sectors, thereby maintaining positive correlations with occupational mortality across most quantiles. The marginal mortality impact of fishery production reaches its peak at upper quantiles and its minimum at lower quantiles. Improved fossil fuel utilization efficiency simultaneously boosts fishery productivity and energy-related fatality risks in high-development regions. Advanced economies within the EU14 possess sufficient capital to finance energy transition costs from traditional fossil fuels to renewable green energy. Nevertheless, both developed and developing European nations remain reliant on conventional energy sources for fishery production. To reduce fishery mortality rates, EU14 economies must accelerate the elimination of fossil fuel-dependent fishing operations and promote clean energy transformation within the marine industry. This study contains identifiable limitations and provides viable directions for future research. Firstly, additional macroeconomic variables, including tourism development, globalization intensity, and industrial commercialization, can be incorporated to enrich the analytical framework of marine environmental degradation and fishery risk research. Secondly, subsequent studies may adopt advanced econometric techniques to conduct disaggregated country-specific evaluations or expand research samples covering ASEAN, BRICS, G7, and N11 economies. Constrained by data accessibility, this research fails to quantify the marginal benefits of marine hazard mitigation, requiring further detailed investigation. Furthermore, the empirical scope can be expanded to diverse geographical regions for comprehensive comparative analysis. Despite adopting contemporary econometric methodologies, this study retains inherent limitations: the empirical dataset spans from 1990 to 2022 with a constrained variable selection scope. For future research, scholars are recommended to investigate the interactive effects of public–private cooperation and ocean energy utilization on fishery productivity based on the existing dataset. Moreover, the current analytical framework centered on EU economies can be extended to high-pollution global economies. Lastly, future academic works should conduct comparative economic cost assessments of marine pollution across distinct economic cycles to capture structural data mutations and improve empirical comprehensiveness.

References

1. Thorvaldsen T, Sønvisen SA, Holmen IM (2022) The impact of fisheries management on fishers' health and safety: a case study from Norway. Mar Policy 140:105066. https://doi.org/10.1016/j.marpol.2022.105066
2. Barrow A et al (2022) Epidemiology of occupational hazards and injuries among fishermen at Tanji fishing site in The Gambia: an analytical cross-sectional study design. Environ Health Insights 16:11786302221088699. https://doi.org/10.1177/11786302221088699
3. Kim JH, Vaughan A, Kincl L (2023) Characterization of musculoskeletal injury risk in dungeness crab fishing. J Agromedicine 28(2):309–320. https://doi.org/10.1080/1059924X.2022.2068715
4. Siame S, Bygvraa DA, Jensen OC (2022) Facilitators and barriers to implementing occupational safety interventions in the fishing industry: a scoping review. Saf Sci 145:105512. https://doi.org/10.1016/j.ssci.2021.105512
5. Guillot-Wright S, Cherryhomes E, Davis L (2022) The impact of economic deregulation for health disparities among Gulf of Mexico commercial fishermen. Mar Policy 141:105099. https://doi.org/10.1016/j.marpol.2022.105099
6. Shan D (2022) Enforcement of fishing Occupational Health and Safety (OHS) standards: challenges in Atlantic Canada. Mar Policy 145:105282. https://doi.org/10.1016/j.marpol.2022.105282
7. Perera F, Nadeau K (2022) Climate change, fossil-fuel pollution, and children's health. N Engl J Med 386(24):2303–2314. https://doi.org/10.1056/NEJMra2117706
8. Arafat WMGM, Haq IU, Mehmed B, Abbas A, Gamage SKN, Gasimli O (2022) The causal nexus among energy consumption, environmental degradation, financial development and health outcome: empirical study for Pakistan. Energies 15(5):1859. https://doi.org/10.3390/en15051859
9. Trowbridge J, Goin DE, Abrahamsson D, Sklar R, Woodruff TJ (2023) Fossil fuel is the common denominator between climate change and petrochemical exposures, and effects on women and children's health. Int J Gynecol Obstet 160(2):368–371. https://doi.org/10.1002/ijgo.14408

10. Safdar N, Liaquat M, Iqbal S, Ambreen A (2021) Industrialization, greenhouse gas emission and its health outcomes in South Asian countries. Multicult Educ 7(10) [Online]. https://www.researchgate.net/profile/Sidra-Iqbal-15/publication/356280369_Multicultural_Education_Industrialization_Greenhouse_Gasemissionand_itsHealth_Outcomes_in_South_Asian_Countries/links/6194538f61f0987720a5dabb/Multicultural-Education-Industrialization-Greenhouse-Gasemissionand-itsHealth-Outcomes-in-South-Asian-Countries.pdf. Accessed 20 May 2026
11. Oyedele O (2022) Carbon dioxide emission and health outcomes: is there really a nexus for the Nigerian case? Environ Sci Pollut Res 29(37):56309–56322. https://doi.org/10.1007/s11356-022-19365-x
12. Romanello M et al (2022) The 2022 report of the Lancet Countdown on health and climate change: health at the mercy of fossil fuels. Lancet 400(10363):1619–1654
13. Weng YH, Hirata Y (2022) Design-centered HRI governance for healthcare robots. J Healthc Eng 2022:1–8. https://doi.org/10.1155/2022/3935316
14. Sheehan MC, Khudairi F, Swaich GS, Hines W, Mehta S, Fox MA (2022) Urban climate-health governance: charting the role of public health in large global city adaptation plans. PLOS Clim 1(3):e0000012. https://doi.org/10.1371/journal.pclm.0000012
15. Zaratin P et al (2022) The MULTI-ACT model: the path forward for participatory and anticipatory governance in health research and care. Health Res Policy Sys 20(1):22. https://doi.org/10.1186/s12961-022-00825-2
16. Kickbusch I, Liu A (2022) Global health diplomacy—reconstructing power and governance. Lancet 399(10341):2156–2166. https://doi.org/10.1016/S0140-6736(22)00583-9
17. Lanford D, Petiwala A, Landers G, Minyard K (2022) Aligning healthcare, public health and social services: a scoping review of the role of purpose, governance, finance and data. Health Soc Care Comm 30(2):432–447. https://doi.org/10.1111/hsc.13374
18. Such E, Smith K, Woods HB, Meier P (2022) Governance of intersectoral collaborations for population health and to reduce health inequalities in high-income countries: a complexity-informed systematic review. Int J Health Policy Manag 11(12):1. https://doi.org/10.34172/ijhpm.2022.6550
19. Pega F, Liu SY, Walter S, Pabayo R, Saith R, Lhachimi SK (2017) Unconditional cash transfers for reducing poverty and vulnerabilities: effect on use of health services and health outcomes in low- and middle-income countries. Cochrane Database Syst Rev 2020(4):CD011135. https://doi.org/10.1002/14651858.CD011135.pub2
20. Woolley KE et al (2021) Effectiveness of interventions to reduce household air pollution from solid biomass fuels and improve maternal and child health outcomes in low- and middle-income countries: a systematic review protocol. Syst Rev 10(1):33. https://doi.org/10.1186/s13643-021-01590-z
21. Bellerose M, Collin L, Daw JR (2022) The ACA medicaid expansion and perinatal insurance, health care use, and health outcomes: a systematic review: systematic review examines the effects of expanding Medicaid on insurance coverage, health care use, and health outcomes during preconception, pregnancy, and postpartum. Health Aff 41(1):60–68. https://doi.org/10.1377/hlthaff.2021.01150
22. Akanni L, Lenhart O, Morton A (2022) Income trajectories and self-rated health status in the UK. SSM Popul Health 17:101035. https://doi.org/10.1016/j.ssmph.2022.101035
23. Văidean VL, Achim MV (2022) When more is less: do information and communication technologies (ICTs) improve health outcomes? An empirical investigation in a non-linear framework. Socio Econ Plan Sci 80:101218. https://doi.org/10.1016/j.seps.2021.101218
24. Badida P, Krishnamurthy A, Jayaprakash J (2023) Meta analysis of health effects of ambient air pollution exposure in low- and middle-income countries. Environ Res 216:114604. https://doi.org/10.1016/j.envres.2022.114604
25. Van Der Zwet A, Connolly J, Huggins C, McAngus C (2023) Network resilience and EU fisheries policy engagement in third countries: lessons for post-Brexit governance. Br J Polit Int Rel 25(1):121–138. https://doi.org/10.1177/13691481211067146

26. Connolly J, Zwet AVD, Huggins C, McAngus C (2022) The governance capacities of Brexit from a Scottish perspective: the case of fisheries policy. Public Policy Adm 37(3):342–362. https://doi.org/10.1177/0952076720936328
27. Drakopulos L (2022) Privatizing the fisheries observer industry: neoliberal science and policy in the U.S. West Coast fisheries. Geoforum 131:116–125. https://doi.org/10.1016/j.geoforum.2022.03.004
28. Apergis N, Payne JE (2010) The emissions, energy consumption, and growth nexus: evidence from the commonwealth of independent states. Energy Policy 38(1):650–655
29. Dogan E, Seker F (2016) Determinants of CO2 emissions in the European Union: the role of renewable and non-renewable energy. Renew Energy 94:429–439. https://doi.org/10.1016/j.renene.2016.03.078
30. Ehigiamusoe KU, Lean HH, Smyth R (2020) The moderating role of energy consumption in the carbon emissions-income nexus in middle-income countries. Appl Energy 261:114215. https://doi.org/10.1016/j.apenergy.2019.114215
31. Anwar A, Siddique M, Dogan E, Sharif A (2021) The moderating role of renewable and non-renewable energy in environment-income nexus for ASEAN countries: evidence from Method of Moments Quantile Regression. Renew Energy 164:956–967
32. World Bank Fish to 2030: prospects for fisheries and aquaculture
33. European Commission. Joint Research Centre (2022) Clean Energy Technology Observatory, Ocean energy in the European Union: status report on technology development, trends, value chains and markets: 2022. LU: Publications Office. [Online]. https://data.europa.eu/doi/10.2760/162254. Accessed 30 Dec 2025
34. Alsaleh M, Xia T, Abdul-Rahim AS (2025) From climate change to occupational risks: the vulnerability of the fisheries sector to climatic accidents. Isr J Aquacult Bamidgeh 77(3):267–285. https://doi.org/10.46989/001c.144566
35. Pedroni P (2004) Panel cointegration: asymptotic and finite sample properties of pooled time series tests with an application to the PPP hypothesis. Econ Theory 20(03):597–625. https://doi.org/10.1017/S0266466604203073
36. Cerrato M, Sarantis N (2002) The cross sectional dependence puzzle. London Guildhall University. [Online]. http://fmwww.bc.edu/RePEc/res2002/Cerrato.pdf. Accessed 20 May 2026
37. Andrews DWK (2005) Cross-section regression with common shocks. Econometrica 73(5):1551–1585. https://doi.org/10.1111/j.1468-0262.2005.00629.x
38. Kao C, Chiang MH (2000) On the estimation and inference of a cointegrated regression in panel data. In: Advances in econometrics, vol 15. Bingley, Emerald (MCB UP), pp 179–222. https://doi.org/10.1016/S0731-9053(00)15007-8
39. Sarkodie SA, Strezov V (2019) Effect of foreign direct investments, economic development and energy consumption on greenhouse gas emissions in developing countries. Sci Total Environ 646:862–871. https://doi.org/10.1016/j.scitotenv.2018.07.365
40. Koenker R, Bassett G (1978) Regression quantiles. Econometrica 46(1):33. https://doi.org/10.2307/1913643
41. Binder M, Coad A (2011) From Average Joe's happiness to Miserable Jane and Cheerful John: using quantile regressions to analyze the full subjective well-being distribution. J Econ Behav Organ 79(3):275–290. https://doi.org/10.1016/j.jebo.2011.02.005
42. Machado JAF, Santos Silva JMC (2019) Quantiles via moments. J Econ 213(1):145–173. https://doi.org/10.1016/j.jeconom.2019.04.009
43. Canay IA (2011) A simple approach to quantile regression for panel data. Econ J 14(3):368–386
44. Phillips PCB, Sul D (2003) Dynamic panel estimation and homogeneity testing under cross-sectional dependence. Econ J 6(1):217–259. https://doi.org/10.1111/1368-423X.00108
45. Pesaran MH (2021) General diagnostic tests for cross-sectional dependence in panels. Empir Econ 60(1):13–50. https://doi.org/10.1007/s00181-020-01875-7
46. Breitung J, Pesaran MH (2008) Unit roots and cointegration in panels. In: Mátyás L, Sevestre P (eds) The econometrics of panel data, Advanced studies in theoretical and applied econometrics, vol 46. Springer Berlin Heidelberg, Berlin, Heidelberg, pp 279–322. https://doi.org/10.1007/978-3-540-75892-1_9

47. Chudik A., Pesaran H (2015) Large panel data models with cross-sectional dependence: a survey. [Online]. https://academic.oup.com/edited-volume/41981/chapter/355335206?searchresult=1&utm_source=TrendMD&utm_medium=cpc&utm_campaign=Oxford_Academic_Books_TrendMD_0. Accessed 20 May 2026
48. Levin A, Lin CF, James Chu CS (2002) Unit root tests in panel data: asymptotic and finite-sample properties. J Econ 108(1):1–24. https://doi.org/10.1016/S0304-4076(01)00098-7
49. Im KS, Pesaran MH, Shin Y (2003) Testing for unit roots in heterogeneous panels. J Econ 115(1):53–74. https://doi.org/10.1016/S0304-4076(03)00092-7
50. Westerlund J (2007) Testing for error correction in panel data*. Oxf Bull Econ Stat 69(6):709–748. https://doi.org/10.1111/j.1468-0084.2007.00477.x
51. Kremers JJM, Ericsson NR, Dolado JJ (1992) The power of cointegration tests. Oxf Bull Econ Stat 54(3):325–348. https://doi.org/10.1111/j.1468-0084.1992.tb00005.x
52. Balsalobre-Lorente D, Shahbaz M, Roubaud D, Farhani S (2018) How economic growth, renewable electricity and natural resources contribute to CO2 emissions? Energy Policy 113:356–367. https://doi.org/10.1016/j.enpol.2017.10.050
53. Bekun FV, Alola AA, Sarkodie SA (2019) Toward a sustainable environment: nexus between CO2 emissions, resource rent, renewable and nonrenewable energy in 16-EU countries. Sci Total Environ 657:1023–1029. https://doi.org/10.1016/j.scitotenv.2018.12.104
54. Leite MC, Weidmann J (1999) Does mother nature corrupt? Natural resources, corruption, and economic growth. International Monetary Fund. [Online]. https://books.google.com/books?hl=en&lr=&id=UFYYEAAAQBAJ&oi=fnd&pg=PA3&dq=Leite,C.,+Weidmann,+J.+(1999).+Does+Mother+Nature+Corrupt:+Natural+Resources,+Corruption,+and+Economic+Growth+IMF+Working+Papers.+International+Monetary+Fund.&ots=o29DeYoDjX&sig=TVLC-xIqLLL9VCD6nJ30s1TCTho. Accessed 20 May 2026
55. Mehlum H, Moene K, Torvik R (2006) Institutions and the resource curse. Econ J 116(508):1–20. https://doi.org/10.1111/j.1468-0297.2006.01045.x
56. Flaten O, Lien G, Koesling M, Løes AK (2010) Norwegian farmers ceasing certified organic production: characteristics and reasons. J Environ Manag 91(12):2717–2726. https://doi.org/10.1016/j.jenvman.2010.07.026
57. Isham J (2005) The varieties of resource experience: natural resource export structures and the political economy of economic growth. World Bank Econ Rev 19(2):141–174. https://doi.org/10.1093/wber/lhi010

Chapter 4
Fisheries Operation and Marine Water Pollution in the EU Region

Acronyms

EU	European Union
CO_2	Carbon dioxide emissions
MMQR	Method of moments quantile regression
EUR	Euro
CD	Cross-sectional dependency
FP	Fisheries production
WDI	World Bank Development Indicators
GDP	Gross domestic product
PF	Fossil fuel
WP	Water pollution
DOLS	Dynamic ordinary least squares
FMOLS	Fully modified ordinary least squares
FEOLS	Fixed-effects ordinary least squares
Europa	European data portal
Eurostat	European Commission''s database
LGDP	Natural logarithms of economic growth
LFP	Natural logarithms of fisheries production
LFF	Natural logarithms of fossil fuel
LLC	Lin & Chu Test
IPPS	Im, Pesaran, and Shin Test
EU27	27 European Union Member States
EU13	13 European Union Member States
BRICS	Brazil, Russia, India, China, South Africa
ASEAN	Association of Southeast Asian Nations
EU14	14 European Union Member States
ETS	Emissions Trading Scheme

M. Alsaleh, *Fisheries Sustainability in the EU*, Urban Sustainability,
https://doi.org/10.1007/978-981-92-4496-6_4

Fisheries-related pollutants, including oil residues, toxic chemical mixtures, aquacultural waste, and plastic debris, gradually dissolve in seawater and deplete dissolved oxygen. Such anthropogenic contamination triggers invisible marine water degradation, severely endangering marine organisms and destabilizing aquatic ecosystems. Based on a panel dataset covering 27 European Union economies from 1990 to 2022, this empirical chapter quantitatively explores the dynamic impacts of fisheries production, fossil fuel consumption, economic growth, and governance quality on marine water pollution. Adopting the fixed-effects method of moments quantile regression (MMQR), the empirical results confirm that fisheries output exerts a statistically significant positive influence on marine water pollution across most conditional quantiles. Specifically, fishery production aggravates water contamination at both lower and upper quantiles, with a stronger marginal impact observed at higher pollution quantiles. Sub-sample evidence further reveals that fisheries activities generate more severe water pollution in developed EU14 economies compared with developing EU13 nations. In contrast, fossil fuel consumption imposes greater water contamination pressure within the EU13 developing group. Economic growth demonstrates consistent pollution-mitigating effects across all quantiles in developed EU14 countries, thereby validating the growth hypothesis for fishery-dependent economies. Within the EU14 sample, governance presents significant negative inhibitory effects on water pollution from the 1st to the 6th quantiles, with statistically significant outcomes recorded at both extreme quantile levels. To alleviate marine water contamination across the European Union, policymakers are recommended to promote eco-friendly fishing apparatus, optimize sustainable harvesting practices, and deploy renewable marine energy technologies such as tidal and wave power across both developed and developing EU member states.

4.1 Fisheries Operation and Water Pollution

The ocean sustains fundamental planetary circulation systems and maintains habitable ecological conditions for human beings. Marine ecosystems provide indispensable natural resources, including precipitation regulation, drinking water provision, climate stabilization, coastal protection, food supply, and atmospheric oxygen [1]. Therefore, rational marine resource governance constitutes a core component of global sustainable development. Nevertheless, progressive ocean acidification and continuous coastal water pollution severely degrade marine biodiversity and damage natural habitats for aquatic organisms [1]. Effective regulatory frameworks, sufficient resource investment, and stringent legal constraints are essential to mitigate unsustainable fishing behaviors, control marine contamination, and establish well-protected marine conservation zones [2]. Anthropogenic carbon emissions elevate atmospheric carbon dioxide concentrations, accelerating ocean carbon absorption. Although the average seawater pH remains weakly alkaline at 8.1, continuous carbon intake gradually reduces pH values, exacerbating ocean acidification. This chemical alteration endangers aquatic biota and weakens the ocean's natural carbon sequestration capacity [2]. Seas, oceans, and coastal territories represent strategically vital natural assets for the European Union. Marine industries, including fisheries, tourism, international trade, and commercial shipping, sustain livelihoods for numerous European residents. The EU fisheries sector ranks fourth globally in production volume, yielding approximately 6.4 million tonnes of aquatic products annually and employing over 350,000 workers [2]. Nevertheless, fisheries-induced water pollution has become a primary bottleneck restricting blue economic sustainability. Water quality, comprehensively determined by salinity, temperature, dissolved oxygen, pH value, and biochemical characteristics, serves as a fundamental indicator of marine ecosystem health [3]. Each aquatic species requires stable water physicochemical conditions for survival and reproduction. An Australian empirical experiment further confirms that bottom trawling disturbs benthic microbial communities, weakening natural nitrogen purification capacity and deteriorating coastal water quality [3]. Additionally, unregulated fisheries activities discharge excessive nutritional additives, antibiotics, methane, hydrogen sulfide, pesticides, and chemical residuals into seawater, threatening marine biotic habitats and ecological balance [3]. Oceans absorb approximately 30% of anthropogenic carbon emissions, substantially alleviating global warming pressures [4]. However, preferential fuel tax subsidies provided to EU fishing fleets generate severe environmental externalities. Statistical evidence indicates that EU fishery vessels consume massive fossil fuels, releasing up to 7.3 million tonnes of carbon dioxide annually [4]. Such fiscal incentives disrupt marine biogeochemical circulation, accelerate climate deterioration, contaminate aquatic habitats, and destabilize marine biological equilibrium.

Marine sustainability is increasingly threatened by irreversible biogeochemical alterations, including rising sea surface and benthic temperatures, disrupted primary productivity, ocean acidification, and coastal hypoxia [5]. Ocean warming is projected to trigger sharp declines in marine biomass by 2050. The ongoing expansion of fisheries-derived waste pollution will further exacerbate marine environmental degradation and water quality deterioration [5]. Eutrophication and anthropogenic contamination collectively induce coastal ecological recession. Without coordinated regulatory intervention, coastal eutrophication severity is predicted to increase by 20% by 2050 [6]. The Black Sea stands as one of Europe's most severely polluted marine basins, suffering from chronic eutrophication caused by nutrient overload, intensive fishery exploitation, industrial chemicals, and phosphate detergent discharge. Severe plastic and mercury contamination are also prevalent within the Mediterranean and Black Sea regions. Accumulated synthetic chemicals in seawater and sediment disrupt marine biochemical balance, constituting one of the most persistent threats to sustainable blue development [6]. Aquatic organisms accumulate toxic substances through direct water contact and indirect food chain transmission, inducing widespread biological hazards.

The fisheries industry imposes multifaceted pressure on marine conservation, manifested in fish stock depletion, water contamination, and habitat degradation [7]. Several scholars argue that global fishery scale contraction is indispensable for marine ecological restoration [8], whereas such industrial shrinkage inevitably conflicts with the livelihood dependence of fishery practitioners. Empirical projections indicate that unsustainable harvesting practices may lead to the depletion of wild seafood resources by 2048. Excessive fishing intensity, water pollution, and ecological degradation collectively trigger continuous marine population recession [8]. This study addresses three core research questions: (1) How does fisheries expansion affect the long-term marine environmental sustainability across European regions from 1990 to 2022? (2) Are developed EU economies more vulnerable to water contamination originating from fishery expansion? (3) Can optimized fishery output achieve ecological benefits and assist the European Union in accomplishing its 2050 sustainable development agenda? Correspondingly, the research objectives are defined as follows: (1) to quantitatively evaluate marine water pollution induced by fishery production expansion across the European Union between 1990 and 2022; (2) to conduct a comparative empirical analysis of fishery-driven pollution disparities between developed and developing EU economies; (3) to explore the heterogeneous pollution impacts of fishery output across differentiated economic development levels.

Most prior EU-focused marine environmental studies contain empirical limitations by ignoring cross-sectional dependence (CD) across panel units. Unobserved common shocks, spatial spillover effects, and compound external disturbances generate cross-sectional correlation among EU economies [9–13]. High economic integration renders EU member states susceptible to synchronous external shocks, including resource crises and seafood price volatility, thereby interfering with economic growth and fishery consumption patterns. According to [9], incorporating cross-sectional dependence correction ensures unbiased and consistent empirical outcomes for energy-income-environment research. Unlike previous literature adopting aggregated natural resource indicators [14, 15], this study selects fishery production as a typical point-source natural resource. Point-source resources generally generate stronger adverse growth externalities than diffuse resources [16, 17]. To capture heterogeneous institutional effects, this research categorizes 27 EU economies into developed and developing subgroups and incorporates governance as a regulatory control variable. The sample encompasses advanced economies (Germany, Italy, Sweden) and emerging European nations (Estonia, Bulgaria, Malta), with an empirical time range spanning from 1990 to 2022. This research adopts the [18] method of moments quantile regression (MMQR) to examine the complex nexus between fishery production, control variables, and water pollution. Equipped with fixed-effect specifications, the MMQR technique identifies distributional heterogeneity across conditional pollution quantiles, which cannot be captured by conventional mean regression models. Therefore, this method enables comprehensive verification of the growth hypothesis within fishery-dependent economies. This paper pioneers the quantitative assessment of income-induced marine water pollution by simultaneously incorporating fishery output, fossil fuel consumption, economic growth, and governance indicators. Quantile regression possesses prominent statistical advantages over mean regression: conditional quantile estimation exhibits higher robustness against outlier interference [19]. Furthermore, quantile regression intuitively categorizes heterogeneous variable distributions, facilitating cross-sectional group classification and detailed heterogeneous impact analysis. Each quantile provides unique empirical information that cannot be extracted from traditional mean-based estimators. The remainder of this thesis is structurally organized as follows: Sect. 4.2 reviews relevant theoretical literature concerning the growth hypothesis; Sect. 4.3 elaborates data specification and econometric methodologies; Sect. 4.4 presents empirical estimation outcomes and detailed discussions; Sect. 4.5 concludes the research findings and proposes targeted policy implications.

4.2 Water Pollution from Fisheries Operation Background

Bottom trawling represents one of the most destructive fishing techniques. Trawling nets dragged across sea beds physically injure marine organisms and permanently destroy benthic habitats. Recent marine ecological research confirms that bottom trawling disrupts sediment microorganism metabolism, weakening natural nutrient purification functions and aggravating coastal water contamination [1, 2] explored the interactive relationship between fisheries productivity, water pollution, and market failure, emphasizing that industrial optimization simultaneously promotes fishery economic expansion and water quality conservation. Through statistical simulation, [3] examined zooplankton community degradation in Egypt's Mariout Lake, confirming that unregulated fishing activities accelerate water contamination and form extensive hypoxic dead zones lethal to aquatic organisms [4] detected microplastic contamination in Turkish freshwater systems, indicating that improper fishery operations elevate surface water pollutant concentrations [5] adopted heavy metal distribution indicators to construct a water pollution evaluation system for Indonesian aquatic environments, proving that fishery misconduct exacerbates heavy metal accumulation. Conversely, [6] identified positive feedback between water quality improvement and fishery stock regeneration in the United States, demonstrating that standardized fishing practices mitigate aquatic contamination [8] further confirmed that overfishing and targeted harvesting constitute primary anthropogenic drivers of aquatic ecosystem pollution. The fisheries industry extensively relies on fossil fuels for vessel propulsion, power generation, plastic manufacturing, and daily operational consumption. Fossil fuel combustion generates persistent environmental externalities, particularly aquatic ecological degradation. Chinese fishery economic research advocates renewable energy substitution to achieve sustainable industrial transformation. Based on American water pollution trend assessments, [6] emphasized that clean energy transition guarantees long-term fishery sustainability. [20] documented continuously intensifying marine ecological pressure caused by fossil fuel dependence in fisheries [21]. proposed fossil fuel taxation policies to accelerate industrial green upgrading. Consistent empirical conclusions have been validated across multiple emerging economies, including Pakistan [22], Malaysia [23], and India [24].

Ocean governance integrates multi-stakeholder institutional forces to conserve marine ecosystems and realize sustainable aquatic resource utilization [25]. evaluated self-governance efficiency in Chinese aquaculture industries, suggesting that individualized pollution treatment fails to resolve commons tragedy dilemmas [26]. Examined marine regulatory frameworks in India, concluding that legalized fishery supervision comprehensively improves ocean health in social, economic, and environmental dimensions [27]. analyzed African small-scale fisheries, highlighting that institutional complexity, overexploitation, and fishery-related crimes collectively exacerbate water contamination. Additional global empirical studies conducted by [9, 27, 28] further corroborate the significant governance-pollution correlation within marine economic systems. A World Bank report reveals the existence of a global invisible water quality crisis, which reduces economic potential in heavily polluted regions by one-third and endangers public and ecological health. A substantial body of Chinese empirical literature explores the economic growth-water pollution nexus [29–33, 34, 35], confirming that rapid economic expansion and population growth hinder water pollution abatement [28]. emphasized that technological advancement optimizes pollution prevention mechanisms [36, 37] analyzed eco-justice issues in Bangladesh and India, demonstrating that inadequate law enforcement and inefficient water management obstruct ecological equity [38, 39] compared fossil fuel consumption, economic growth, and water pollution across developed and developing nations, highlighting socioeconomic determinants of water quality conservation. Distinct from prior literature focusing narrowly on fishery industrial characteristics, this study comprehensively evaluates marine resource consumption, water contamination, and macroeconomic development. It systematically examines the fishery-pollution-growth nexus within the EU to provide empirical evidence for blue sustainability optimization. Static panel regression and advanced quantile techniques are synchronously adopted to capture heterogeneous influencing mechanisms. The primary research novelties are summarized as follows: Firstly, this paper systematically investigates the interactive relationship between fishery production, fossil fuel usage, governance quality, economic growth, and water pollution across the European Union from 1990 to 2022. Secondly, it comparatively analyzes pollution disparities between developed and developing EU economies. Thirdly, unit root and cointegration tests are applied to verify data stationarity considering stochastic trending characteristics. Fourthly, MMQR estimation is utilized to capture distributional heterogeneity across pollution quantiles. Finally, this research predicts future pollution evolution trends driven by fishery expansion.

4.3 Marine Water Pollution Modeling

Referring to the empirical model specifications established by [40–43], this study constructs the functional econometric formula for marine water pollution assessment:

$$WP_{it} = f\left(FP_{it}, FF_{it}, GDP_{it}, GV_{it}\right)$$

In this formula, WP denotes water pollution intensity, while FP represents fisheries production. Specifically, fisheries production (FP) is quantified as total fishery output measured in metric tons. Water pollution (WP) acts as a comprehensive proxy indicator for aquacultural and fishery contamination, calculated in million cubic meters. Both indicators were extracted from the World Development Indicators [44]. Gross domestic product (GDP), representing economic growth at constant US dollar prices, was sourced from the Eurostat database [45]. Fossil fuel consumption (FF), proxied by petroleum consumption measured in tons of oil equivalent, was also obtained from Eurostat [46]. Governance (GV), reflecting governmental institutional effectiveness, was collected from the official Europa statistical platform [46]. The empirical sample covers EU economies from 1990 to 2022, with all variables processed in per capita real terms. Balanced annual panel data were compiled from three authoritative databases: the World Bank WDI database, Eurostat, and the Europa official statistics website. Detailed variable definitions, measuring units, data sources, and data transformation procedures are systematically summarized in Table 4.1.

In this part, we discuss the findings. To start, Table 4.2 displays the chosen components" statistical characteristics (i.e., maximum, minimum, mean, and standard deviation). The relative mean values for WP, FP, GDP, FF, and GV are 1.178, 4.955, 1.274, 3.771, and 1.857. The outstanding standard deviation for each of the study"s components is 0.143, 0.833, 0.196, 1.034, and 0.107, respectively. Thanks to the factors" descriptive properties, we may proceed to the unit root test.

This study employs three mainstream panel cointegration estimation techniques for comparative empirical analysis, namely fixed-effects ordinary least squares (FEOLS), fully modified ordinary least squares (FMOLS), and dynamic ordinary least squares (DOLS). Within the FEOLS framework, Driscoll–Kraay standard errors are incorporated to effectively eliminate statistical biases induced by cross-sectional dependence and residual autocorrelation within constrained lag intervals. According to [49], heterogeneous cross-sectional adjustments and divergent

Table 4.1 Data, units, transformation, and source

Data	Measurement unit	Transformation	Source
Water pollution	Million cubic meters	Natural logarithm transformation	WDI
Fishery production	Metric tons	Natural logarithm transformation	WDI
Fossil fuel	Metric tons of oil equivalent	Natural logarithm transformation	Eurostat
Economic growth	Constant US dollars	Natural logarithm transformation	Eurostat
Governance	% of confidence for governance	Natural logarithm transformation	Europa

Source: [47, 48]

Table 4.2 Results of descriptive statistics

Variable	Observations	Mean	Std. dev.	Min	Max
WP	864	1.178	0.143	0.425	1.419
FP	864	4.955	0.833	2.902	6.310
GDP	864	1.274	0.196	0.436	5.713
FF	864	3.771	1.034	0.185	1.612
GV	864	1.857	0.107	1.205	1.989

Source: [47, 48]

convergence speeds toward long-run cointegrated equilibrium constitute the primary empirical challenges in dynamic panel estimation. To mitigate such heterogeneous disturbances, the FMOLS model proposed by Pedroni incorporates individual-specific intercepts, allowing serial correlation optimization for residual sequences across distinct cross-sectional units. Monte Carlo simulation evidence confirms that DOLS estimators generate unbiased finite-sample outputs compared with FMOLS and conventional OLS techniques. Motivated by such statistical superiority, [50] extended the DOLS estimator to panel data contexts. By introducing lead and lag difference terms, the DOLS approach effectively suppresses endogenous feedback and alleviates endogenous statistical interference. Conventional mean-based regression techniques exhibit inherent empirical limitations; therefore, panel quantile regression is adopted to capture distributional heterogeneous impacts across diverse conditional quantiles [51]. Originating from the foundational work of [52], quantile regression differs fundamentally from traditional least-squares regression. Instead of merely estimating conditional mean coefficients, this technique precisely calculates conditional median values and multiple quantile distributions of dependent variables. Quantile regression possesses strong robustness against outlier interference and maintains reliable estimation performance even when the conditional mean correlation between variables is weak or insignificant [53].

Nevertheless, conventional quantile regression fails to capture unobserved individual heterogeneity within panel datasets. To address this methodological defect, the current research adopts the fixed-effects method of moments quantile regression (MMQR) developed by [18]. Different from traditional quantile regression models proposed by [19, 54], MMQR permits individual-level fixed effects and comprehensively identifies distributional alterations rather than simple mean shifts. This methodological advantage enables the precise detection of conditional covariance heterogeneity among water pollution determinants. The MMQR technique is particularly applicable for integrated panel data models containing endogenous explanatory variables and individual fixed effects. Additionally, MMQR generates non-crossing quantile estimations with high computational simplicity. Under the location-scale variation model, the conditional quantile expression Q_{Y}(\tau|X) is specified as follows:

$$Y_{it} = \alpha_{i+} X_{it}'{}' \beta + \left(\delta_i + Z_{it} \gamma' \right) U_{it} \tag{4.1}$$

Where the probability, $P(\delta_i + Z_{it}\gamma' > 0) = 1.(\alpha, \beta', \delta, \gamma')$ are parameters to be estimated. $(\alpha_i, \delta_i), i = 1, \ldots, n$, designates the individual i fixed effects, and Z is a k-vector of identified components of X which are differentiable transformations with element l given by:

$$Z_l = Z_l(X), l = 1, \ldots, k \tag{4.2}$$

For each fixed individual i, the explanatory variable X_{it}follows an independent and identical distribution across time t. The moment conditions constructed by [18] eliminate stringent homogeneity assumptions. Within this framework, the error termU_{it} remains orthogonal to X_{it}and obeys an identical and independent distribution across both individuals (i) and time (t), satisfying standard normalized constraint conditions. The mathematical expression presented in Eq. (4.1) is structured as follows;

$$Q_Y(\tau|X_{it}) = (\alpha_{i+}\delta_i q(\tau)) + X_{it}\beta' + Z_{it}\gamma' q(\tau) \tag{4.3}$$

Examining it from the third equation, the natural logarithm of FP (LFP), the economic growth logarithm, which is also natural (LGDP), the fossil fuel production natural logarithm (LFF), and the natural logarithm of governance make up the independent variables vectors X_{it}' in this current research. (LGV). The scalar coefficient, $Q_Y(\tau|X_{it})$, represents the quantile -τ fixed effect for individual $X_{it} - \alpha_i(\tau) \equiv \alpha_i$ $_+$ $\delta_i q(\tau)$ and displays the distributional quantile of the outcome variable Y_{it} (water pollution natural logarithm), which is dependent on the independent variable location X_{it}. Unlike the typically fixed effects of least squares, the individual impact does not pinpoint a shift in the intercepts. Their heterogeneous effects are allowed to make variance crosswise the conditioned distribution quantile of the variable, categorized as endogenous Y. These parameters, time-wise, are invariant parameters. The τ-th quantile sample, denoted as $Y.\ q(\tau)$, is calculated by resolving the following optimization problem.

$$\min_q \sum_i \sum_t P_\tau \left(R_{it} - \left(\delta_i + Z_{it}\gamma' \right) q \right) \tag{4.4}$$

Where $P_\tau(A) = (\tau - 1)AI\{A \le 0\} + TAI\{A > 0\}$ denotes the check function.

4.4 The Finding of Water Pollution from Fisheries Activities

Before estimating the unknown parameters, a series of standard preliminary tests are conducted to determine the time-series characteristics of the variables. The first step involves assessing whether cross-sectional dependency (CD) exists within the

Table 4.3 Cross-sectional dependence results

Variables	WP	FP	GDP	FF	GV
Pesaran (2004)	(4.26)***	(18.82)***	(83.55)***	(65.35)***	(31.30)***

– Remark: *** refer importance at the 1%, scale
– Source: [47, 48]

panel. Accurate coefficient estimates for the variables may be subject to distortions arising from cross-sectional dependency. If disregarded, cross-sectional dependency can result in common variables going unobserved, which could noticeably diminish the potential advantages of panel data [55]. It is therefore essential to account for this issue in order to obtain reliable coefficient estimates. The [56] CD testing analysis was adopted to evaluate cross-sectional dependency within the panel. As shown in Table 4.3, the remaining quantitative parameters indicate a considerable degree of cross-sectional dependency among the nations under consideration, with the exception of fisheries production. To minimize the risk of size distortions, it is necessary to adopt techniques for the cointegration testing analysis and unit root test analysis that incorporate panel estimation procedures demonstrating sufficient resistance to the effects of cross-sectional dependency.

To reliably assess the integration properties of the variables under investigation, this study employs the panel unit root tests proposed by Levin, Lin, and Chu (2002) (LLC) and Im, Pesaran, and Shin (2003) (IPS). Panel unit root tests are designed to enhance testing efficiency by combining both cross-sectional and time-series data. Several panel unit root frameworks have been developed to test the unit root null hypothesis against various competing alternatives. Levin, Lin, and Chu (2002) test the unit root null hypothesis against a homogeneous alternative, under which each series in the panel is stationary and shares the same rate of mean reversion. Im, Pesaran, and Shin (2003), on the other hand, test the unit root null hypothesis against the alternative that at least one series in the panel is stationary. The test introduced in this study places emphasis on panel stationarity, which naturally draws attention to the LLC framework. This restrictive assumption is relaxed in the Im-Pesaran-Shin (IPS) (2003) test, which permits each individual panel member to have its own autoregressive parameter. For moderately small panel datasets such as the one used in the present study, the Levin, Lin, and Chu (2002) (LLC) panel unit root testing analysis has been found to outperform other comparable unit root tests in terms of statistical power. Both testing analyses are applied here to determine how cross-sectional dependency influences the panel unit root test results. As revealed in Table 4.4, all variable parameters across all unit root testing specifications are non-stationary at levels but stationary at first differences, confirming that each variable in the estimation is integrated of order one, I(1).

To determine whether a genuine long-run relationship exists among the variables under consideration, the panel cointegration testing analysis of [49] was adopted alongside the bootstrapped panel cointegration testing of [57], as presented in Table 4.5 [49] offers a comprehensive panel cointegration testing framework grounded in the principles of the Granger two-step and Engle technique. The method

Table 4.4 Panel unit root test results

	Difference		First difference	
Variable	LLC	IPS	LLC	IPS
WP	−0.711	2.658	−8.721***	−17.947***
FP	−3.076	−3.700	−24.136***	−22.737***
GDP	−1.571	−0.210	−18.74***	−68.362***
FF	−9.255	−4.970	−20.610***	−21.409***
GR	−1.090	−0.254	−23.471***	−23.479***

– Remark: *** refer importance at the 1%, scale
– Levin, Lin, & Chu test (LLC), and Im, Pesaran, and Shin W-stat test (IPS)
– Source: Source: [47, 48]

Table 4.5 Panel cointegration test for EU countries

Test	Without trend		With trend	
A. Pedroni residual cointegration test:				
Alternative hypothesis: common AR coefficients (within dimension):				
Panel v-statistic	1.277 (0.100)		0.969 (0.833)	
Panel rho-statistic	−2.948*** (0.001)		−3.229*** (0.000)	
Panel PP-statistic	−10.579*** (0.000)		−11.225*** (0.000)	
Panel ADF-statistic	−3.662*** (0.000)		−2.182** (0.014)	
Alternative hypothesis: common AR coefficients (between dimensions)				
Group rho-statistic	−2.196**		(0.014)	
Group PP-statistic	−14.165***		(0.000)	
Group ADF-statistic	−1.069***		(0.142)	
B). Westerlund (2007) Bootstrap panel cointegration:				
Statistics	Value	Z-value	p-value	Robust p-value
Gt	−3.415	−3.218	0.001	0.000**
Ga	−9.377	−4.889	1.023	0.145
Pt	−16.956	−3.045	0.001	0.000***
Pa	−11.328	−1.574	0.942	0.236

– Remark: ***, **, and * refer importance at the 1%, 5%, and 10% scales, respectively
– Values in parentheses are *p*-values
– Source: [47, 48]

controls for heterogeneity by removing short-run factors and person-specific deterministic patterns in the first stage of the process. Seven alternative test statistics are derived from the estimated residuals, falling into two categories: those that assume a single common process, generally referred to as “”pooled”” or “”within-dimension”” tests, and those that account for individual processes, commonly referred to as “”grouped”” or “”between-dimension”” tests, as presented in Table 4.5. The [57] method introduces four novel tests under the null hypothesis that no cointegration exists. Since the underlying dynamics are structural rather than purely residual in nature, the test removes the standard common factor restrictions typically imposed on residual-based dynamic tests. The failure to satisfy common factor restrictions can substantially undermine the effectiveness of residual-based cointegration analyses, making the incorporation of structural dynamics necessary [58].

With this restriction lifted, it is no longer necessary for the short-run and long-run adjustment procedures to produce the same outcomes. By applying [57] bootstrap methodology, it becomes possible to reduce the distortionary effects of cross-sectional dependency and obtain more reliable, robust critical values. The findings presented in Table 4.5 confirm that cointegration is strongly supported by both [49] testing technique and [57] bootstrapped cointegration testing analysis.

Table 4.6 presents the outcomes of the FMOLS, DOLS, and FEOLS estimation procedures. Although there are variations in statistical significance across the three estimators, the coefficient estimates produced for Model 1 in Table 4.6 are broadly comparable on average. In terms of both statistical significance and coefficient magnitude, fish output and fossil fuel use in the EU region demonstrate the greatest consistency among the three estimators. According to the FEOLS, FMOLS, and DOLS estimators, a one percentage point increase in fisheries production is associated with reductions in water pollution of 0.03%, 0.02%, and 0.19%, respectively. These findings are consistent with the main conclusions of [1–3, 8]. The introduction of hazardous substances and dangerous contaminants into ocean waters—including waste from fisheries operations, chemicals used on vessels, oil consumption, and plastic debris—is a primary driver of marine water pollution. Irresponsible fishing practices have posed a serious threat to the ocean environment and water quality, affecting numerous marine animal and plant species as well as the communities that depend on them. As anticipated, a statistically significant positive relationship is also observed between fossil fuel use and marine water pollution across the EU region. According to the FEOLS, FMOLS, and DOLS estimators, a one percentage point increase in fossil fuel use is associated with increases in marine water pollution of 0.02%, 0.09%, and 0.12%, respectively. These results are consistent with earlier findings by [21, 22, 24, 39]. Although diesel and petrol are primarily used in wild fisheries operations to power vessel equipment, this generates significant CO2 emissions, with approximately 70% of these emissions

Table 4.6 Model 1. Panel estimation for the EU-27 region from 1990 to 2022

Long-run coefficient	DOLS		FMOLS		FEOLS	
	Coefficient	Std. error	Coefficient	Std. error	Coefficient	Std. error
FP	0.387	(0.192)**	0.071	(0.022)***	0.038	(0.097)***
GDP	0.236	(0.067)*	0.021	(0.012)*	0.022	(0.017)
FF	0.121	(0.114)**	0.093	(0.022)***	0.022	(0.093)**
GV	−0.504	(0.431)	−0.109	(0.022)***	−1.100	(0.047)***

– Note: ***, **, and * indicate significance at the 1%, 5%, and 10% levels, respectively
– Values in parentheses are P-values
– Source: [47, 48]

absorbed and dissolved in the ocean, degrading water quality and contributing to pollution. Notably, the EU"s approved Emissions Trading Scheme (ETS) currently excludes fisheries operations, effectively absolving the industry of responsibility under climate policy. Economic expansion also significantly reduces the degree of marine water pollution across the EU region, though the effect varies across FEOLS specifications. For the FMOLS and DOLS estimators, a one percentage point increase in economic growth is associated with reductions in marine water pollution of 0.01% and 0.23%, respectively, lending support to the growth theory for the panel of fishery-producing nations. These findings echo the main conclusions of [29, 31], who identified an inelastic and positive influence of economic development on marine water pollution. An unaddressed water quality crisis is affecting species, ocean ecosystem health, and up to one-third of potential economic development in the most heavily contaminated areas of the planet. Governance in the EU region, on the other hand, is found to have a detrimental impact on marine water contamination. While this result is broadly consistent with the modernization hypothesis, it also conflicts with the perspectives advanced by [9, 26]. The growth hypothesis receives strong statistical support in the FMOLS and FEOLS specifications, while the DOLS specification provides only moderate statistical support. To maintain high marine water quality and a healthy ocean, a coalition of businesses, civil society organizations, and local and regional governments has entered into a mutual agreement under private law aimed at measuring and reducing ship-generated marine waste, promoting clean marine fisheries, and preserving pristine beaches across European countries.

As shown in Table 4.7, the panel quantile regression estimates indicate that the effect of fisheries production on marine water pollution is statistically significant at the first and ninth quantiles—specifically between the first and third quantiles and between the eighth and ninth quantiles—but not at the median quantiles. This finding supports the effectiveness of marine water regulations for the fisheries sector in nations with higher levels of marine water pollution. The results further demonstrate that the growth hypothesis is consistent with only the first and last quantiles of marine water pollution in the EU27, suggesting that fisheries growth takes precedence over marine water quality in nations falling somewhere in the middle range. Furthermore, nations at the median level with lower water pollution levels may find it comparatively easier to transition the fisheries sector from non-recycled to recycled sources due to the associated investment costs. In nations with the worst levels of marine pollution, the fisheries sector has no meaningful positive impact on ocean water quality, which may be attributable to the differing extraction techniques available in the established EU14 countries and the growing EU13 economies. These findings could have significant consequences for the policy approach supported by [1–3], which is associated with extremely high levels of marine water pollution. The results also align with recent empirical research by [4], which supported the growth hypothesis as another important contributor to marine water quality, as well as with the findings of [5], who demonstrated the validity of the growth hypothesis for the primary source of marine water quality deterioration. At the median quantiles—specifically the fifth, sixth, and seventh—the effects of economic expansion as

Table 4.7 Panel quantile estimation results for EU27 countries

Variables	Method of moments quantile regression with fixed effects										
	Location	Scale	Quantiles								
			0.1	0.2	0.3	0.4	0.5	0.6	0.7	0.8	0.9
FP	(0.097)***	(0.090)*	(0.050)*	(0.027)**	(0.031)**	(0.059)	(0.064)	(0.071)	(0.069)	(0.026)**	(0.062)***
GDP	(0.017)	(0.318)	(0.049)	(0.077)	(0.043)	(0.035)	(0.065)***	(0.061)***	(0.061)***	(0.043)	(0.066)
FF	(0.093)**	(0.064)**	(0.089)***	(0.073)***	(0.057)***	(0.049)***	(0.055)***	(0.059)***	(0.061)***	(0.018)	(0.021)
GR	(0.047)***	(0.012)***	(−0.530)*	(−0.261)	(−0.589)***	(−0.509)***	(−0.553)***	(−0.461)***	(−0.403)***	(−0.546)***	(−0.423)***

Notes: *, **, and *** indicate significance at 10%, 5%, and 1% levels, respectively
Source: [47, 48]

Table 4.8 Model 2. Panel estimation for the EU-14 region from 1990 to 2022

Long-run coefficient	DOLS		FMOLS		FEOLS	
	Coefficient	Std. error	Coefficient	Std. error	Coefficient	Std. error
FP	0.132	(0.077)*	0.169	(0.047)***	0.116	(0.036)***
GDP	0. 023	(0.049)	0.072	(0.036)**	0.142	(0.038)***
FF	0.023	(0.026)	0.079	(0.017)	0.040	(0.013)
GV	−0.913	(0.270)***	−0.326	(0.169)*	−0.334	(0.129)**

– Note: ***, **, and * indicate significance at the 1%, 5%, and 10% levels respectively
– Values in parentheses are P-values
– Source: [47, 48]

measured by GDP are strongly and positively associated with marine water pollution, with the magnitude of this influence increasing at higher quantiles. Fossil fuel use has a major impact on marine water pollution from the first through the seventh quantiles, while its effect becomes negligible at the eighth and ninth quantiles, supporting the effectiveness of energy use measures that promote marine ecology in more heavily polluted nations. Regarding governance, the findings indicate that its impact on marine water pollution is negative across all quantiles and statistically significant in nations with high levels of water pollution from the first through the ninth quantiles. This negative relationship between governance and water pollution—a cornerstone of the modernization hypothesis—may account for this outcome.

Table 4.8 presents the findings from the FMOLS, DOLS, and FEOLS estimation procedures. Although differences are observed in their levels of statistical significance, the coefficient estimates produced by the three specifications for Model 2 in Table 4.8 are broadly comparable on average. In terms of both coefficient magnitude and statistical significance, fishery production and governance in the EU14 developed nations demonstrate the greatest consistency across all three estimators. According to the FEOLS, FMOLS, and DOLS estimators, a one percentage point increase in fisheries production is associated with reductions in marine water pollution of 0.03%, 0.04%, and 0.07%, respectively. These findings are consistent with the main conclusions of [4, 5, 8, 6]. The challenge of hidden seawater contamination illustrates how a combination of bacteria, waste, chemicals, and plastics can deplete water of oxygen and render it toxic to marine life and ecosystems. Economic expansion significantly reduces the degree of marine water pollution in the EU14 developed nations, with the FEOLS and FMOLS estimators showing reductions of 0.14% and 0.07%, respectively, for a one percentage point increase in economic growth, though the coefficient appears inconsistent across DOLS specifications.

These findings lend support to the growth theory for the panel of EU14 developed nations. This outcome contradicts [38, 39], who were unable to identify a significant positive and elastic impact of economic growth on marine water pollution in developed countries, but echoes the main conclusions of [30, 35, 37]. Fisheries-related economic activities—including shipping, aquaculture, and recreational uses—can contribute to marine water pollution, and ineffective fisheries practices may result in billions of dollars in costs associated with cleaning up contaminated waterways and preventing marine environmental disasters. Governance in the EU14 developed nations is found to have a considerable negative impact across all estimated parameters, a result that conflicts with [9] but supports the modernization theories advanced by [27, 28, 59]. According to the FEOLS, FMOLS, and DOLS estimators, a one percentage point increase in governance is associated with reductions in marine water pollution of 0.33%, 0.32%, and 0.91%, respectively. The growth hypothesis receives strong statistical support across the FEOLS, FMOLS, and DOLS specifications. As shown in Table 4.9, the panel quantile regression results indicate that the effects of the fisheries sector on marine water pollution are statistically significant from the first through the sixth quantiles, supporting the effectiveness of marine water policy in enhancing fisheries production in the EU14 developed nations. These results confirm that the growth hypothesis is valid for EU14 developed countries at lower and median levels of marine water pollution, suggesting that marine water policy takes precedence over fisheries production in the earlier and median quantiles but not at the higher quantiles. Furthermore, EU14 developed countries currently at high-demand stages of water pollution may find it more challenging to transition from conventional to recycled and environmentally friendly fisheries techniques due to the associated investment costs. In EU14 nations with the most severe levels of marine water pollution, the fisheries industry has a less favorable impact on water pollution, which may be attributable to the varying extraction techniques available within EU14 developed fisheries economies. These findings carry significant policy implications, as they support the validation of the growth hypothesis by [4] for several affluent nations generating very high levels of marine water pollution. The growth hypothesis was also supported for another primary source of marine pollution in recent empirical research by [5], consistent with the present findings. The results further align with those of [6, 8], who demonstrated the existence of the growth hypothesis for the primary source of marine water pollution. From the first through the ninth quantiles, the impact of economic expansion on marine water pollution as measured by GDP is considerable, supporting the notion that environmental regulations may contribute to economic development in more heavily polluted nations and suggesting that economic growth takes precedence over marine water quality in the EU14 developed nations. Fossil fuel use also shows a significant correlation with marine water contamination from the first through the ninth quantiles, with the magnitude of its influence diminishing at higher quantiles. Regarding governance, the findings reveal a negative and statistically significant impact on marine water pollution across all quantiles, with the exception of the seventh and ninth quantiles, and the magnitude of governance''s influence similarly decreases at higher quantiles. The effectiveness of governance compels government officials and their managers to promote environmentally friendly policies within the fisheries sector.

Table 4.10 in Model 3 presents the outcomes of the various panel estimation approaches (FMOLS, DOLS, and FEOLS), which yield broadly comparable coefficient values. The results demonstrate that fisheries production has a considerable and positive influence on marine water pollution in the EU13 developing nations. For FEOLS, FMOLS, and DOLS, respectively, a 1% increase in fisheries

Table 4.9 Panel quantile estimation results for EU14 countries

Variables	Method of moments quantile regression with fixed effects										
	Location	Scale	Quantiles								
			0.1	0.2	0.3	0.4	0.5	0.6	0.7	0.8	0.9
FP	(0.036)***	(0.026)***	(0.057)***	(0.042)***	(0.036)***	(0.037)***	(0.031)***	(0.019)**	(0.003)	(0.001)	(0.001)
GDP	(0.038)***	(0.096)	(0.213)***	(0.233)***	(0.237)***	(0.249)***	(0.241)***	(0.249)***	(0.197)***	(0.125)**	(0.094)*
FF	(0.013)	(0.068)***	(0.217)	(0.302)	(0.073)	(0.002)	(0.078)	(0.031)	(0.062)	(0.137)	(0.074)
GR	(0.129)**	(0.567)	(−0.057)***	(−0.026)***	(−0.013)***	(−0.020)***	(−0.004)***	(−0.005)**	(0.018)	(−0.008)*	(−0.022)

Notes: *, **, and *** indicate significance at 10%, 5%, and 1% levels, respectively
Source: [47, 48]

Table 4.10 Model 3. Panel estimation for the EU-13 region from 1990 to 2022

Long-run coefficient	DOLS		FMOLS		FEOLS	
	Coefficient	Std. error	Coefficient	Std. error	Coefficient	Std. error
FP	0.127	(0.023)***	0.081	(0.030)***	0.032	(0.013)**
GDP	0.173	(0.034)***	0.021	(0.019)	0.030	(0.019)
FF	0.067	(0.011)***	0.096	(0.028)***	0.039	(0.011)***
GR	−0.670	(0.082)	−0.109	(0.025)***	−0.178	(0.065)***

– Note: ***, **, and * indicate significance at the 1%, 5%, and 10% levels respectively
– Values in parentheses are P-values
Source: [47, 48]

production is associated with increases in marine water pollution of 0.03%, 0.8%, and 0.12%. The positive and statistically significant coefficients of fisheries production across all three techniques confirm the existence of the growth hypothesis in the EU13 developing nations. This result is consistent with the findings of [4, 5], who indicate that marine water pollution rises during the early phases of development as fisheries production increases. Fossil fuel use is found to be directly associated with marine water pollution across the FEOLS, FMOLS, and DOLS methods, with a 1% increase in fossil fuel use associated with increases in marine water pollution of 0.03% for FEOLS, 0.09% for FMOLS, and 10.06% for DOLS, confirming that fossil fuel consumption worsens water pollution in developing nations. The results further demonstrate that GDP has a considerable and positive effect on marine water pollution in the EU13 developing nations. According to DOLS, a 1% increase in GDP leads to a 0.17% increase in marine water pollution, though the coefficient appears inconsistent across the FEOLS and FMOLS estimation parameters. The positive and statistically significant GDP coefficient in the DOLS technique further confirms the existence of the growth hypothesis in the EU13 developing nations, consistent with the findings of [35, 36]. The empirical findings for the EU13 developing nations also reveal a negative relationship between governance and marine water pollution. Reductions in marine water pollution of 0.17% and 0.10% are found for FEOLS and FMOLS, respectively, for a 1% increase in governance, though the coefficient appears inconsistent across DOLS estimation settings, suggesting that regulated governance contributes to reductions in marine water pollution.

Table 4.11 presents the results of the method of moments quantile regression. The findings confirm that fisheries production has a positive impact on marine water pollution across all quantiles, with the magnitude of variation in marine water pollution attributable to fisheries production decreasing from 0.216 to 0.124 as the quantile rises from the first to the ninth. A substantial and positive relationship between fossil fuel use and marine water contamination is also observed across all quantiles, with the magnitude of the fossil fuel effect increasing at lower quantiles and subsequently declining at higher quantiles toward the ninth. Governance is found to have an adverse influence on marine water pollution at both the lowest and highest quantiles, with this relationship being statistically significant at the first, second, and ninth quantiles. Unlike in earlier instances, the magnitude of

Table 4.11 Panel quantile estimation results for EU13 countries

Variables	Method of moments quantile regression with fixed effects										
	Location	Scale	Quantiles								
			0.1	0.2	0.3	0.4	0.5	0.6	0.7	0.8	0.9
FP	(0.013)**	(0.021)***	(0.216)***	(0.182)***	(0.134)***	(0.125)***	(0.091)***	(0.088)***	(0.085)***	(0.113)***	(0.124)***
GDP	(0.019)	(0.048)	(0.086)***	(0.091)***	(0.097)***	(0.114)***	(0.112)***	(0.122)***	(0.143)***	(0.127)***	(0.171)***
FF	(0.011)***	(0.067)***	(0.071)***	(0.084)***	(0.087)***	(0.081)***	(0.082)***	(0.076)***	(0.080)***	(0.059)***	(0.048)**
GR	(0.065)***	(0.506)***	(−0.302)***	(−0.224)***	(−0.090)	(−0.078)	(−0.030)	(−0.036)	(−0.044)	(−0.038)	(−0.182)***

– Notes: *, **, and *** indicate significance at 10%, 5%, and 1% levels, respectively
– Values in parentheses are P-values
Source: [47, 48]

governance''s influence diminishes as higher quantiles are reached. Economic expansion is also confirmed to have a positive impact on marine water contamination, with Table 4.11 showing that the increase in marine water pollution attributable to economic expansion is significant and rises from 0.086 to 0.172 as the quantile increases from the first to the ninth across all quantiles.

Panel FMOLS and DOLS estimations confirmed the robustness of the FEOLS panel results. The coefficients obtained from the FEOLS model exhibited consistent signs and similar levels of statistical significance compared to those derived from the FMOLS and DOLS approaches. This consistency indicates that the FMOLS and DOLS estimations are dependable and suitable for drawing valid conclusions. Although the FEOLS coefficients differed slightly in significance thresholds, their directional effects remained identical to those of FMOLS and DOLS. Moreover, the DOLS estimations are considered robust as they effectively address issues related to endogeneity and serial correlation. To further investigate the impact of the fisheries sector on marine water pollution, the EU27 countries were categorized into two groups: 14 developed nations and 13 developing nations, according to their marine pollution levels. Tables 4.8 and 4.9 present the estimated influence of the fisheries sector on hazardous marine waste within the developed EU14 countries, while Tables 4.10 and 4.11 illustrate the corresponding effects for the developing EU13 countries. The findings from Tables 4.8 and 4.10 reveal that the fisheries sector exerts a statistically significant influence on marine water pollution. The evidence further indicates that the developed EU14 countries experience a stronger adverse impact on marine pollution compared to the developing EU13 nations, with estimated effect magnitudes of 0.116 and 0.032, respectively. These results imply that developed EU14 countries possess greater potential to mitigate marine water pollution through stricter fisheries regulations and marine taxation policies targeting the fisheries sector. In addition, the results demonstrate that fossil fuel consumption significantly contributes to marine water pollution in the EU13 developing countries, whereas no substantial effect is observed in the EU14 developed nations. For the EU13 countries, the estimated coefficient is 0.039, suggesting that developing countries could substantially reduce marine pollution through effective energy policies and taxation on fossil fuel usage within the fisheries industry. The findings also indicate that economic development in the EU14 developed countries has a more pronounced and statistically significant effect on marine water pollution than in the EU13 nations, with an estimated magnitude of 0.142. This suggests that investments in environmentally sustainable and green fisheries practices may contribute more effectively to reducing marine water pollution in the developed EU14 countries compared to the developing EU13 countries.

Furthermore, governance-related effects on marine water pollution appear stronger in the EU13 developing countries than in the EU14 developed nations. The estimated coefficients for governance are −0.178 for the EU14 countries and −0.334 for the EU13 countries. These results suggest that adopting effective ocean governance strategies within the fisheries sector may enable the EU13 developing countries to reduce marine water pollution more rapidly than the EU14 developed countries. One possible explanation is that, under existing tax policy modifications, the fisheries sector in several EU14 countries remains relatively exempt from taxation.

4.5 Conclusion and Solutions for Marine Water Pollution from Fisheries Activities

This study empirically explored the relationships among fishery production, fossil fuel consumption, economic growth, governance, and marine water pollution within the framework of the growth hypothesis. The research contributes to the expanding literature on environmental pollution associated with fisheries activities and sustainable development. Using panel data covering 27 EU countries over a 31-year period, the study employed panel cointegration analysis, panel unit root tests, and panel estimation techniques. Governance, measured as the percentage of confidence in governance, was incorporated into the model to evaluate sustainable development and adjust for marine water pollution effects. The empirical findings confirmed that all variables followed an I(1) process and shared a stable long-run relationship under the panel cointegration and unit root testing frameworks. Although coefficient significance varied across different estimation methods, the magnitudes remained relatively consistent. To provide a more comprehensive empirical assessment, the study further adopted the method of moments quantile regression (MMQR), which enabled the analysis of heterogeneous effects of explanatory variables across different conditional quantiles of marine water pollution. The coefficient estimates obtained from DOLS, FMOLS, and FEOLS strongly supported the growth hypothesis for fisheries-producing EU27 countries. The MMQR findings revealed that fisheries production exerted a positive and statistically significant effect on marine water pollution at both the lower and upper conditional quantiles, while the relationship weakened around the median quantile. Fossil fuel consumption showed a positive and significant relationship from the first through the seventh quantiles, indicating that fossil fuel utilization in fisheries contributes substantially to marine pollution across multiple distribution levels. Governance, on the other hand, significantly reduced marine water pollution across nearly all quantiles except the second, implying that governance quality plays an important role in environmental improvement throughout different development stages in EU27 countries. The results also suggest that fisheries-producing countries with higher pollution levels tend to be more economically diversified than emerging economies, which rely more heavily on seafood imports. Consequently, the fisheries sector may constitute a larger proportion of marine pollution in wealthier nations. The graphical analysis presented in Table 4.7 further demonstrated that fisheries production and fossil fuel consumption exhibited similar MMQR dynamic patterns. The coefficients associated with fisheries production and fossil fuel use were highest at the lower quantiles and gradually declined toward the upper quantiles. This finding indicates that the effects of fisheries activities and fossil fuel usage on marine pollution are strongest at the lower quantile levels. In this context, resource dependence does not necessarily imply economic abundance, thereby reinforcing the resource curse hypothesis associated with deindustrialization. The negative governance effects observed at the lower and median quantiles may reflect the impact of recent environmental protection measures introduced to combat marine pollution. Residents in the highly polluted EU14 developed countries generally place greater value on environmental quality compared to those in the EU13 developing nations. Consequently, governments seeking public support are more likely to implement policies aimed at reducing marine water pollution.

At higher pollution quantiles, fisheries production in EU14 developed countries may involve substantial trade-offs between economic growth and environmental sustainability. This finding highlights the need for balancing fisheries expansion with pollution control strategies. Environmental regulations may receive stronger public support in countries where environmental degradation is immediately observable, thereby reducing political emphasis on pollution-intensive fisheries-led economic growth. Governance effectiveness may also become increasingly important at higher quantiles due to the wider adoption of sustainable fisheries practices and sustainability-oriented policy frameworks in both EU13 and EU14 countries. The upper quantile estimates in Table 4.9 further indicate that fisheries production exerts a positive and significant effect in developed EU14 nations, suggesting that fisheries practices in these countries have become increasingly advanced and industrialized. Empirical evidence from countries such as Egypt [3], Turkey [4], Indonesia [5], and the USA [6] supports the argument that stronger governance frameworks can substantially reduce marine water pollution. Accordingly, this study emphasizes the necessity of improving governance systems in fisheries-producing countries, particularly in the developed EU14 nations where marine pollution indicators remain high. The study also found that fisheries production maintained a positive association with marine water pollution at both lower and upper quantiles, while fossil fuel consumption remained positively significant across nearly all quantiles except the eighth and ninth. These findings are consistent with the growth hypothesis documented in earlier literature. Several possible explanations emerge for the differing quantile effects within EU27 fisheries-producing countries. At the median quantiles, deindustrialization may occur when fisheries production replaces other tradable sectors dependent on marine environmental quality. Consequently, these economies become increasingly reliant on seafood imports and exports as major components of economic growth. The relocation of labor-intensive fisheries industries from EU13 developing countries to EU14 developed markets in pursuit of higher profit margins may further influence economic growth patterns at higher quantiles while simultaneously reducing pollution levels in EU13 nations. In developed EU14 countries, strict environmental regulations affecting export-oriented and environmentally dependent industries may encourage fisheries operators to minimize marine waste generation. Conversely, firms facing high compliance costs associated with energy-efficient technologies and fuel switching may relocate operations to countries with less stringent environmental regulations and taxation systems. Given that fossil fuel consumption influences nearly every productive sector of the economy, its positive relationship with marine water pollution across all quantiles is unsurprising. The study further observed that the impact of fisheries production on marine pollution was strongest at the lower quantiles and weakest at the highest quantiles. This does not necessarily imply that increasing energy consumption automatically reduces marine pollution; rather, it may reflect greater use of renewable and sustainable energy sources at higher quantiles. These findings suggest that fisheries industries become more efficient as they transition from lower to higher quantile levels. They also reflect the ability of developed EU14 countries to invest in modern, environmentally sustainable fisheries technologies and equipment.

Despite fishery production remaining relatively high compared to pollution levels, the fisheries gear utilized in both EU14 developed and EU13 developing countries may still largely rely on conventional technologies. Therefore, greater efforts are required to transition fisheries operations toward environmentally sustainable practices to reduce marine pollution in developed EU14 nations. Promoting eco-friendly fisheries methods—including biodegradable fishing lures, environmentally friendly fishing lines, and sustainable angling practices—would contribute significantly to marine environmental protection. Where immediate adoption of eco-friendly fisheries technologies is not feasible, EU13 developing countries should implement fair environmental taxation systems targeting pollution-intensive fisheries activities. Revenue generated from such taxes could then support long-term technological innovation and infrastructure development for sustainable fisheries management. Finally, future studies could broaden this analysis by incorporating additional variables such as tourism, globalization, and financial development, all of which influence the sustainability of fisheries and marine development within the EU region. Similar analyses could also be conducted at the national level or extended to other regional groupings such as ASEAN, BRICS, the Group of Seven, or the Next 11 countries using advanced econometric techniques. Like all empirical models, this study is subject to several limitations. Data constraints prevented the analysis from examining the potential benefits associated with marine water pollution, which remain an area for future investigation. Additionally, the model was limited to a specific set of variables and focused solely on EU countries between 1990 and 2022. Future research could expand the framework by exploring the influence of public-private investment in fisheries and ocean energy on fishery production using broader datasets and additional indicators. The model could also be extended to include a wider range of polluted countries to enhance generalizability. Furthermore, future researchers may account for structural breaks in the data and compare the economic complexity of marine water pollution across different time periods and regions.

References

1. Datta S, Sinha D, Chaudhary V, Kar S, Singh A (2022) Water pollution of wetlands: a global threat to inland, wetland, and aquatic phytodiversity. In: Handbook of research on monitoring and evaluating the ecological health of wetlands. IGI Global Scientific Publishing, pp 27–50. [Online]. https://www.igi-global.com/chapter/water-pollution-of-wetlands/295739. Accessed 20 May 2026
2. Kroetz K, Nøstbakken L, Quaas M (2022) The future of wild-caught fisheries: expanding the scope of management. Rev Environ Econ Policy 16(2):241–261. https://doi.org/10.1086/721097
3. Heneash AM, Alprol AE, Abd El-Hamid HT, Khater M, El Damhogy KA (2021) Assessment of water pollution induced by anthropogenic activities on zooplankton community in Mariout Lake using statistical simulation. Arab J Geosci 14(7):641. https://doi.org/10.1007/s12517-021-06977-9
4. Özhan Turhan D (2021) Evaluation of microplastics in the surface water, sediment and fish of Sürgü Dam Reservoir (Malatya) in Turkey. Turk J Fish Aquat Sci 22(Special Issue):TRJFAS20157. https://doi.org/10.4194/TRJFAS20157
5. Hamuna B, Tanjung RHR (2021) Heavy metal content and spatial distribution to determine the water pollution index in depapre waters, Papua, Indonesia. Curr Appl Sci Technol 21(1):1–11
6. Keiser DA, Shapiro JS (2019) Consequences of the clean water act and the demand for water quality*. Q J Econ 134(1):349–396. https://doi.org/10.1093/qje/qjy019
7. Palomino E, Pardue J (2021) Alutiiq fish skin traditions: connecting communities in the COVID-19 era. Heritage 4(4):4249–4263. https://doi.org/10.3390/heritage4040234
8. Prakash S, Verma AK (2019) Effect of arsenic on lipid metabolism of a freshwater catfish, Mystus vittatus. Liver 10(20):30
9. Andrews DWK (2005) Cross-section regression with common shocks. Econometrica 73(5):1551–1585. https://doi.org/10.1111/j.1468-0262.2005.00629.x
10. Breitung J, Pesaran MH (2008) Unit roots and cointegration in panels. In: Mátyás L, Sevestre P (eds) The econometrics of panel data, Advanced studies in theoretical and applied econometrics, vol 46. Springer Berlin Heidelberg, Berlin, Heidelberg, pp 279–322. https://doi.org/10.1007/978-3-540-75892-1_9
11. Cerrato M, Sarantis N (2002) The cross sectional dependence puzzle. London Guildhall University. [Online]. http://fmwww.bc.edu/RePEc/res2002/Cerrato.pdf. Accessed 20 May 2026
12. Chudik A, Pesaran H (2015) Large panel data models with cross-sectional dependence: a survey. [Online]. https://academic.oup.com/edited-volume/41981/chapter/355335206?searchresult=1&utm_source=TrendMD&utm_medium=cpc&utm_campaign=Oxford_Academic_Books_TrendMD_0. Accessed 20 May 2026
13. Flaten O, Lien G, Koesling M, Løes AK (2010) Norwegian farmers ceasing certified organic production: characteristics and reasons. J Environ Manag 91(12):2717–2726. https://doi.org/10.1016/j.jenvman.2010.07.026
14. Balsalobre-Lorente D, Shahbaz M, Roubaud D, Farhani S (2018) How economic growth, renewable electricity and natural resources contribute to CO2 emissions? Energy Policy 113:356–367. https://doi.org/10.1016/j.enpol.2017.10.050
15. Bekun FV, Alola AA, Sarkodie SA (2019) Toward a sustainable environment: nexus between CO2 emissions, resource rent, renewable and nonrenewable energy in 16-EU countries. Sci Total Environ 657:1023–1029. https://doi.org/10.1016/j.scitotenv.2018.12.104
16. Isham J (2005) The varieties of resource experience: natural resource export structures and the political economy of economic growth. World Bank Econ Rev 19(2):141–174. https://doi.org/10.1093/wber/lhi010
17. Mehlum H, Moene K, Torvik R (2006) Institutions and the resource curse. Econ J 116(508):1–20. https://doi.org/10.1111/j.1468-0297.2006.01045.x

18. Machado JAF, Santos Silva JMC (2019) Quantiles via moments. J Econ 213(1):145–173. https://doi.org/10.1016/j.jeconom.2019.04.009
19. Koenker R (2004) Quantile regression for longitudinal data. J Multivar Anal 91(1):74–89. https://doi.org/10.1016/j.jmva.2004.05.006
20. Mondal S, Singh SP, Lahir YK (2022) Emerging trends in environmental biotechnology, 1st edn. CRC Press, Boca Raton. https://doi.org/10.1201/9781003186304
21. Moosavian SF, Borzuei D, Zahedi R, Ahmadi A (2022) Evaluation of research and development subsidies and fossil energy tax for sustainable development using computable general equilibrium model. Energy Sci Eng 10(9):3267–3280. https://doi.org/10.1002/ese3.1217
22. Shafeeque M et al (2021) Understanding temporary reduction in atmospheric pollution and its impacts on coastal aquatic system during COVID-19 lockdown: a case study of South Asia. Geomat Nat Haz Risk 12(1):560–580. https://doi.org/10.1080/19475705.2021.1885503
23. Chia SR, Nomanbhay S, Ong MY, Shamsuddin AHB, Chew KW, Show PL (2022) Renewable diesel as fossil fuel substitution in Malaysia: a review. Fuel 314:123137. https://doi.org/10.1016/j.fuel.2022.123137
24. Kalair A, Abas N, Saleem MS, Kalair AR, Khan N (2021) Role of energy storage systems in energy transition from fossil fuels to renewables. Energy Storage 3(1):e135
25. Luo Y, Wu J, Xu Y (2022) Can self-governance tackle the water commons? — Causal evidence of the effect of rural water pollution treatment on farmers' health in China. Ecol Econ 198:107471. https://doi.org/10.1016/j.ecolecon.2022.107471
26. Kelkar N, Arthur RI (2022) A review of governance and tenure in inland capture fisheries and aquaculture systems of India. [Online]. https://books.google.com/books?hl=en&lr=&id=wL1qEAAAQBAJ&oi=fnd&pg=PP1&dq=Kelkar,+N.,+%26+Arthur,+R.+I.+(2022).+A+review+of+governance+and+tenure+in+inland+capture+fisheries+and+aquaculture+systems+of+India.+https://doi.org/10.4060/cb8615en&ots=43o6zSOa5g&sig=lUJVzKdVu4wfxr1-PtME1opbuaU. Accessed 20 May 2026
27. Okafor-Yarwood I, Kadagi NI, Belhabib D, Allison EH (2022) Survival of the richest, not the fittest: how attempts to improve governance impact African small-scale marine fisheries. Mar Policy 135:104847. https://doi.org/10.1016/j.marpol.2021.104847
28. He M et al (2022) Waste-derived biochar for water pollution control and sustainable development. Nat Rev Earth Environ 3(7):444–460. https://doi.org/10.1038/s43017-022-00306-8
29. Huang Y, Mi F, Wang J, Yang X, Yu T (2022) Water pollution incidents and their influencing factors in China during the past 20 years. Environ Monit Assess 194(3):182. https://doi.org/10.1007/s10661-022-09838-4
30. Pan D, Tang J (2021) The effects of heterogeneous environmental regulations on water pollution control: quasi-natural experimental evidence from China. Sci Total Environ 751:141550. https://doi.org/10.1016/j.scitotenv.2020.141550
31. Tang W, Pei Y, Zheng H, Zhao Y, Shu L, Zhang H (2022) Twenty years of China's water pollution control: experiences and challenges. Chemosphere 295:133875. https://doi.org/10.1016/j.chemosphere.2022.133875
32. Wang Y, Wei H, Wang Y, Peng C, Dai J (2021) Chinese industrial water pollution and the prevention trends: an assessment based on environmental complaint reporting system (ECRS). Alex Eng J 60(6):5803–5812. https://doi.org/10.1016/j.aej.2021.04.015
33. Wang M, Janssen ABG, Bazin J, Strokal M, Ma L, Kroeze C (2022) Accounting for interactions between Sustainable Development Goals is essential for water pollution control in China. Nat Commun 13(1):730. https://doi.org/10.1038/s41467-022-28351-3
34. Zhang Y, Liu W, Cai Y, Khan SU, Zhao M (2021) Decoupling analysis of water use and economic development in arid region of China—based on quantity and quality of water use. Sci Total Environ 761:143275. https://doi.org/10.1016/j.scitotenv.2020.143275
35. Zhou Z, Liu J, Zhou N, Zhang T, Zeng H (2021) Does the '10-Point Water Plan' reduce the intensity of industrial water pollution? Quasi-experimental evidence from China. J Environ Manag 295:113048. https://doi.org/10.1016/j.jenvman.2021.113048

36. Faroque S, South N (2021) Water pollution and environmental injustices in Bangladesh. Int J Crime, Justice Soc Democr 10(2):1–13. https://doi.org/10.5204/ijcjsd.2006
37. Mumbi AW, Watanabe T (2022) Cost estimations of water pollution for the adoption of suitable water treatment technology. Sustainability 14(2):649. https://doi.org/10.3390/su14020649
38. Radelyuk I, Tussupova K, Klemeš JJ, Persson KM (2021) Oil refinery and water pollution in the context of sustainable development: developing and developed countries. J Clean Prod 302:126987. https://doi.org/10.1016/j.jclepro.2021.126987
39. Shapiro JS (2022) Pollution trends and US environmental policy: lessons from the past half century. Rev Environ Econ Policy 16(1):42–61. https://doi.org/10.1086/718054
40. Apergis N, Payne JE (2010) The emissions, energy consumption, and growth nexus: evidence from the commonwealth of independent states. Energy Policy 38(1):650–655. https://doi.org/10.1016/j.enpol.2009.08.029
41. Dogan E, Seker F (2016) Determinants of CO2 emissions in the European Union: the role of renewable and non-renewable energy. Renew Energy 94:429–439. https://doi.org/10.1016/j.renene.2016.03.078
42. Ehigiamusoe KU, Lean HH, Smyth R (2020) The moderating role of energy consumption in the carbon emissions-income nexus in middle-income countries. Appl Energy 261:114215. https://doi.org/10.1016/j.apenergy.2019.114215
43. Anwar A, Siddique M, Dogan E, Sharif A (2021) The moderating role of renewable and non-renewable energy in environment-income nexus for ASEAN countries: evidence from method of moments quantile regression. Renew Energy 164:956–967. https://doi.org/10.1016/j.renene.2020.09.128
44. World Bank, Fish to 2030: prospects for fisheries and aquaculture
45. European Commission. Joint Research Centre (2022) Clean Energy Technology Observatory, Ocean energy in the European Union: status report on technology development, trends, value chains and markets: 2022. LU: Publications Office. [Online]. https://data.europa.eu/doi/10.2760/162254. Accessed 30 Dec 2025
46. European Commission, The European Green Deal – European Commission. [Online]. https://commission.europa.eu/strategy-and-policy/priorities-2019-2024/european-green-deal_en. Accessed 08 Jun 2025
47. Alsaleh M, Abdul-Rahim AS (2024) What are the influence of fishery activities and their implication on marine water pollution? An empirical analysis. Environ Dev Sustain 28(6):15063–15094. https://doi.org/10.1007/s10668-024-05561-x
48. Alsaleh M, Abdul-Rahim AS, Bai K, Berhane G (2025) Using method of moments quantile regression to examine the influences of fishery operations and their ramifications for marine water contamination. Turk J Fish Aquat Sci 26(6):TRJFAS26217. https://doi.org/10.4194/TRJFAS26217
49. Pedroni P (2004) Panel cointegration: asymptotic and finite sample properties of pooled time series tests with an application to the PPP hypothesis. Economic Theory 20(03):597–625. https://doi.org/10.1017/S0266466604203073
50. Kao C, Chiang MH (2000) On the estimation and inference of a cointegrated regression in panel data. In: Advances in econometrics, vol 15. Emerald (MCB UP), Bingley, pp 179–222. https://doi.org/10.1016/S0731-9053(00)15007-8
51. Sarkodie SA, Strezov V (2019) Effect of foreign direct investments, economic development and energy consumption on greenhouse gas emissions in developing countries. Sci Total Environ 646:862–871. https://doi.org/10.1016/j.scitotenv.2018.07.365
52. Koenker R, Bassett G (1978) Regression quantiles. Econometrica 46(1):33. https://doi.org/10.2307/1913643
53. Binder M, Coad A (2011) From Average Joe's happiness to Miserable Jane and Cheerful John: using quantile regressions to analyze the full subjective well-being distribution. J Econ Behav Organ 79(3):275–290. https://doi.org/10.1016/j.jebo.2011.02.005
54. Canay IA (2011) A simple approach to quantile regression for panel data. Econ J 14(3):368–386

55. Phillips PCB, Sul D (2003) Dynamic panel estimation and homogeneity testing under cross section dependence. Econ J 6(1):217–259. https://doi.org/10.1111/1368-423X.00108
56. Pesaran MH (2021) General diagnostic tests for cross-sectional dependence in panels. Empir Econ 60(1):13–50. https://doi.org/10.1007/s00181-020-01875-7
57. Westerlund J (2007) Testing for error correction in panel data*. Oxf Bull Econ Stat 69(6):709–748. https://doi.org/10.1111/j.1468-0084.2007.00477.x
58. Kremers JJM, Ericsson NR, Dolado JJ (1992) The power of cointegration tests. Oxf Bull Econ Stat 54(3):325–348. https://doi.org/10.1111/j.1468-0084.1992.tb00005.x
59. Bender M, Bustamante R, Leonard K (2022) Living in relationship with the ocean to transform governance in the UN ocean decade. PLoS Biol 20(10):e3001828. https://doi.org/10.1371/journal.pbio.3001828

Chapter 5
Fisheries Production Toward Food Security in the EU Region

Acronyms

EU	European Union
CO_2	Carbon dioxide emissions
MMQR	Method of moments quantile regression
EUR	Euro
CD	Cross-sectional dependency
FP	Fisheries production
WDI	World Bank Development Indicators
GDP	Gross domestic product
PF	Fossil fuel
WP	Water pollution
DOLS	Dynamic ordinary least squares
FMOLS	Fully modified ordinary least squares
FEOLS	Fixed-effects ordinary least squares
Europa	European data portal
Eurostat	European Commission's database
LFP	Natural logarithms of food production
LLC	Lin & Chu Test
IPPS	Im, Pesaran, and Shin Test
EU27	27 European Union Member States
EU13	13 European Union Member States
BRICS	Brazil, Russia, India, China, South Africa
ASEAN	Association of Southeast Asian Nations
EU14	14 European Union Member States
CSA	Climate-smart agriculture
IUU	Illegal, unreported, and unregulated
EG	Economic growth
OLS	Ordinary least square
LAP	Natural logarithm of agricultural production
LGV	Natural logarithm of governance
LEP	Natural logarithm of energy production

M. Alsaleh, *Fisheries Sustainability in the EU*, Urban Sustainability,
https://doi.org/10.1007/978-981-92-4496-6_5

Foods derived from marine ecosystems play a vital role in promoting food security, eliminating hunger, and supporting sustainable, resilient, and environmentally friendly food systems. Despite their importance, blue foods are often overlooked in global food security discussions. In 2017, blue food exports generated greater revenues for developing economies than all other agricultural commodities combined. This study investigates the dynamic influence of the fisheries sector on sustainable food security in 27 European countries while accounting for agricultural production, energy consumption, economic growth, and governance from 1990 to 2022. Using the innovative method of moments quantile regression (MMQR) with fixed effects, the findings reveal a significant positive relationship between fisheries production and sustainable food security across all quantiles except the fifth quantile. Fisheries output enhances sustainable food security from the first to the ninth quantiles, with the strongest effect observed at the upper quantiles and the weakest effect at the lower quantiles. The results further indicate that fisheries production contributes more positively to sustainable food security in the EU13 developing countries than in the EU14 developed countries. Conversely, agricultural production demonstrates a stronger positive contribution to sustainable food security in the EU14 developed nations compared to the EU13 developing economies. The findings also show a substantial positive relationship between agricultural output and sustainable food security, particularly at the median and upper quantiles. Across the EU27 region, sustainable food security improved significantly at the middle and higher quantiles, thereby supporting the growth hypothesis in fisheries-producing economies. Governance was also found to have a positive and statistically significant effect on sustainable food security across both lower and upper quantiles. Furthermore, governance exerted a stronger beneficial influence on food security in the EU14 developed nations than in the EU13 developing countries. The study concludes that effective fisheries sector policies can enhance food security levels in both EU14 and EU13 nations while promoting long-term sustainable food systems.

5.1 Fisheries Toward Food Security

Global concern over food security continues to mount, as world food production needs to increase by some 60–70% in order to feed an estimated 10 billion people by 2050 [1]. In Europe, the challenges to ensuring an adequate food supply for the world's growing population highlight the central role of the EU agri-food sector to improve global food security through increased productivity and efficiency [1]. More specifically, the EU agri-food industry is expected to improve its global competitiveness to feed an expanding, wealthier world population. However, a focus on food availability alone fails to address the broader dimensions of food insecurity. Food security challenges differ widely between low- and high-income countries, and are highly context-specific [1]. Thus, sustainable food security in Europe needs to be addressed through food availability, accessibility, utilization, and stability [1]. EU agricultural exports are currently facing several challenges due to the

Russia-Ukraine conflict, increasing the pressures on food inflation and putting the European food supply at risk [2]. This conflict is undermining food systems sustainability, affecting the price and stability of grains in both EU and international markets. Developing countries, dependent on imports, become more exposed, and this creates humanitarian concerns and additional burdens for European economies [2]. Geopolitical instability and climate change remain the principal threats to food security for many European countries [3]. The European Union is a world-leading supporter of development aid, and it is expanding its response to food security needs to provide relief and address chronic food shortages [3]. In June 2022, EU countries agreed on an €8 billion food security investment package for the period 2020–2024, on top of the average of €1–1.4 billion annual support from the EU that was in place between 2016–2021 prior to the crisis [4]. Additionally, an extra €600 million was allocated in June 2022; €150 million for emergency assistance and €350 million for medium and long-term measures, aimed at boosting food production in European regions [4]. While adapting agriculture is one response to increasing global demand for food, food systems also interact strongly with energy systems, and their interrelations affect EU sustainable food security [3].

The high pressure in fertilizer prices is expected to put more stress on food security and inflation in the next few years. A decrease of 1% in the use of fertilizers in the world may decrease availability of calories by 0.4%, and a 5% reduction could result in 2% fewer calories available in food products [3]. Some countries in the EU could experience a food shortage of around 3% in case of a 5% reduction in the use of fertilizers [3]. The EU, among its roles as the world's largest producer and trader in the agri-food sector (in 2021) [4], exports a significant volume of food to regions in developing countries. More than half of the EU's wheat exports, for instance, go to Asia and Africa, targeting markets like Pakistan, Algeria, Morocco, Egypt, and Nigeria [4]. However, in 2021, disruptions in agricultural production due to geopolitical uncertainty and climate change, combined with soaring inflation and food prices, have presented new obstacles to EU sustainable food security. Indeed, EU food prices, in particular, rose by around 9% in July 2022 compared to the same month last year, according to Eurostat [4]. One of the most resource-intensive areas of the EU's agri-food sector and one responsible for many externalities globally and in Europe, the meat industry has become an interesting focus for this research. As the EU's system for producing meat is considered relatively efficient, reducing production in the EU without altering consumption would negatively affect global food security [5]. Such changes may even lead to increased demand for imports of meat from other parts of the world, putting more strain on land resources and undermining local food production opportunities globally [5]. A move toward diets based on less protein-intensive sources of meat, like poultry or fish (replacing red meat), is being encouraged to ensure healthy diets, and potentially freeing up more land to increase global food availability [5]. Around one-third of all food lost and discarded globally originates from fisheries, impacting around 10–12% of the world's population who are directly dependent on seafood for their livelihoods [6]. Fish provide up to 20% of animal protein per capita for the world's 3.3 billion people, along with essential nutrients. A decline in wild-caught seafood production may put over 800 million people at risk of nutrient deficiency [6]. In 2017, the export revenue from blue foods for developing countries exceeded that of all other agricultural commodities [6].

Around 91% of employment globally occurs in small-scale fisheries [6]. In the EU, in 2016, over 20% of individuals did not have meat, fish, or a vegetarian alternative to include in their diets every 2 days [7]. Low-income households, in particular, increasingly depend on social assistance schemes and subsidized meals at school to provide enough food for their households [7]. By 2020, nearly 9% of the EU population had difficulties affording nutritious food every 2 days [7].

Due to rising prices of energy and raw materials, the geopolitical crisis of 2022 placed significant pressure on EU fisheries operations, making operations economically unviable when fuel prices rose above 0.60 EUR/liter (vs just above 1 EUR/liter in several countries), thus forcing many vessels to cease activity [8]. Fishermen operating in Black Sea countries also suffered, with those formerly fishing in close proximity to Ukraine's waters now actively avoiding conflict areas [8]. Increasing operating costs related to fuel and raw materials, as well as rising logistical costs, also affected the fish farming and processing industries, and while not directly targeting Russian imports broadly, trade flows with the country remained limited, particularly affecting processing companies that rely on Russian whitefish imports [8]. The present study will investigate the following research questions:

Has the growth of fisheries sector led to long-run sustainability of food security in European regions over the period of 1990 to 2022? Are EU countries more vulnerable to threats to food security from agricultural production or fisheries production? Can fisheries sector growth facilitate the achievement of EU sustainable development goals by 2050? Following this stream, this study has three main objectives: (i) to investigate how sustainable fisheries production evolved across EU regions over 1990–2022; (ii) to compare the relative impacts of fisheries and agricultural productions on sustainable food security in the EU14 developed and EU13 developing countries; and (iii) to assess the extent of fisheries sector impacts on food security across developed and developing EU countries over the same period. A considerable weakness of studies on EU economy data is the fact that they are based on the assumption that exogenous disturbance factors are independent and there is no cross-sectional dependence (CD) across country variables. However, countries within the EU are highly integrated and exposed to common shocks, spillover effects, and external crises (resource insecurity, oil price shocks), etc. [9–12]. Ignoring cross-sectional dependence can produce biased and inconsistent evidence about the role of energy use and income in the context of resource insecurity. In this context, this study aims to check whether fisheries-producing economies are equally exposed to the growth-moderating effect of resource dependence.

Rather than other works that utilize wider measures of resource insecurity ([13, 14]), this research makes use of fisheries production as the point source resource variable. Point source resources have relatively stronger impact on growth-reducing effects than diffuse resources ([15, 16]). The analysis sample consists of developed and developing fisheries-producing countries and uses governance as the regulatory variable that can attenuate the resource dependence effect [17]. This research analyzes the case of EU fisheries-producing countries such as Germany, Italy, Sweden, Estonia, Bulgaria, and Malta for the period 1990–2022. To analyze the role of fisheries production on sustainable food security, this study develops a novel method of moments quantile regression (MMQR) following [18]. MMQR allows analysis of differential effects of fisheries production at different quantiles of the conditional distribution of sustainable food security across developed and developing economies, thereby providing richer insights into distributional heterogeneous effects than mean regression methods with the help of fixed effects. Therefore, this study provides a richer account of the growth hypothesis in fisheries-producing countries.

This is among the first studies examining food security in fisheries-producing nations while considering fisheries production, agricultural production, fossil fuel consumption, governance, and economic development simultaneously. The comparative analysis of growth hypothesis across quantiles is important as quantile estimations are relatively less sensitive to outliers compared to those conditional mean estimators [19]. In addition, quantile regression analyses of fisheries production impact on food security give a more comprehensive understanding of the distributional effects which cannot be captured by the mean regression analysis techniques. The remainder of this chapter is organized as follows. Section 5.2 provides a survey of the relevant literature; Sect. 5.3 discusses the data and econometric methodology employed; Sect. 5.4 presents and discusses the empirical results; and finally Sect. 5.5 provides a summary of the findings and policy implications.

5.2 Fisheries and Food Security: Historical Development

Blue food protein is a sustainable and important element of securing the global food supply. Aquaculture-based food systems play a significant role in supporting the diversity and functions of ecosystems and the safe, nourishing, and adequate supply of foods within broader food and nutrition security objectives. For instance, studies [20, 21] investigated the contribution of fisheries in framing agendas for food security in the context of developing countries and highlighted some critical perspectives on the importance of the role of aquaculture in ensuring food security worldwide and stressed that further expansion of aquaculture in regions with poor nutrition and food insecurity is highly necessary.

Likewise, [6, 22] examined the food security challenges associated with the decline of technical efficiency in the fisheries sector of developing economies.

They highlighted the significant threat posed by the inefficiency of fisheries to food security and livelihoods, particularly under conditions of climate change [23–26] investigated the contribution of small-scale fisheries to food security and livelihood development in Africa, identifying the critical role they play in food provisioning, employment, and income generation. However, factors such as climate change, an increase in large-scale industrial fishing activities, and decreasing fish stocks negatively impact their long-term profitability and sustainability [7] looked at how adaptation strategies related to climate change among fishing communities in Indonesia affected their household food security. This study found that greater adaptation measures significantly improve food security indicators in households [8] studied the vulnerability of the socio-ecological system of the Solomon Islands tuna fisheries and stressed the importance of cross-fisheries governance collaboration and cooperation between stakeholders. The researchers emphasized that further international and national cooperation and integration are essential, suggesting stronger partnerships among Pacific nations and coordinated national and subnational management systems.

For effective and equitable global development to be attained, the supply of stable, nutritious, and readily available food sources is vital for the welfare of human beings.

Given their significant role in the food supply and revenue of a country, agricultural systems are central for the economies of developing countries [27, 28]., for instance, implemented agent-based models to analyze the impact of climate-smart agriculture (CSA) on food security, finding that its adoption improves welfare of rural households, especially when coupled with access to capital, networks, and developed food markets. In their systematic review of agricultural research, [29, 30] identified the factors driving sustainable food systems such as land-use change, agricultural productivity, climate change, farmer motivation, urban and peri-urban agriculture, and land suitability [31, 32] looked at the effect of climate variations on agricultural production and food security of rural communities in India and suggested that while climate change impacts agricultural productivity and food grain availability in these areas (threatening food security and livelihood of marginal and small farming households), effective and timely administration of the agricultural sector can lead to more stable long-term food security, as pointed out by [33] for Africa.

Along with food security, the security of energy supply is also fundamental in sustainable development. Energy security, while highly dependent on the local context, typically revolves around the availability, affordability, access, and acceptability of energy in society. For example, [34, 35] assessed the connection between food systems and energy systems within the framework of sustainable food production.

They revealed that technologies in the field of renewable energy are central for enhancing food production processes.

In their assessment of the effects of energy technologies on the Colombian community using the nexus of energy, water, and food (EWF) security, [36] determined that the application of energy technologies boosts food and water security for people, helping achieve Sustainable Development Goals (SDGs). Likewise, [37] studied the interplay of renewable energy and food security in the US using statistical methods and found that even though renewable energy policies increase energy independence, they could negatively impact the accessibility, availability, and stability of foods. Finally, with reference to Africa and based on ecological footprint, biocapacity, and eco-balance data, [38–41] carried out a research study with the objective of understanding the implications for ecological development of implementing the EWF nexus concept. They suggested the utility of the ecological footprint and biocapacity measures for assessing the sustainability impact of the system of energy-food provision.

Also in South Asia, [42] carried out a study aimed at identifying drivers that impact sustainable food systems and highlighted the significance of creating an integrated, environmentally friendly food-energy system for sustaining food production and food security.

Protein obtained from blue foods is a sustainable and vital food source to satisfy the global food demand. Aquatic food systems contribute to the delivery of biodiversity and ecological services, as well as the provision of safe, nutritious, and sufficient food as part of wider food and nutrition security strategies. For example, [20, 21] and others provided insights into fisheries contribution to food security agendas of developing countries, their implications for global food security, and the expansion of aquaculture in highly malnourished and food-insecure regions.

Similarly, [6, 22] showed concerns on the food security challenges from dwindling fisheries technical efficiency in developing economies, emphasizing the dire impact on long-term food security and vulnerable livelihoods in a changing climate.

In the case of Africa, [23–26] studied the role of small-scale fisheries in food security and livelihood outcomes, indicating the importance of small-scale fisheries for food provisioning, employment and income, although at risk from climate change, industrialization of large-scale fisheries and overfishing [7] in Indonesia assessed the contribution of adaptation strategies to climate change to household food security, reporting a higher resilience of household food security for fishing communities with higher climate change adaptations [8]. In their study on the resilience and vulnerability of the Solomon Islands' socio-ecological tuna fisheries system emphasized strong regional cooperation, co-ordination, and cooperation between different levels of fisheries governance frameworks and strongly recommended strong regional partnerships through organizations like PNA as well as greater national and local level cooperation. Providing a secure, nutritious, and adequate food supply for all is paramount for human well-being and sustainable development. For developing economies, agriculture plays a particularly important role, contributing to overall food security and national income.

For instance, [27, 28] employed agent-based modeling to study climate-smart agriculture and its effects on food security.

Their results show how climate-smart agriculture (CSA) significantly improves rural household welfare when coupled with access to capital, social networks, and well-integrated food markets. Similarly, [29, 30] completed systematic literature reviews of agriculture research, identifying key drivers for sustainable food systems. Those drivers included: land-use change, agriculture production, climate change, farmers motivation, urban and peri-urban agriculture, and land suitability.

For India, [31, 32] used models to investigate impacts of climate variability on agricultural productivity and food security for rural communities, showing that negative climate change impacts on crop and grain availability resulted in food insecurity for small and marginal farmers and threats to their livelihood [33] in the African context found evidence that good governance and efficient administration of the agriculture sector are vital for securing food security and farm incomes in the long run. Secure food supply is fundamental to the survival, health, and well-being of people, as is energy for human activities and sustainable development. Energy security (availability, affordability, acceptability, accessibility) has long been understood as a necessary condition for development, even if its definition is very context-dependent.

For example, [34, 35] explored the relationship between the energy sector and food security in the context of sustainable food systems and recommended sustainable use of energy by utilizing renewable energy technologies in food systems [36] analyzed the impact of the energy systems on the community in Colombia via an energy-water-food security nexus approach and provided evidence of how renewable energy access improved both food and water security.

Likewise, [37] studied the interplay of renewable energy and food security in the United States utilizing Partial Wavelet Coherence and Spillover Index analysis and found that while the US strengthened its energy independence due to energy-renewable policy, food availability, stability, and accessibility were likely compromised. By using indicators of ecological footprint, biocapacity, and eco-balance, [38–41] analyzed the environmental dimensions of implementing the water-energy-food security nexus in Africa. Among many findings, they highlighted ecological footprint and biocapacity indicators for the evaluation of ecological sustainability of food and energy systems. The study by [42] identified integrated and sustainable food-energy system can enhance the provision of food for all in the long term by considering a wide range of exogenous factors in the South Asia region.

Sustainable fisheries and ocean governance; sustainable management of fish resources and seafood; ocean governance; marine cooperation; management systems in coastal zones [43–45]. The effect of fishery management on food security and sustainable development was analyzed. Their study revealed that good governance practices in agriculture and fishery could ensure sustainable development and food and nutrition security outcomes.

Moreover, [46–49] fisheries and aquaculture management systems in Africa have been examined as tools for ensuring food security and economic growth. They showed that strong institutions for regulation, a secure system of property rights, and effective tenure structures are needed in fisheries and aquaculture for achieving food security and economic development in African regions, but this situation is only partially fulfilled for most African fisheries. The food sector's economic growth contributes more to reducing poverty than that of many other economic sectors. As one of the factors that can lead to food insecurity, poverty must be reduced by improving food systems, thereby stimulating the broader economy.

[50, 51] assessed the relationships between food security and economic development for several countries and concluded that economic growth in production, transportation, and food loss reduction can have positive effects on sustainable development and global food security.

[52] studied the impact of agricultural production on sustainable economic growth in Asia, pointing out the role of technological innovation to increase food demand and promote long-term economic growth. Similarly, [53] in terms of food security sustainability in Africa, they advocated for the role of renewable energy in rural development policies in ensuring sustainable food security. The role of

economic growth was discussed by [54]. In their work analyzing the various factors determining the food security of different regions in China, i.e., socio-economic, agricultural, and climatic aspects, they established that the role of the economy in ensuring sustainable food production and food security is fundamental.

Instead of most previous studies, which focus on the relationship between food production and food security, this study aims to contribute by analyzing how consumption of fisheries can lead to more food-secure and sustainable food systems in general, and how sustainable food production, driven by Fisheries production, contributes to sustainable food security in the European Union (EU). To ensure blue sustainability, this chapter analyses how sustainable development goals, regarding food security in the EU region, can be ensured through efficient and sustainable fishery production. Furthermore, instead of using traditional time-series methods, this study applies both static and panel regression analysis. Several new and interesting contributions distinguish the current study from existing literature.

Firstly, it comprehensively analyzes the interaction between food security, fisheries production, agriculture production, fossil fuel consumption, governance, and economic development for EU countries for the period between 1990 and 2022.

Secondly, the effects of fisheries production on sustainable food systems and food security for sustainability goal can vary across regions, and thus this chapter investigates the heterogeneous effects. Finally, to address potential differences between developed and developing regions within the EU, the full sample of EU27 is divided into two sub-groups: the EU14 developed countries and the EU13 developing countries. Considering that the dataset ranges from 1990 to 2022 and there are potential unit roots in variables, the study utilizes unit root and cointegration tests to analyze the stationarity and long-run relationship between variables.

In addition, this chapter employs the method of moments quantile regression (MMQR) to examine the impact of fisheries production, economic development, and food security on each other at different conditional quantiles. Static and panel regressions are employed in the static analysis and panel analysis, respectively, to reveal the determinants of food security and sustainable ocean development. Lastly, this chapter uses the estimated parameters of the panel regression model to analyze potential futures of the impact of fisheries production on food security.

5.3 Food Security and Fisheries Modeling

The following model was developed in part as a result of recent investigations by [55–58].

$$FS_{it} = f\left(FP_{it}, AP_{it}, EP_{it}, EG_{it}, GV_{it}\right)$$

Variables include FS—food supply security, FP—fisheries production, AP—agricultural production, EP—energy consumption, EG—economic growth, and GV–governance. An adequate assessment of the amount of food production relies critically on the availability of data. In the context of the estimation of the amount of food and nutrition production, average values for food supply level are utilized. FS is for food supply sustainability [59], and FP is for fisheries production growth by taking the volume of all fisheries production in tons of output in weight; AP is for agricultural production growth by land, taking the proportion of land dedicated to arable and horticultural crops, permanent and temporary pastures. All data series on these variables are from the World Development Indicators database [60]. Gross domestic product (GDP) constant US$ represents EG; the data were collected from the European Commission statistical database [61]. Energy consumption (EP) data are derived by taking the energy consumption of petrol in thousand tons of oil equivalent from Eurostat [61]. Official European data portal (Europa) provided data on the quality of government (GV) as proxied by governmental effectiveness. In this study, we focus on the EU economies and years from 1990 to 2022, and all variables are converted to a per capita base. To obtain the findings of this study, we have selected countries in the European region and used their annual information from 1990 to 2022, from the World Bank's World Development Indicators (WDI) and the European Commission database [61] and from the official European data portal (Europa) covering 1990 to 2022 years. Table 5.1 shows the data, units, sources, and transformations of the variables.

Descriptive statistics: This sub-section shows and comments upon the empirical results of the research. Table 5.2 reveals the descriptive statistical measures of selected variables like maximum, minimum, mean, and standard deviation values. The means of the variables are 2.001, 4.927, 1.584, 4.377, 2.753, and 1.867, while the standard deviation values of the selected variables are 0.044, 0.825, 0.255, 0.395, 0.761, and 0.070. Thus, the means of the selected variables were presented as it gives us first idea about the distribution of selected variable; this is also a guide in unit root test for this study.

We used fully modified ordinary least squares (FMOLS), dynamic ordinary least squares (DOLS), and fixed-effects ordinary least squares (FEOLS) for comparisons. In FEOLS, we used Driscoll and Kraay standard errors, which allow common types

Table 5.1 Data, units, transformation, and source

Data	Measurement unit	Transformation	Source
Food supply	(2004–2006 = 100)	Natural logarithm transformation	WDI
Fishery production	Metric tons	Natural logarithm transformation	WDI
Agriculture production	Agricultural land (% of land area)	Natural logarithm transformation	WDI
Energy production	Metric tons of oil equivalent	Natural logarithm transformation	Eurostat
Economic growth	Constant US dollars	Natural logarithm transformation	Eurostat
Governance	% of confidence for governance	Natural logarithm transformation	Europa

Source: [62]

Table 5.2 Results of descriptive statistics

Variable	Observations	Mean	Std. dev.	Min	Max
FS	864	2.001	0.044	1.848	2.227
FP	864	4.927	0.825	3.211	6.310
AP	864	1.584	0.255	0.848	1.867
EG	864	4.377	0.395	3.129	5.253
FF	864	2.753	0.761	0.011	4.600
GV	864	1.867	0.070	1.482	1.979

Source: [62]

of cross-sectional dependence and autocorrelation up to a lag (up to the second lag in this analysis). One source of heterogeneity for estimating dynamic cointegrated panels, according to [63] (as discussed by some other papers), concerns heterogeneity from variations of cross-sectional means, while another concerns it related to heterogeneous adjustment toward equilibrium across the cross-sectional units.

To solve the heterogeneity concerns caused by differential cointegrating parameters and common serial correlation properties, the FMOLS of [60] model allows each panel member its own intercept and allows differences in the serial correlation properties of the error processes among panel members.

Results of a Monte Carlo simulation show that the DOLS estimator is unbiased with respect to OLS and FMOLS estimators in finite samples, which motivated [64] to adapt the DOLS estimators to the panel context. Furthermore, DOLS estimation further solves the problem of endogenous feedback through incorporating leads and lags of the differences in estimating a dynamic equation and then corrects the problem of endogeneity. In the face of limitations of previous estimation techniques, a panel quantile regression was employed to examine the effects that could occur on conditional quantiles [65]. The work of [66] was the first to introduce panel quantile regression estimation methods.

As different from the least-squares estimation, which provides estimations of the conditional mean of the dependent variable as a function of the independent variables, quantile regression typically is used to estimate conditional medians or other specific quantiles of the distribution of the response variable.

Compared to standard least-squares estimates, quantiles are less sensitive to outliers in the estimation. They are especially useful when there is weak or non-linear association between the conditional means of the two variables [67]. Nevertheless, in this chapter we adopted the method of moments quantile regression (MMQR) of [18] with fixed effects.

Unlike quantile regression, which is robust to outliers, the MMQR approach handles the unobserved heterogeneity issue among individuals in panel settings, and allows for individual effects impacting the full distribution, not just the mean. That is why this methodology was successfully used in order to analyze heterogeneity in many cases like [19, 68], to name just a few, thus making the identification of the conditional heterogenous covariance effects of determinants of water pollution possible. It has been proved that MMQR approach is suitable if a panel data model with individual-level effects has endogenous variables.

The MMQR estimator is also relatively simple to compute and provide non-crossing quantile estimates.

If we assume a location-scale model specification, then the computation of the conditional quantiles QY(τ|X) is given by:

$$Y_{it} = \alpha_{i+} X_{it}^{\prime} \beta^{\prime} + \left(\delta_i + Z_{it} \gamma^{\prime} \right) U_{it} \tag{5.1}$$

Where parameters that need to be evaluated and the probability, $P(\delta_i + Z_{it}\gamma' > 0) = 1.(\alpha, \beta', \delta, \gamma')$. Z is a k-vector of identifiable components of X that are differentiable transformations with element l provided by: $(\alpha_i, \delta_i), i = 1, \ldots, n$, represents the individual i fixed effects.

$$Z_l = Z_l\left(X\right), l = 1, \ldots, k \tag{5.2}$$

Xit is separately identifiable over time and over i, for each fixed i. Uit is uncorrelated with Xit (the orthogonality condition), and is identically and independently distributed over individuals (i) and over time (t); furthermore, Uit satisfies the specified moment conditions which require, inter alia, only weak exogeneity. From Eq. (5.1), it follows that

$$Q_Y\left(\tau | X_{it}\right) = \left(\alpha_{i+} \delta_i q\left(\tau\right)\right) + X_{it} \beta^{\prime} + Z_{it} \gamma^{\prime} q\left(\tau\right) \tag{5.3}$$

In our current paper, the independent variables are the natural logarithm of food production (LFP), natural logarithm of agricultural production (LAP), natural logarithm of economic growth (LEG), natural logarithm of energy production (LEP), and natural logarithm of governance (LGV). The term Xit − αi(τ) + βi q(τ) indicates the quantile-specific fixed effect for individual i and quantile. Yit is the dependent variable (natural logarithm of food supply sustainability); the conditional quantile distribution is QY(τ|Xit), depends on the placement of the independent variable Xit, and q(τ) indicates the -th sample quantile of the distribution that we estimate by solving the optimization problem. It is a time-invariant term and not a simple shift of intercept like the individual effects under conventional least-squares fixed effects.

$$\min_q \sum_i \sum_t P_\tau \left(R_{it} - \left(\delta_i + Z_{it} \gamma^{\prime} \right) q \right) \tag{5.4}$$

Where P_τ (A) = $(\tau - 1)$ AI {A ≤ 0} + TAI {A > 0} denotes the check function.

5.4 Food Security from Fisheries Resources

In this study, we employ several standard preliminary tests to evaluate time-series properties of the variables prior to estimate the unknown coefficients. Our first and foremost task is to ascertain whether there is cross-sectional dependence (CD) in the panel. Cross-sectional dependence can lead to spurious values of the true parameters of coefficient estimates. Unreported cross-sectional dependence from unobserved factors can significantly reduce the potential benefits of using panel data [69] if it goes unchecked. Hence, this has to be controlled to get valid coefficient estimates. The test of cross-sectional dependence for the panel, CD, developed by [70], is used. We can observe from other quantitative variables in Table 5.3 that, except for fisheries production, significant CD exists among countries. This requires us to conduct unit root and cointegration tests as well as use a panel estimation technique which takes care of the impact of CD to prevent size distortions [71].

We use both [71] (LLC) and [72] (IPS) unit root tests to evaluate unbiasedly the integration properties of the variables in question. The main goal of panel unit root tests is to increase test power by pooling cross-sectional and time-series information. A number of panel unit root tests have been constructed to test the unit root null hypothesis against an alternative. Tests are generally based on combining cross-sectional information into single test statistics that may have greater power than individual, standard, non-panel tests. Tests by [71] test the unit root null hypothesis against a homogeneous alternative, where each series in the panel is stationary with the same speed of mean reversion [72] test the unit root null hypothesis against the alternative that at least one series in the panel is stationary. As the new test that this study suggests focuses on panel stationarity, it can reasonably point toward the LLC framework. This premise is, however, let go by the Im-Pesaran-Shin (IPS) (2003) panel test, which allows for each individual to have a separate autoregressive coefficient. When it comes to fairly small panels, like in this case, the Levin, Lin, and Chu (2002) (LLC) panel unit root test provides better power performance against other unit root tests in similar settings. Here we use these two tests, which analyze how cross-sectional dependence impacts the panel unit root tests. As shown in Table 5.4, all the variables are non-stationary at levels, but they become stationary after taking first differences in any of the unit root test specifications. Thus, we can assume that all variables in the estimations are integrated of order one, I(1).

We then use the [63] Pedroni (2004) panel cointegration test and the [73] bootstrapped panel cointegration test in Table 5.5 to establish whether the true long-run relationships exist among variables [63] developed a complete approach to testing panel cointegration based on the framework of the Engle and Granger two-step

Table 5.3 Cross-sectional dependence results

Variables	FS	FP	AP	EG	EP	GV
Pesaran (2004)	(3.12)***	(18.82)***	(45.65)***	(83.55)***	(65.35)***	(31.30)***

– Remark: *** refer importance at the 1%, scale
– Source: [62]

Table 5.4 Panel unit root test results

	Difference		First difference	
Variable	LLC	IPS	LLC	IPS
FS	0.263	−4.857	−21.00***	−25.58***
FP	−3.076	−3.700	−24.13***	−22.73***
AP	−1.279	−25.13	−1.730***	−24.03***
EG	−1.571	−0.210	−18.74***	−68.36***
EP	−9.255	−4.970	−20.61***	−21.40***
GV	−1.090	−0.254	−23.47***	−23.47***

– Remark: *** refer importance at the 1%, scale
– Levin, Lin, & Chu test (LLC), and Im, Pesaran, and Shin W-stat test (IPS)
– Source: [62]

Table 5.5 Panel cointegration test for EU countries

Test	Without trend		With trend	
A). Pedroni residual cointegration test:				
Alternative hypothesis: common AR coefficients (within dimension):				
Panel v-statistic	−0.611 (0.729)		−1.884 (0.970)	
Panel rho-statistic	0.631 (0.736)		0.829 (0.796)	
Panel PP-statistic	−7.646*** (0.000)		−9.669*** (0.000)	
Panel ADF-statistic	−1.319*** (0.093)		−3.490*** (0.000)	
Alternative hypothesis: common AR coefficients (between dimensions)				
Group rho-statistic	−2.715		(0.996)	
Group PP-statistic	−12.747***		(0.000)	
Group ADF-statistic	−2.811***		(0.002)	
B). Westerlund (2007) Bootstrap panel cointegration:				
Statistics	Value	Z-value	p-value	Robust p-value
Gt	−3.648	−3.643	0.000	0.000**
Ga	−9.401	−5.661	1.000	0.145
Pt	−16.568	−1.835	0.033	0.045***
Pa	−10.446	3.071	0.999	0.849

– Remark: ***, **, and * refer importance at the 1%, 5%, and 10% scales respectively
– Values in parentheses are *p*-values
Source: [62]

procedure. The main advantage of Pedroni test is to take heterogeneity into account by detrending variables with individual deterministic trends and short-run dynamics during the first processing. With the help of estimated residuals, Pedroni produces seven different statistics, which are separated into common process—so-called "pooled" or "within-dimension" tests, and individual processes called "grouped" or "between-dimension" tests in Table 5.5. There are four new tests in the approach of

Table 5.6 Model 1. Panel estimation for the EU-27 region from 1990 to 2022

Long-run coefficient	DOLS		FMOLS		FEOLS	
	Coefficient	Std. error	Coefficient	Std. error	Coefficient	Std. error
FP	0.088	(0.027)***	0.112	(0.030)***	0.017	(0.067)***
AP	0.336	(0.119)***	0.532	(0.126)***	0.053	(0.022)
EP	0.023	(0.035)	0.112	(0.033)***	0.016	(0.055)***
EG	0.191	(0.063)***	0.174	(0.051)***	0.131	(0.068)**
GV	0.154	(0.130)	0.264	(0.136)***	0.152	(0.051)***

– Note: ***, **, and * indicate significance at the 1%, 5%, and 10% levels, respectively
– Values in parentheses are P-values
Source: [62]

[63] with no cointegration being tested on the null hypothesis. The test specification has the advantages of allowing the dynamics of the series to be structural, rather than residual-based, thereby dropping the common factor restrictions imposed in residual-based tests, since the restriction can severely limit the power of residual-based cointegration tests [74].

Thus, the short-run and long-run adjustment parameters do not have to be equal [73]. The bootstrap procedure helps eliminate the distorting effect of the cross-section dependence, and the resulting critical values are therefore robust. From Table 5.5, we notice that the bootstrapped cointegration test by [73] and the test by [63] strongly support cointegration.

The results of the estimation using FMOLS, DOLS, and FEOLS methods are shown in Table 5.6. As demonstrated in Model 1 in Table 5.6, the coefficient estimates of the three specifications on average tend to be similar in general but vary in degree and statistical significance. Fisheries production in the EU region turns out to be the most stable of all three estimators with respect to coefficient values and statistical significance. For FMOLS, FEOLS, and DOLS, a 1% increase in fisheries production increases sustainable food security by 0.11%, 0.11%, and 0.08%, respectively. The finding accords with the principal findings of [20–22] with respect to sustainable consumption of fisheries resources in the long run, habitat conservation, increased food security, and reduced poverty in fisheries-dependent areas.

The results on the impact of agricultural production show that it leads to increased food security in the EU region by 0.31% and 0.12% in DOLS and FMOLS, respectively, as we expected. The coefficient, however, turned out to be inconsistent across the FEOLS specifications, which are consistent with those of [27, 29]. It should therefore be emphasized that, to ensure that comprehensive agricultural sector development is crucial for sustainable food security and nutrition. It includes expanding availability and variety of foods, facilitating economic transformation and the creation of primary income generation for the residents in the rural EU region.

The results of energy production show a significant impact on food security in the EU region, as it turns out to improve food security by 0.01% and 0.11% from FEOLS and FMOLS, respectively; however, it is found to be inconsistent in the DOLS specifications, which corroborates with the growth hypothesis in energy-producing economies. The finding corroborates with the principal findings of [34–36], where energy production was found to have a positive and an inelastic impact on food security.

Similarly, results from all specifications revealed that economic growth has a significant positive impact on food security in the EU region, unlike [62], but similar to the modernization hypothesis of [50–52]. The FMOLS, DOLS, and FEOLS specifications provided highly significant evidence to support the growth hypothesis. Adequate, safe and nutritious food is a human right, which deserves to be given top priority especially for the most vulnerable people. Economic security goes along with development, sustainability, and appropriate nutrition.

In a similar fashion, the coefficient for governance is found to have a positive and significant influence across the estimation techniques, which is in line with the modernization hypothesis of [43], but inconsistent with the results of [44]. Both FMOLS and FEOLS are found to provide a high degree of significant evidence for the growth hypothesis whereas the DOLS model offered a moderately significant one. Therefore, strategic intervention on rural development and food security would play a significant role in permanently eradicating hunger in the region.

The results of the panel quantile regression presented in Table 5.7 are significant for most quantiles except for the fifth and median ones; in particular, fisheries production has a significant long-term effect on food security. These findings also suggest that the performance of the fisheries production strategies is more effective for those countries with a higher food security level.

Results showed that a growth hypothesis exists in all food security levels, except for the fifth quantile in EU27. This could suggest that fish production's growth has priority compared to the other food-producing sectors in all of the countries. Development of renewable resources from nonrenewable resources in fisheries sector may not be that easy at the median level of developed countries due to climate change concerns and related investment costs. In median-level countries, no significantly positive effect of fisheries' production on food security was revealed.

One of the reasons for such a result may be the different level of technology used for catching fisheries in the established EU14 countries versus newly admitted ones, the EU13 countries. From [20] and [21] confirmed the growth hypothesis for several countries and documented extraordinarily high food security level; the implications of our results for future policies could be very large. Also, the results we reached corroborate recent empirical research by [75], which affirmed the growth hypothesis and another fundamental conceptualization for food security. Our findings are also in agreement with [22], which confirmed the growth hypothesis for top leading food production and nutrition producers. In more detail, our results reveal a strong positive relationship between agricultural production and food security for the fourth to the ninth quintiles, while the influence of agricultural production is quite low for the first and third quintiles. A similar increasing trend can be observed for the influence of agricultural production, as it happened for fish production. On the other side, energy production is positively associated with food security in the bottom and top quantiles, but not for the intermediate quantiles. As with the others, the influence of energy production tends to increase as the quantile gets higher. All the

Table 5.7 Panel quantile estimation results for EU27 countries

Variables	Method of moments quantile regression with fixed effects										
	Location	Scale	Quantiles								
			0.1	0.2	0.3	0.4	0.5	0.6	0.7	0.8	0.9
FP	(0.067)***	(0.048)**	(0.010)***	(0.088)***	(0.057)**	(0.040)**	(0.031)	(0.036)*	(0.066)***	(0.092)***	(0.021)***
AP	(0.022)	(0.043)	(0.084)	(0.056)	(0.030)	(0.070)*	(0.093)**	(0.113)**	(0.114)**	(0.168)***	(0.362)***
EP	(0.055)***	(0.091)***	(0.072)***	(0.069)***	(0.043)	(0.018)	(0.014)	(0.044)	(0.088)***	(0.099)***	(0.251)***
EG	(0.068)**	(0.013)**	(0.075)***	(0.060)***	(0.035)*	(0.028)	(0.038)***	(0.025)*	(0.030)	(0.030)	(0.280)***
GV	(0.051)***	(0.169)***	(0.055)**	(0.061)***	(0.054)***	(0.037)***	(0.015)	(0.034)	(0.112)***	(0.197)***	(0.445)***

Notes: *, **, and *** indicate significance at 10%, 5%, and 1% levels, respectively
Source: [62]

Table 5.8 Model 2. Panel estimation for the EU-14 region from 1990 to 2022

Long-run coefficient	DOLS		FMOLS		FEOLS	
	Coefficient	Std. error	Coefficient	Std. error	Coefficient	Std. error
FP	0.105	(0.013)***	0.042	(0.024)*	0.032	(0.020)
AP	0.196	(0.127)	0.809	(0.111)***	0.180	(0.048)***
EP	0. 021	(0.082)**	0.081	(0.015)***	0.012	(0.024)
EG	0.084	(0.031)**	0.053	(0.019)***	0.048	(0.029)*
GV	0.243	(0.114)**	0.343	(0.056)***	0.119	(0.034)***

– Note: ***, **, and * indicate significance at the 1%, 5%, and 10% levels, respectively
– Values in parentheses are P-values
Source: [62]

variables, in particular economic growth, had a significant positive influence on food security in general, but it is not significantly present for the fourth, seventh, and eighth quantiles. Based on these, food security and economic development strategies can promote fisheries development in more established fish-producing nations. However, according to governance results, the influence of governance on food security is positive in all quantiles and statistically significant in the case of strong fish production (except for the fourth and fifth quantiles), because it confirms the positive link between governance and food security.

Table 5.8 provides estimates of DOLS, FMOLS, and FEOLS estimators. Though statistical significance differs across the estimates, Model 2 in Table 5.8 shows that estimates from the three specifications appear relatively close to each other. Fisheries development in the EU14 developed countries appears to provide consistent results across two specifications as far as magnitudes of estimated coefficients and statistical significance are concerned. Based on the DOLS estimator, an increase of 1% in fisheries production positively affects sustainable food security by 0.04%; FMOLS yields a higher positive impact of 0.10%, whereas the FEOLS estimation specification yields weakly significant results. These results corroborate the principal results of [6, 25, 26]. Food security can be obtained via the direct pathway (nutrition) as well as the indirect path (income to purchase food) through fisheries.

With regard to coefficients' magnitudes and statistical significance level, agricultural development in the EU14 developed countries seems to confirm similar patterns across two specifications. An increase of 1% in agricultural output enhances sustainable food security by 0.18% according to the FEOLS estimator and by 0.80% according to the FMOLS estimator, whereas the DOLS estimation specification presents statistically marginal results. These findings are consistent with the main results in [30–32]. Food security can be obtained directly (as nutritional intake) or indirectly (income effect on consumption) via agricultural activities.

The impact of energy production in the EU14 developed countries on the sustainability of food security appears to positively and significantly affect food security, as expected. However, the coefficient of energy appears not to show any consistent result across the FEOLS estimates but not across FMOLS and DOLS estimates with agricultural production estimated at 0.08% and 0.02%, respectively, as a result of 1% increase in energy production based on FMOLS and DOLS estimators, respectively. These are not different from [37–39], who report similar findings and evidence. As observed, promoting energy security simultaneously promotes food security. The direct connection lies in the fact that the production and supply of food generally requires high energy input; hence, energy security is of critical concern for food security as well as for fighting food insecurity.

Furthermore, the level of sustainable food security in the EU14 developed countries seems significantly influenced by economic development. With an increase in economic growth by 1%, sustainable food security increases by 0.04%, 0.05%, and 0.08% in FEOLS, FMOLS, and DOLS estimators, respectively. This confirms the growth hypothesis in the panel of the EU14 developed countries. The results align with the key findings of [53, 54] on the inelastic and positive impact of economic development on sustainable food security. Sustainable development is not possible without a country achieving food security across the EU14. A country-owned and country-driven food security policy is an essential prerequisite for a world, regional, and national economy to develop, especially if food insecurity still prevails globally, regionally, and nationally.

The governance aspect in the EU14 developed countries reveals a positive and substantial impact, regardless of the estimation specification (FEOLS, FMOLS, DOLS). This result goes contrary to the evidence of [46] but aligns with the modernization theory presented by [46, 47]. An increase of 1% in governance boosts sustainable food security by 0.11% according to the FEOLS estimation specification, 0.34% by FMOLS, and 0.24% by DOLS. In sum, FEOLS, FMOLS, and DOLS specifications present strong and statistically significant evidence supporting the growth hypothesis.

Table 5.9 shows that the effect of fisheries on food security is statistically insignificant across quantiles 1 to 8 but significant at the top of the food security spectrum (9th quantile), suggesting policy interventions targeting fisheries-related aspects of food security are effective in the developed EU14 economies. The finding also indicates that the growth hypothesis holds in developed EU14 only at the top quantiles of food security. This could imply that for food security purposes in developed EU14 countries, fisheries is not given precedence over other agricultural subsectors at lower quantiles. In developing EU14 countries at lower quantiles and lower development stage, lack of resource may also make it difficult to support the fisheries subsector. In countries with a high level of food security (represented by top quantile) in EU14, fisheries has a larger, significant impact on food security, which may be related to a high level of diversity in fisheries' production and processing techniques used in developed EU14 economies. Such findings auger well for food policy in these countries, which currently enjoy a high level of food security in recently year according to [6] who confirm the growth hypothesis in many developed economies. It appears our findings conform to findings reported in [26], which again has confirmed the growth hypothesis at the top quantile level. Similarly, our findings concur with the work of [25], who demonstrate that the growth hypothesis in main fisheries exporting country holds at the top quantile for recent years.

For agricultural production, a positive and significant impact is detected across quantiles, suggesting policy measures aiming to support agricultural food security

Table 5.9 Panel quantile estimation results for EU14 countries

Variables	Method of moments quantile regression with fixed effects										
	Location	Scale	Quantiles								
			0.1	0.2	0.3	0.4	0.5	0.6	0.7	0.8	0.9
FP	(0.020)	(0.017)	(0.022)	(0.022)	(0.013)	(0.019)	(0.021)	(0.021)	(0.006)	(0.005)	(0.053)**
AP	(0.048)***	(0.040)**	(0.011)	(0.017)***	(0.014)***	(0.015)***	(0.016)***	(0.015)***	(0.020)***	(0.025)***	(0.035)***
EP	(0.024)	(0.032)	(0.068)	(0.035)	(0.035)	(0.074)	(0.076)	(0.180)	(0.173)	(0.152)	(0.577)
EG	(0.029)*	(0.015)**	(0.024)	(0.018)	(0.028)*	(0.010)	(0.020)	(0.029)	(0.038)*	(0.041)*	(0.038)**
GV	(0.034)***	(0.039)**	(0.215)*	(0.169)	(0.136)	(0.096)	(0.086)	(0.084)	(0.083)	(0.105)	(0.228)***

Notes: *, **, and *** indicate significance at 10%, 5%, and 1% levels, respectively
Source: [62]

are effective in countries with higher levels of food security. Thus, in the EU14 developed countries, the agricultural subsector has received preference over the fisheries subsector.

Interestingly, for the governance effect on food security in developed EU14, the positive impact observed at both low and high quantiles appears to be statistically weak at higher quantiles, suggesting the limitation on the scope of governance, regulatory actions, or specific capacity in facilitating and strengthening food production security in the highly productive food-producing countries in EU14. The impact of energy production on the sustainability of food security is positive and statistically significant at all quantiles, except for the quantiles 3, 7, 8, and 9, where it is substantially correlated. In turn, at the lowest and highest quantiles of food security, the growth hypothesis indicates that economic development significantly affects food security in developed EU14, while there is an insignificant result at the low quantiles (1, 2, 4th to 6th). These results reflect the greater policy attention on production and economic growth aspects than food security in the developed EU14 countries.

The coefficients obtained from the various panel estimation techniques (FEOLS, DOLS, FMOLS) employed in Model 3 are shown in Table 5.10. In Table 5.10, the various estimation techniques used to obtain results are nearly similar in magnitude. Thus, it is evident that fishery production has a strong and positive influence on the long-run food security of the EU13 developing countries.

A 1% increase in fishery production results in 0.02%, 0.06%, and 0.03% change in sustainable food security under the FEOLS, FMOLS, and DOLS estimators, respectively.

The positive and statistically significant fishery production coefficients obtained through these three methodologies support the presence of the growth hypothesis in the EU13 developing countries. This result aligns with that of [23, 24, 8] who confirm that the levels of food security increase as fisheries production increases in the developing stages. Furthermore, it indicates that agricultural production has a positive and significant effect on the long-run food security of EU13 developing countries. A 1% change in agricultural production leads to 0.33% and 0.20% change in sustainable food security according to the FEOLS and FMOLS estimators, respectively.

Table 5.10 Model 3. Panel estimation for the EU-13 region from 1990 to 2022

Long-run coefficient	DOLS		FMOLS		FEOLS	
	Coefficient	Std. error	Coefficient	Std. error	Coefficient	Std. error
FP	0.039	(0.014)***	0.065	(0.026)**	0.025	(0.011)**
AP	0.202	(0.069)***	0.333	(0.115)***	0.059	(0.027)
EP	0.025	(0.017)	0.026	(0.023)	0.027	(0.005)
EG	0.038	(0.130)	0.067	(0.038)*	0.086	(0.032)**
GV	0.894	(0.089)***	0.357	(0.141)**	0.060	(0.031)*

– Note: ***, **, and * indicate significance at the 1%, 5%, and 10% levels, respectively
– Values in parentheses are P-values
Source: [62]

Positive and statistically significant coefficients for agricultural production based on both the FEOLS and FMOLS methodologies prove the existence of the growth hypothesis in EU13 developing countries.

However, the results from the DOLS coefficient appear not consistent with its specifications [33, 28, 29] also confirm that the levels of food security increase as agricultural output increases in the initial stages of development. In addition, it further reveals that economic development has a strong and positive effect on the long-run food security of EU13 developing countries. A 1% rise in the rate of economic growth has resulted in a 0.08% and 0.06% rise in sustainable food security for FEOLS and FMOLS estimators, respectively.

Nevertheless, the DOLS estimation specification appears to show a limited statistical significance. Positive and significant economic growth coefficients according to the FEOLS and FMOLS estimation procedures show support for the growth hypothesis in the EU13 developing countries [51, 52, 54] provide similar results. The findings of these regressions also imply a positive relationship between governance and long-run food security of the EU13 developing countries.

A 1% increase in the levels of governance results in an increased level of sustainable food security of 0.06%, 0.35%, and 0.89% under the FEOLS, FMOLS, and DOLS estimators, respectively.

The analysis strongly supports the assertion that better governance could bring more improvements in sustainable food security in developing and EU countries. The method of moments quantile regression results are reported in Table 5.11. Firstly, we can observe a positive effect of fishery production on food security. It can be seen from Table 5.11 that for all the quantiles, the variations in food security resulting from fishery production have a significant positive influence, and its effect on food security increases as we move from the first to the ninth quantiles (0.015 to 0.082).

Likewise, the table displays the results of method of moments quantile regression; it can be clearly observed that agriculture production has a positive influence on food security.

The result of agricultural production also has an impact across the different quantiles, and its impact on food security increases from the first to the ninth quantiles (0.099 to 0.320). It can also be seen that the impact of energy production has a significant and positive effect on food security across the quantiles, with the impact increasing as we move toward higher quantiles. Similar to this, governance has a positive impact across all quantiles, but from the first to the ninth quantiles the effect is statistically significant.

Conversely, to the other variable, governance impacts negatively as we move toward higher quantiles. Moreover, the results show that economic development also shows a positive effect on food security for most of the higher quantiles; particularly the sixth to the ninth quantiles have a statistically significant effect, with coefficients rising from 0.080 to 0.115 as we move toward higher quantiles.

FEOLS panel estimations were supported by FMOLS and DOLS panel estimators. As observed, the panel FEOLS coefficients are not only similar to panel FMOLS and DOLS in signs but also in significance level. Therefore, it implies that panel FMOLS

Table 5.11 Panel quantile estimation results for EU13 countries

Variables	Method of moments quantile regression with fixed effects										
	Location	Scale	Quantiles								
			0.1	0.2	0.3	0.4	0.5	0.6	0.7	0.8	0.9
FP	(0.011)**	(0.021)***	(0.015)	(0.023)**	(0.022)***	(0.037)***	(0.044)***	(0.053)***	(0.058)***	(0.071)***	(0.082)***
AP	(0.027)	(0.025)	(0.099)***	(0.075)**	(0.107)***	(0.150)***	(0.178)***	(0.178)***	(0.213)***	(0.258)***	(0.320)***
EP	(0.005)	(0.048)	(0.006)***	(0.004)***	(0.007)***	(0.019)***	(0.023)***	(0.026)***	(0.030)***	(0.038)***	(0.037)***
EG	(0.032)**	(0.067)***	(0.011)	(0.007)	(0.006)	(0.059)	(0.063)	(0.080)**	(0.114)**	(0.136)***	(0.115)**
GV	(0.031)*	(0.506)***	(0.934)***	(0.944)***	(0.934)***	(0.856)***	(0.827)***	(0.808)***	(0.759)***	(0.698)***	(0.652)***

Notes: *, **, and *** indicate significance at 10%, 5%, and 1% levels, respectively
Source: [62]

and DOLS results are quite reliable and conclusions can be drawn. Same as panel FMOLS and DOLS, the FEOLS coefficients are similar in signs while varying significantly in significance level, but panel DOLS estimate results seem reliable enough and avoid serial correlation and endogeneity concerns. The EU27 countries were grouped into 14 developed countries and 13 developing countries in order to identify the impacts of the fisheries sector on long-term food security in each country based on the level of food security of countries. The expected influence of fisheries sector on long-term food security in the EU14 developed countries has been presented in Tables 5.8 and 5.9, and that of fisheries sector on long-term food security in the EU13 developing countries has been presented in Tables 5.10 and 5.11. The evidence of the EU14 developed and EU13 developing countries provided in Tables 5.8 and 5.10 reveals clearly that the fisheries sector greatly improves the sustainability of food security and also provides evidence that the impact of fisheries sector on long-term food security is greater in EU13 developing countries than in EU14 developed countries, of about 0.025 in magnitude. This may imply that the effective way to improve long-term food security in the EU13 developing countries would be through fishing policy and marine resource subsidy policy rather than in EU14 developed countries.

The evidence provided by the data indicates also that the impact of agriculture sector on the sustainability of food security is more significantly important to developed EU14 countries than in developing EU13 countries by magnitudes of 0.180 and 0.059, respectively. In other words, developed countries in the EU14 will enhance food security far more significantly through agriculture policy and food crop subsidy policies than developing countries in the EU13. Moreover, data suggest that the impact of economic development on long-term food security is greater in EU13 developing countries than in EU14 developed countries of magnitudes 0.086 and 0.048, respectively. Investment in green food and environment-sustainable food industry in EU13 developing countries would achieve food security to a far greater extent and much more quickly than in EU14 developed countries. Again, evidence also implies that developed EU14 countries have a higher positive impact of governance than developing EU13 countries by magnitudes of 0.119 and 0.060, respectively. The impact implies that developed countries in EU14 can develop a far higher degree of food security and much more quickly than developed countries in EU13 by implementing food governance policy, food policy, and marine food subsidy policy in agriculture sector. This is also explainable by the reason that fisheries sector may be exempted from tax by the current law amendment of fiscal policy in EU13 developing countries.

5.5 Conclusion and Condition of Food Security Through Fisheries Resources

In this study, we will evaluate the empirical relationships of fishery production, agricultural production, energy production, economic growth, governance, and food security while allowing for the growth hypothesis. It adds to literature by building on sustainable food security literature and with a focus on the development of fisheries sector. We use panel cointegration, panel unit root, and panel estimation techniques, drawing upon more than 31 years of data for 27 EU countries. Governance is measured quantitatively (% confidence in governance) and is added to the model to gauge the level of sustainable development and to control for food security.

The panel unit root and panel cointegration analysis demonstrated that all series under consideration follow I(1) process and have non-spurious long-run relationship. While the estimates from different specifications of the models are quite similar in magnitude, traditional panel cointegration estimation techniques provide varying levels of coefficient significance. In order to undertake robust empirical analysis of the study relationships, the researchers have used the MMQR technique, which allows to estimate effects of exogenous factors across different quantiles of the conditional distribution of food security in a heterogeneous way. The FMOLS, DOLS, and FEOLS estimation of coefficients have supported the growth hypothesis for fishery-producing economies of the EU27 member countries.

Our MMQR estimates suggest that fisheries sector is significant and positive in all quantiles; however, it does not show up at the median quantile of the conditional distribution of sustainable food security. It must be noted that in richer countries, where reliance on fish imports is higher, economic diversification tends to be greater, and this tends to mean that less of food security is reliant on domestic fish production when compared with developing countries that generally produce higher levels of fisheries output and generally experience a higher level of food security. This could mean that the role played by the fisheries sector is of much more importance for developing countries for their food security profiles. Our fisheries production and economic growth MMQR results seem to be quite similar in the general tendency (shown in Table 5.7).

The estimates are strong in the upper quantiles for fisheries production and economic development. We can note an increase in the coefficient from the lowest to the highest quantile for fisheries production and economic development, and thus we observe an increase in the impact on food security. From these quantiles of food security, it becomes visible that at the median quantile of food security the concern on the Fisheries sector can be of lesser importance to the people of the EU countries where the negative impacts on food production insecurity are immediate, whereas there is a positive and considerable effect of the fishery sector on food security for those countries at the highest quantiles indicating a more efficient Fisheries production among the EU13 developing economies.

Similarly, the results of our MMQR analysis show that the agricultural sector is positive and significant in the upper quantiles above the median of the conditional distribution of sustainable food security but not significant in the lower quantiles below the median. Since the richer developed countries that are more food secure generally have more diversified economies than developing countries, which are more import dependent on agriculture, the agricultural sector could be playing a relatively larger role in the food security profile. For example, the positive significance of governance at the upper quantiles might be related to governments at this stage that are seeking to ensure food security and prevent continuing increases in unsustainability and insecurity. For the less productive EU14 developing countries, food security appears more linked to food production, when compared with their developed counterpart countries of EU14 developing countries that had the greatest amounts of fisheries production.

It is important to understand the trade-offs and that as well as well-being, which is always maximized at higher food security. In this instance, governance efficiency that responds to voter demands would attempt to provide more food security, providing governance efficiency leads to this demand, so it is important that voter demands can and do put pressure on governments; otherwise well-being will be low at higher food security for given amounts of governance, which makes people want fewer governments for food security. The importance of fisheries production in the highest quantile of food security might be explained by food security concerns acting as a constraint on overall development and that the EU14 developing countries that have the highest fisheries production are subject to the largest trade-offs in terms of overall growth that could be enhanced by fisheries sector production that depends on food security. For example, the EU13 developing and EU14 developed countries that have to put the most emphasis on efficient governance to overcome the impacts of the resource curse that would otherwise prevent development will have to employ this most effectively for efficient allocation and resource management when it is associated with resource constraints and development limitations for all sectors but in particular resource-abundant fisheries production.

From evidence relating to the Solomon Islands, as examined in [8], Indonesia, as presented in [7], and Africa, as reported in [22], it is concluded that the level of governance would indeed be the determining factor of increasing food security for these countries.

Henceforth, we therefore recommend strengthening governance in fisheries-producing countries, especially those in the EU13 developing regions that show high levels of sustainable food security. We observe from the MMQR results that energy production is positive and significant in the lower and upper quantiles. This suggests a positive link between energy production and food security at all levels of food security and suggests that although the effect of energy production is not statistically significant in the central quantiles (third to sixth), it is positive in the early and late stages of the transition and it also is the strongest at the highest quantile in the EU27 country regions.

In other words, more of this production would lead to a healthier overall ecosystem and greater overall good food production if, at lower stages of food security, this energy production is not coming from traditional energy sources. This implies countries that depend less on fossil fuels in the fisheries sector will produce more energy; this does not imply abundance, and gives more support for de-industrialization-induced resource curse. All quantiles showed positive and strong results for energy production but for the third through sixth quantiles.

However, with its all-pervasive impact on all producing sectors of an economy, a food security outcome should surely benefit from increased energy production across all quantiles.

Fisheries production had positive but relatively stronger effects for the highest quantile and smallest effects at the lowest quantile, which could suggest an increase in energy consumption and a subsequent increase in production in the fisheries sector from lower to upper quantiles, leading to improvements in food security per unit of energy. Alternatively, this reflects that developing EU13 countries and developed EU14 countries have adequate resources to shift from traditional to more clean forms of energy production, which directly or indirectly contribute to the welfare gained from fisheries production and food security, so perhaps increased fisheries production can and will go hand-in-hand with increasing cleaner and more sustainable forms of energy production including but not limited to the renewable ocean energy resources in our developed economies which could be achieved if some form of reasonable environmental taxes for marine industries were established for developed economies like those within the EU14 who are unable to establish renewable ocean energy resources and thus marine fisheries producing food. Economic growth was correlated positively with food security in all quantiles but the seventh, eighth, and fourth. It seems the growth hypothesis applies here.

There appear to be various reasons for this based on the context for fisheries-producing countries in the EU27 area.

The fourth, seventh, and eighth quantiles are likely to be countries experiencing a phenomenon such as de-industrialization that is characterized by fish production substituting for all other sectors of the economy, in which they traded all other sources of food depended on them for imports into this area from the outside. Thus, much of this trade was concentrated on the fisheries and fish import and export. Alternatively, higher growth rates in some regions due to the influx of EU14 capital-intensive, outward-oriented fisheries industries for maximization of profits could distort the process of development for the benefit of the EU13.

This trend could significantly depress food security levels within those developed economies, and is in line with the growth hypothesis. Another potential factor affecting economic growth, in a manner that may not result in improved sustainable food security levels at lower quantiles and may well act to depress food security in the developed countries with higher food security, may be attributed to the stringent regulations for marine-resource-reliant industries that force EU13 developing countries, which rely on exporting such products, to reduce waste at onshore fishing firms and force other companies away from expensive technology due to high levels of carbon tax and fuel substitution taxes, moving toward lower production areas for tax reasons, including other such taxes that make high production uneconomical. We will consider adding more factors into the models in future analyses, including, amongst other variables, those such as tourism, globalization, and financial development, among other variables, all of which influence both the productivity and sustainability of fisheries sector and sustainable food security within the EU regions.

Additionally, research could be conducted on other regions such as the People's Republic of China, ASEAN, BRICS, the group of seven, and the next 11 countries using more appropriate and advanced economic modeling techniques, perhaps country-based models.

As with any form of modeling, it is unavoidable that the analysis will have some shortcomings. Based on data availability, it was not possible to analyze the effects of fisheries sector saving. This needs to be studied and verified with additional information. As per the design of this study, there is no additional data available for similar national cases as provided.

This implies that there is a need to look into this aspect for further research with additional country examples and more appropriate econometric techniques.

Our models may only be used for selected countries and for only the periods of 1990–2022 based on available data and parameters. Other areas that we suggest for further analysis could involve public and private investment in fisheries sectors and offshore renewable energy in the fisheries production in EU27, with additional parameter metrics from additional sources; however the same data and metrics were unavailable to us due to the study limitations; we have restricted ourselves to the parameters already available to enable completion of this study; however we anticipate this information may be available for another study and in turn contribute toward enhancing our understanding of fisheries food security for the benefit of future generations. Other potential limitations may also involve applying a range of econometric techniques for other country regions such as the developing world, such as ASEAN, Africa, the Middle East, and BRICS countries, or perhaps regions within some of these areas, such as sub-regions of Africa.

In future studies, it is possible to account for changes in parameters over time such as by considering time-varying parameters or by conducting analysis in different historical periods to account for periods of structural breaks; these are considerations for the more comprehensive modeling of sustainable fisheries production and food security in future work.

References

1. Beyer RM, Hua F, Martin PA, Manica A, Rademacher T (2022) Relocating croplands could drastically reduce the environmental impacts of global food production. Commun Earth Environ 3(1):49. https://doi.org/10.1038/s43247-022-00360-6
2. Glauber JW, Laborde Debucquet D, Mamun A (2023) From bad to worse: how Russia-Ukraine war-related export restrictions exacerbate global food insecurity. [Online]. https://cgspace.cgiar.org/items/a4d641c2-2bdd-4c27-9595-754be33ee7b0. Accessed 20 May 2026
3. Caldeira C, De Laurentiis V, Corrado S, Van Holsteijn F, Sala S (2019) Quantification of food waste per product group along the food supply chain in the European Union: a mass flow analysis. Resour Conserv Recycl 149:479–488. https://doi.org/10.1016/j.resconrec.2019.06.011
4. Hebebrand C, Laborde Debucquet D (2023) High fertilizer prices contribute to rising global food security concerns. [Online]. https://cgspace.cgiar.org/items/e988b2b7-6ef4-4721-a4d2-089066fdf1e8. Accessed 20 May 2026
5. Hirschmugl M, Sobe C, Khawaja C, Janssen R, Traverso L (2021) Pan-European mapping of underutilized land for bioenergy production. Land 10(2):102
6. Tidd AN, Rousseau Y, Ojea E, Watson RA, Blanchard JL (2022) Food security challenged by declining efficiencies of artisanal fishing fleets: a global country-level analysis. Glob Food Sec 32:100598. https://doi.org/10.1016/j.gfs.2021.100598

7. Rahman MS, Huang WC, Toiba H, Efani A (2022) Does adaptation to climate change promote household food security? Insights from Indonesian fishermen. Int J Sustain Dev World Ecol 29(7):611–624. https://doi.org/10.1080/13504509.2022.2063433
8. Syddall VM, Fisher K, Thrush S (2022) Collaboration a solution for small island developing states to address food security and economic development in the face of climate change. Ocean Coast Manage 221:106132. https://doi.org/10.1016/j.ocecoaman.2022.106132
9. Chudik A, Pesaran H (2015) Large panel data models with cross-sectional dependence: a survey. [Online]. https://academic.oup.com/edited-volume/41981/chapter/355335206?searchresult=1&utm_source=TrendMD&utm_medium=cpc&utm_campaign=Oxford_Academic_Books_TrendMD_0. Accessed 20 May 2026
10. Breitung J, Pesaran MH (2008) Unit roots and cointegration in panels. In: Mátyás L, Sevestre P (eds) The econometrics of panel data, Advanced studies in theoretical and applied econometrics, vol 46. Springer Berlin Heidelberg, Berlin, Heidelberg, pp 279–322. https://doi.org/10.1007/978-3-540-75892-1_9
11. Andrews DWK (2005) Cross-section regression with common shocks. Econometrica 73(5):1551–1585. https://doi.org/10.1111/j.1468-0262.2005.00629.x
12. Cerrato M, Sarantis N (2002) The cross sectional dependence puzzle. London Guildhall University. [Online]. http://fmwww.bc.edu/RePEc/res2002/Cerrato.pdf. Accessed 20 May 2026
13. Balsalobre-Lorente D, Shahbaz M, Roubaud D, Farhani S (2018) How economic growth, renewable electricity and natural resources contribute to CO2 emissions? Energy Policy 113:356–367. https://doi.org/10.1016/j.enpol.2017.10.050
14. Bekun FV, Alola AA, Sarkodie SA (2019) Toward a sustainable environment: nexus between CO2 emissions, resource rent, renewable and nonrenewable energy in 16-EU countries. Sci Total Environ 657:1023–1029. https://doi.org/10.1016/j.scitotenv.2018.12.104
15. Isham J (2005) The varieties of resource experience: natural resource export structures and the political economy of economic growth. World Bank Econ Rev 19(2):141–174. https://doi.org/10.1093/wber/lhi010
16. Mehlum H, Moene K, Torvik R (2006) Institutions and the resource curse. Econ J 116(508):1–20. https://doi.org/10.1111/j.1468-0297.2006.01045.x
17. Leite MC, Weidmann J (1999) Does mother nature corrupt? Natural resources, corruption, and economic growth. International Monetary Fund. [Online]. https://books.google.com/books?hl=en&lr=&id=UFYYEAAAQBAJ&oi=fnd&pg=PA3&dq=Leite,C.,+Weidmann,+J.+(1999).+Does+Mother+Nature+Corrupt:+Natural+Resources,+Corruption,+and+Economic+Growth+IMF+Working+Papers.+International+Monetary+Fund.&ots=o29DeZnChU&sig=cQgr9PJf7fX7d18hNwAwf2FL4go. Accessed 20 May 2026
18. Machado JAF, Santos Silva JMC (2019) Quantiles via moments. J Econ 213(1):145–173. https://doi.org/10.1016/j.jeconom.2019.04.009
19. Koenker R (2004) Quantile regression for longitudinal data. J Multivar Anal 91(1):74–89. https://doi.org/10.1016/j.jmva.2004.05.006
20. Garlock T et al (2022) Aquaculture: the missing contributor in the food security agenda. Glob Food Sec 32:100620. https://doi.org/10.1016/j.gfs.2022.100620
21. Canty SWJ, Deichmann JL (2022) Do small-scale fisheries have the capacity to provide food security to coastal populations? Fish Fish 23(3):708–718. https://doi.org/10.1111/faf.12643
22. Akbari M et al (2022) The evolution of food security: where are we now, where should we go next? Sustainability 14(6):3634. https://doi.org/10.3390/su14063634
23. Aura CM et al (2022) Aligning small indigenous fish species (SIS) in policy and management for enhanced food security and nutrition: the case of the Kenyan Lake Victoria Omena fishery. Lakes Reserv 27(1):e12399. https://doi.org/10.1111/lre.12399
24. March A, Failler P (2022) Small-scale fisheries development in Africa: lessons learned and best practices for enhancing food security and livelihoods. Mar Policy 136:104925. https://doi.org/10.1016/j.marpol.2021.104925

25. Onyenekwe C, Okpara U, Opata P, Egyir I, Sarpong D (2022) The triple challenge: food security and vulnerabilities of fishing and farming households in situations characterized by increasing conflict, climate shock, and environmental degradation. Land 11(11):1982. https://doi.org/10.3390/land11111982
26. Olaifa ES, Osabuohien ES, Issahaku H (2022) Enhancing fish production for food security in Nigeria. Mater Today Proc 65:2208–2214. https://doi.org/10.1016/j.matpr.2022.06.243
27. Fujimori S et al (2022) Land-based climate change mitigation measures can affect agricultural markets and food security. Nat Food 3(2):110–121. https://doi.org/10.1038/s43016-022-00464-4
28. Bazzana D, Foltz J, Zhang Y (2022) Impact of climate smart agriculture on food security: an agent-based analysis. Food Policy 111:102304. https://doi.org/10.1016/j.foodpol.2022.102304
29. Verma KK et al (2022) Recent trends in nano-fertilizers for sustainable agriculture under climate change for global food security. Nano 12(1):173. https://doi.org/10.3390/nano12010173
30. Viana CM, Freire D, Abrantes P, Rocha J, Pereira P (2022) Agricultural land systems importance for supporting food security and sustainable development goals: a systematic review. Sci Total Environ 806:150718. https://doi.org/10.1016/j.scitotenv.2021.150718
31. Pandey DK, Momin KC, Dubey SK, Adhiguru P (2022) Biodiversity in agricultural and food systems of jhum landscape in the West Garo Hills, North-eastern India. Food Sec 14(3):791–804. https://doi.org/10.1007/s12571-021-01251-y
32. Kumar A, Sharma P (2022) Impact of climate variation on agricultural productivity and food security in rural India. SSRN J. https://doi.org/10.2139/ssrn.4144089
33. Setsoafia ED, Ma W, Renwick A (2022) Effects of sustainable agricultural practices on farm income and food security in northern Ghana. Agric Econ 10(1):9. https://doi.org/10.1186/s40100-022-00216-9
34. Hunter P (2022) Harnessing the sun's power: hybrid photosynthesis might become a key technology to address the energy crisis and food security challenges. EMBO Rep 23(11):EMBR202256149. https://doi.org/10.15252/embr.202256149
35. Luqman M, Al-Ansari T (2022) Energy, water and food security through a waste-driven polygeneration system for sustainable dairy production. Int J Hydrog Energy 47(5):3185–3210. https://doi.org/10.1016/j.ijhydene.2021.10.214
36. Granit I (2022) Microgrids through the Energy-Water-Food Security Nexus in La Guajira, Colombia: increasing water and food security or jeopardizing groundwater levels? Energy Res Soc Sci 93:102814
37. Guo J, Tanaka T (2022) Energy security versus food security: an analysis of fuel ethanol-related markets using the spillover index and partial wavelet coherence approaches. Energy Econ 112:106142. https://doi.org/10.1016/j.eneco.2022.106142
38. Muhirwa F et al (2022) Ecological balance emerges in implementing the water-energy-food security nexus in well-developed countries in Africa. Sci Total Environ 833:154999. https://doi.org/10.1016/j.scitotenv.2022.154999
39. Kipkoech R, Takase M, Amankwa Afrifa EK (2022) Renewable energies in Ghana in relation to market condition, the environment, and food security. J Renew Energy 2022:1–8. https://doi.org/10.1155/2022/8243904
40. Bhave AG, Conway D, Dessai S, Dougill AJ, Mkwambisi D (2022) Stress-testing development pathways under a changing climate: water-energy-food security in the lake Malawi-Shire river system. Phil Trans R Soc A 380(2221):20210134. https://doi.org/10.1098/rsta.2021.0134
41. Shupler M et al (2021) COVID-19 impacts on household energy & food security in a Kenyan informal settlement: the need for integrated approaches to the SDGs. Renew Sust Energ Rev 144:111018. https://doi.org/10.1016/j.rser.2021.111018
42. Rasul G, Neupane N, Hussain A, Pasakhala B (2021) Beyond hydropower: towards an integrated solution for water, energy and food security in South Asia. Int J Water Resour Dev 37(3):466–490. https://doi.org/10.1080/07900627.2019.1579705

43. Collins AM (2022) Empowerment, rights, and global food governance: gender in the UN Committee for world food security. Globalizations 19(2):220–237. https://doi.org/10.1080/14747731.2021.1877006
44. Jones EC, Ong C, Haynes J (2022) Disaster-related food security and past general governance strategies in a worldwide sample. Weather Clim Soc 14(1):3–18. https://doi.org/10.1175/WCAS-D-20-0138.1
45. Leeuwis C, Boogaard BK, Atta-Krah K (2021) How food systems change (or not): governance implications for system transformation processes. Food Sec 13(4):761–780. https://doi.org/10.1007/s12571-021-01178-4
46. Iitembu JA, Gabriel NN, Tjipute M, Asino H, Hamukwaya J (2022) The governance of aquaculture in namibia as a vehicle for food security and economic growth. In: Mupambwa HA, Nciizah AD, Nyambo P, Muchara B, Gabriel NN (eds) Food security for African smallholder farmers, Sustainability sciences in Asia and Africa. Springer Nature Singapore, Singapore, pp 391–403. https://doi.org/10.1007/978-981-16-6771-8_23
47. Ahwireng AK (2022) Small pelagic fish for food: governance and performance of small pelagic fish value chains for food security and nutrition in Ghana. Anderson Kwasi Ahwireng
48. Anser MK et al (2021) Impact of ICT adoption and governance interaction on food security in West Africa. Sustainability 13(10):5570. https://doi.org/10.3390/su13105570
49. Haysom G (2021) Integrating food sensitive planning and urban design into urban governance actions. Urban Forum 32(3):289–310. https://doi.org/10.1007/s12132-021-09417-9
50. Panghal A, Mor RS, Kamble SS, Khan SAR, Kumar D, Soni G (2022) Global food security post COVID-19: dearth or dwell in the developing world? Agron J 114(1):878–884. https://doi.org/10.1002/agj2.20932
51. Cheng P, Tang H, Lin F, Kong X (2022) Bibliometrics of the nexus between food security and carbon emissions: hotspots and trends. Environ Sci Pollut Res 30(10):25981–25998. https://doi.org/10.1007/s11356-022-23970-1
52. Yaqoob N et al (2022) The effects of agriculture productivity, land intensification, on sustainable economic growth: a panel analysis from Bangladesh, India, and Pakistan economies. Environ Sci Pollut Res 30(55):116440–116448. https://doi.org/10.1007/s11356-021-18471-6
53. Ajayi OO, Mokryani G, Edun BM (2022) Sustainable energy for national climate change, food security and employment opportunities: implications for Nigeria. Fuel Commun 10:100045. https://doi.org/10.1016/j.jfueco.2021.100045
54. Qi W et al (2022) Economic growth dominates rising potential flood risk in the Yangtze River and benefits of raising dikes from 1991 to 2015. Environ Res Lett 17(3):034046. https://doi.org/10.1088/1748-9326/ac5561
55. Apergis N, Payne JE (2010) The emissions, energy consumption, and growth nexus: evidence from the commonwealth of independent states. Energy Policy 38(1):650–655. https://doi.org/10.1016/j.enpol.2009.08.029
56. Dogan E, Seker F (2016) Determinants of CO2 emissions in the European Union: the role of renewable and non-renewable energy. Renew Energy 94:429–439. https://doi.org/10.1016/j.renene.2016.03.078
57. Ehigiamusoe KU, Lean HH, Smyth R (2020) The moderating role of energy consumption in the carbon emissions-income nexus in middle-income countries. Appl Energy 261:114215. https://doi.org/10.1016/j.apenergy.2019.114215
58. Anwar A, Siddique M, Dogan E, Sharif A (2021) The moderating role of renewable and non-renewable energy in environment-income nexus for ASEAN countries: evidence from Method of Moments Quantile Regression. Renew Energy 164:956–967. https://doi.org/10.1016/j.renene.2020.09.128
59. FAO (2022) The state of world fisheries and aquaculture 2022. FAO. https://doi.org/10.4060/cc0461en
60. World Bank, Fish to 2030: prospects for fisheries and aquaculture
61. European Commission. Joint Research Centre (2022) Clean Energy Technology Observatory, Ocean energy in the European Union: status report on technology development, trends,

value chains and markets: 2022. LU: Publications Office. [Online]. https://data.europa.eu/doi/10.2760/162254. Accessed 30 Dec 2025
62. Alsaleh M (2023) The role of the fishery industry in the shift towards sustainable food security: a critical study of blue food. Environ Sci Pollut Res 30(48):105575–105594. https://doi.org/10.1007/s11356-023-29747-4
63. Pedroni P (2004) Panel cointegration: asymptotic and finite sample properties of pooled time series tests with an application to the PPP hypothesis. Econ Theory 20(03):597–625. https://doi.org/10.1017/S0266466604203073
64. Kao C, Chiang MH (2000) On the estimation and inference of a cointegrated regression in panel data. In: Advances in econometrics, vol 15. Emerald (MCB UP), Bingley, pp 179–222. https://doi.org/10.1016/S0731-9053(00)15007-8
65. Sarkodie SA, Strezov V (2019) Effect of foreign direct investments, economic development and energy consumption on greenhouse gas emissions in developing countries. Sci Total Environ 646:862–871. https://doi.org/10.1016/j.scitotenv.2018.07.365
66. Koenker R, Bassett G (1978) Regression quantiles. Econometrica 46(1):33. https://doi.org/10.2307/1913643
67. Binder M, Coad A (2011) From Average Joe's happiness to Miserable Jane and Cheerful John: using quantile regressions to analyze the full subjective well-being distribution. J Econ Behav Organ 79(3):275–290. https://doi.org/10.1016/j.jebo.2011.02.005
68. Canay IA (2011) A simple approach to quantile regression for panel data. Econ J 14(3):368–386
69. Phillips PCB, Sul D (2003) Dynamic panel estimation and homogeneity testing under cross section dependence. Econ J 6(1):217–259. https://doi.org/10.1111/1368-423X.00108
70. Pesaran MH (2021) General diagnostic tests for cross-sectional dependence in panels. Empir Econ 60(1):13–50. https://doi.org/10.1007/s00181-020-01875-7
71. Levin A, Lin CF, James Chu CS (2002) Unit root tests in panel data: asymptotic and finite-sample properties. J Econ 108(1):1–24. https://doi.org/10.1016/S0304-4076(01)00098-7
72. Im KS, Pesaran MH, Shin Y (2003) Testing for unit roots in heterogeneous panels. J Econ 115(1):53–74. https://doi.org/10.1016/S0304-4076(03)00092-7
73. Westerlund J (2007) Testing for error correction in panel data*. Oxf Bull Econ Stat 69(6):709–748. https://doi.org/10.1111/j.1468-0084.2007.00477.x
74. Kremers JJM, Ericsson NR, Dolado JJ (1992) The power of cointegration tests. Oxf Bull Econ Stat 54(3):325–348. https://doi.org/10.1111/j.1468-0084.1992.tb00005.x
75. Campanati C, Willer D, Schubert J, Aldridge DC (2022) Sustainable intensification of aquaculture through nutrient recycling and circular economies: more fish, less waste, blue growth. Rev Fish Sci Aquacult 30(2):143–169. https://doi.org/10.1080/23308249.2021.1897520

Chapter 6
Shaping Sustainable Fisheries and Tourism Industry in EU27

Acronyms

EU	European Union
CO_2	Carbon Dioxide Emissions
MMQR	Method of Moments Quantile Regression
EUR	Euro
CD	Cross-sectional Dependency
FP	Fisheries Production
WDI	World Bank Development Indicators
GDP	Gross domestic Product
TD	Tourism Development
PD	Population Density
DOLS	Dynamic Ordinary Least Squares
FMOLS	Fully Modified Ordinary Least Squares
FEOLS	Fixed-Effects Ordinary Least Squares
Europa	European Data Portal
Eurostat	European Commission's Database
LFP	Natural logarithms Food Production
LLC	Lin & Chu Test
IPPS	Im, Pesaran, and Shin Test
EU27	27 European Union Member States
EU13	13 European Union Member States
BRICS	Brazil, Russia, India, China, South Africa
ASEAN	Association of Southeast Asian Nations
EU14	14 European Union Member States
OLS	Ordinary Least Square
LEG	Natural Logarithm of Economic Growth
LEN	Natural Logarithm of Energy Input
LED	Natural Logarithm of Tourism development
LGV	natural Logarithm of Governance

M. Alsaleh, *Fisheries Sustainability in the EU*, Urban Sustainability,
https://doi.org/10.1007/978-981-92-4496-6_6

Intensive use of coastal resources and construction of new marinas to expand the tourism industry may result in deteriorating water and marine environmental quality due to pollution, noise, and species disturbance, thereby harming aquatic ecosystems and fisheries. By examining the roles of fisheries production, tourism development, energy consumption, economic growth, governance, and population density during 1990–2022, this research seeks to explore the dynamic influence of tourism development on the sustainability of the fisheries sector in 27 European countries. The findings of this study, which employed the novel method of moments quantile regression (MMQR) with fixed effects, revealed a substantial adverse impact of energy consumption and tourism development on fisheries sustainability across the EU27 region. Specifically, the findings demonstrate that across all quantiles, the considerable adverse impact of tourism development on fisheries sustainability is more pronounced in EU13 developing countries than in EU14 developed countries. Simultaneously, the findings indicate that the EU14 developed countries experienced a more significant negative impact on fisheries sustainability than the EU13 developing countries at the uppermost quantiles. The relationship between governance (across all quantiles) and economic growth in the EU27 was also positively significant. Notably, the findings reveal that the EU14 developed countries exert a more substantial positive impact on fisheries sustainability than the EU13 developing countries. At median quantiles, it was discovered that population density enhanced fisheries sustainability. Results demonstrate that population density exerts a more significant positive impact on fisheries sustainability in EU13 developing countries than in EU14 developed countries. To safeguard marine ecosystems and achieve sustainable blue growth, our results provide policymakers with insights into supporting coastal tourism development in EU14 and EU13 countries using effective and environmentally sustainable resources and technologies.

6.1 Fisheries Impacts from Tourism Development

With nearly 2.127 million people directly employed in European countries, coastal tourism constitutes the most significant component of the blue economy on the continent. It accounts for 40% of gross value added (GVA) and 60% of employment in the blue economy [1]. Depending on the subsector, marine and coastal tourism may be linear and area based. Most marine activities occur along the coast and between the beach and on-water tourist destinations, although activities such as diving, snorkeling, and exploring underwater cultural sites are location-specific [1]. The spatial implications of mass tourism will persist even if it is projected to stabilize or decline in the future. Among the elements warranting special consideration are the direct use of sea space near the shore, effects on the marine environment and water quality, and environmental pressure on land [1]. The diversification of a fisherman's activities through tourism may be beneficial. It may be conducted year-round as a source of supplementary income [1]. The annual €2 billion contribution of angling tourism to the European economy could be crucial in assisting the EU region's tourism industry in recovering from the COVID-19 pandemic [1].

The Baltic Sea, Black Sea, Mediterranean Sea, North-East Atlantic, and, more specifically, the Arctic Ocean all exhibit degraded marine and coastal ecosystems [2]. Meeting the immediate needs of populations impacts the environment. However, this has a potentially irreversible effect on fish species and ecosystems that have evolved over thousands, if not millions, of years [2]. These impacts on fish species and habitats are attributable to the substantial and expanding tourism industry along Europe's coasts and associated supporting sectors, such as fisheries, agricultural and industrial chemical pollution, tourism developments, shipping, energy infrastructure, and other marine activities [2]. Despite being productive, Europe's waters cannot be regarded as healthy, clean, or pristine for fish species and ecosystems [2]. Due to air, water, and waste pollution and direct and indirect changes in land use in urban, rural, and natural areas, coastal and marine tourism increasingly generates adverse environmental externalities [3]. While heavily relying on the health of natural ecosystems that contribute significantly to the region's attractiveness, this tourism simultaneously places additional strain on fish species and habitats [3]. The development and utilization of infrastructure and marine or coastal leisure activities that have disrupted the fisheries sector constitute the primary causes of tourism's detrimental environmental effects on the Black Sea, Baltic Sea, Arctic Ocean, and Mediterranean coastal regions [3]. These adverse externalities include water and energy consumption for tourism-related activities, particularly in locations with limited access to clean water, where they also cause land transformation, artificialization of the shoreline, pollution, and biodiversity loss. This has disrupted fish species and habitats, thereby increasing pressure on the fisheries sector [3]. From another perspective, marine debris constitutes a serious problem in several tourism hotspots in the Mediterranean; more than 75% of annual waste generation occurs during the summer months and is directly correlated with visitor numbers [4]. In reality, coastal tourism harms ecosystems by subjecting them to multiple stresses, including waste, water and air pollution, light and noise pollution, invasive species, urbanization, transportation, and resource exploitation. Such pressures affect both ecosystems proximate to the source of the stresses and ecosystems distant from them [4].

The cruise industry exerts a significant adverse impact on the environment. As they transit between ports, cruise ships pollute the air, adversely affect public health, and damage fish species' natural habitats. A single company's fleet of 47 vessels in the Mediterranean generates nearly ten times more SOX emissions than the 260+ million passenger cars in Europe [5]. Cruise ships navigate close to shore, where fish habitats and biodiversity are most vulnerable to pollution. Consequently, both cruise travel and shipping contribute to the decline of fish and marine species [5]. Given the projected sectoral expansion, these environmental and social pressures on fish species and habitats will likely intensify if appropriate regulations are not implemented at the municipal and national levels [5]. These pressures are projected to intensify, particularly in Southern Mediterranean countries with weaker marine regulations and limited fisheries enforcement capabilities, as Northern Mediterranean countries are relatively developed tourism destinations [5].

The research questions addressed are: (1) How do tourism-related activities influence the sustainability of the fisheries sector and associated marine environment in European seas between 1990 and 2022? (2) Are EU member states more susceptible to the adverse effects of coastal and marine tourism on fisheries sustainability? (3) Can tourism activities become more environmentally sustainable to meet the EU's 2030 goals for blue growth and sustainable fisheries? The primary objective of this research is to examine the sustainability of fisheries production in relation to tourism in the European Union seas between 1990 and 2022. (2) To comprehensively analyze the effects of coastal and marine tourism between EU13 developing and EU14 developed countries from 1990 to 2022 on fisheries production sustainability and marine ecosystems. (3) To examine the magnitude and scope of differences between developing and developed countries' sustainable fisheries production arising from tourism activities between 1990 and 2022. Many contend that the recent growth of the tourism industry has become a critical component in determining the prospects and dynamics of sustainable development. Significant policy advancements such as the European Tourism Indicators System, revisions to the common agricultural and fisheries policy, and the most recent EU enlargement have helped substantially reduce and, in some instances, eliminate marine environmental concerns. Due to these profound transformations in the tourism industry and policies, coastal and marine tourism is becoming increasingly popular among tourist destinations. The competitive environment in which the fisheries sector operates presents new risks and challenges. The sustainability of tourism, the health of the marine environment, and the sustainability of ocean resources are also major global concerns. Thus, fisheries sustainability is becoming an increasingly complex issue, as blue food products must be competitive while simultaneously addressing domestic and international organizations' concerns about marine resource sustainability, marine ecology, and tourism. Due to intense competition in the tourism sector, the sustainability of the EU's fisheries industry is no exception. The fisheries industry significantly influences the EU coastal tourism sector, and over the past decade, blue food imports have grown substantially. Because there is a shortage of pertinent literature that does not adequately explain the observed changes in the sustainable development of the fisheries industry, research examining how tourism activities' growth affects fisheries industry sustainability may help develop future tourism development strategies and inform policy. The assumption that disturbance factors are independent and that no cross-sectional dependence (CD) exists among the stated variables across countries constitutes a critical limitation in earlier research on EU states. These factors might represent common or idiosyncratic shocks, spillovers, or a combination thereof [6–10] Incorporating CD into the analysis is crucial since a cross-sectional group of EU economies will likely exhibit mutual dependence due to further economic integration. EU countries are especially susceptible to common shocks affecting governance and economic growth, such as tourism development.

The current study aims to determine whether the growth-mitigating influence of natural resources similarly affects the tourism development profile of fisheries-producing countries. Previously published research [11, 12] employed more aggregate exogenous variables that did not distinguish between point-source and diffuse-source resources. In contrast, this research utilizes a highly specialized natural resource indicator (fisheries production sustainability). Because point-source resources typically generate more detrimental consequences than diffuse ones, we focused on fisheries in EU economies [13, 14]. Because sound policy has been demonstrated to mitigate or even reverse the resource dependence-induced growth-mitigating effect [15], our sample includes both developed and developing economies and regulatory control variables such as governance. The EU27 fisheries-producing countries include developed nations such as Germany, Italy, and Sweden, as well as developing countries such as Estonia, Bulgaria, and Malta. The study's timeframe spans from 1990 to 2022. Because the selected countries exhibit different levels of economic development, our study employs the [16] method of moments quantile Regression, which is unique and novel in its application among previous studies, to examine the economic growth-induced resource curse's sustainability relationship in the fisheries industry's context along with other control parameters across diverse quantiles of the conditional distribution of fisheries sustainability. By incorporating fixed effects, this approach provides unique empirical insights into distributional heterogeneity relationships. By allowing for heterogeneous economic growth-fisheries production relationships at various blue resource sustainability conditional quantiles, the technique captures patterns that conventional mean regression may fail to detect. Consequently, the growth hypothesis is thoroughly investigated in fisheries-producing countries. To our knowledge, this paper is the first to introduce distributional heterogeneity when examining resource sustainability induced by economic growth in the EU27 countries by considering governance, energy consumption, population density, fisheries production, tourism development, and economic growth. Evaluating the growth hypothesis at different quantiles of the fisheries sustainability conditional distribution is necessary for several reasons. First, conditional quantile estimations demonstrate greater robustness to outliers emanating from the dependent variable than conditional mean estimates, which exhibit higher susceptibility to outliers' distorting effects [17]. Conditional mean estimations, conversely, do not accurately capture the full distributional effect of tourism development on fisheries sustainability. Quantile regression, on the other hand, offers greater intuitive appeal, particularly in panel regression contexts, as it disaggregates the independent variable's distributional influences on the dependent variable across quantile ranges in diverse forms. Consequently, it facilitates easier categorization of the heterogeneous impacts across different cross-sectional groups. As a result, conditional quantile estimates provide information rarely found in conditional mean estimations. The remaining sections of this chapter are structured as follows. Section 6.2 briefly reviews the literature related to the growth hypothesis. Section 6.3 outlines the data and methodologies. Section 6.4 presents and discusses the estimation results, and Sect. 6.5 provides conclusions and policy recommendations based on the empirical findings.

6.2 Fisheries and Tourism Historical Background

The fishery regions of Europe possess beautiful scenery, charming fishing ports, and numerous other characteristics that make them attractive tourist destinations. Coastal and marine tourism constitutes the principal nautical activity in Europe, and several regions are already established as tourism hubs. While most communities may benefit from this expanding market, advantages often bypass the fisheries industry, and developing managed tourism operations may even generate detrimental effects. For instance, [1] examined how tourism affected Norway's fisheries resources and concluded that angling marine tourism could exert positive, substantial effects on the economy and marine environment. In Indonesia, [2] investigated the impact of fisheries tourism on fisheries resource availability and the economic health of coastal communities, asserting that fisheries tourism development may significantly enhance the marine and ocean environment. In Ireland, [3] investigated the relationship between coastal tourism and fisheries resource accessibility, demonstrating that regulated fisheries tourism may positively influence Ireland's fisheries expansion. Similarly, [4] investigated the factors determining the value of anglers' fisheries tourism experiences and concluded that these factors could be used to assess whether fisheries tourism is developing sustainably or experiencing degradation. The importance-performance analysis of the fisheries tourism service structure was examined in Finland by [5], who noted that fisheries tourism posed a threat to the salmon population, which has been steadily declining [5]. Findings also assist stakeholders in fisheries tourism in identifying prospective priority areas for development and in evaluating the potential of non-fisheries-related activities that may enhance the region's future sustainability. Coastal regions, the transitional zones between land and water, are characterized by high biodiversity and include some of the most diverse and vulnerable ecosystems on the planet. Massive visitor influxes, often concentrated in confined areas, exert significant impacts. They increase local populations' demands for water and waste management and generate pollution, thereby placing tremendous strain on the environment and its ecosystems. For instance, [18] examined the effects of coastal tourism on marine and coastal pollution in Indonesia, demonstrating that coastal plastic waste constitutes one of the most significant threats to marine life and species [19, 20]. investigated coastal tourism and pollution in coastal environments in India and Bangladesh, respectively, noting that the ecological condition of marine life faces tremendous pressure due to extensive coastal tourism activities in both countries. Accordingly, [21] investigated how tourism affected coastal regions, emphasizing the possibility that marine tourism and travel could exert both positive and negative impacts on the region's economy and ecology, respectively [22]. studied the effects of tourism development on marine ecosystems and resources in the Republic of Palau and found that absent intervention, tourism and associated coastal development may become self-destructive, ultimately deteriorating or destroying the natural resources that serve as the primary attractions. Numerous studies, including [23–28], have examined how poor governance and enforcement encourage overfishing, the significance of sound policy for fisheries management, and various strategies for effective and sustained enforcement [29, 30]., among others, investigated the legal and policy status of the fisheries industry in Indonesia and India, respectively, and asserted that community partnership and co-management constitute the most suitable approaches to achieve an enabling environment for fisheries industry sustainability in those countries. According to evaluations of fisheries management plans by [31, 32], based on environmental governance, there has been notable improvement in fish abundance in Europe, rising from 1229 (in 2008) to 1142 (in 2013) and 1781 (in 2018), with the decline in 2013 being attributed to a nearby wildfire. While local knowledge structures are theoretically acknowledged, [33] investigated alternatives to sustainable seafood certifications in Japan and noted that although influential political and economic actors advocate for sustainable seafood certification to address external shocks, they do not draw upon local knowledge. In the USA, [34, 35] investigated the fundamental design of fisheries management regulations to conserve and sustain marine resources and found that the enabling environment constitutes a significant factor outside fishermen's direct control and that, when combined with unpredictable weather, access to reliable crew, and rapid shifts in market prices, it exerts severe consequences on their general well-being.

According to research, marine fisheries vessels alone contributed 207 million tons of CO_2 to the atmosphere in 2019, nearly equivalent to CO_2 emissions from 51 coal-fired power plants during the same period [36]. For instance, [37] also investigated resource use efficiency in warm-water fish culture across different pond sizes for environmental progress and sustainable energy. In their studies of fossil fuel input and carbon footprint in trawl fisheries, [38, 39]. Noted that efficiency improvements may be achieved by systematically monitoring the fossil fuel profile of fishing vessels through energy audits. Fifteen percent of the animal protein consumed by humans derives from marine fish. Global oceans are failing under the extreme strain of providing for 15% of 7.5 billion people (and increasing) annually, year after year. Numerous earlier studies, including those by [40–42], confirmed this condition. However, several studies, including those by [43–45], have demonstrated that greater investments in enhancing human capital may partially offset the effects of decommissioning in the fisheries sector [46]. examined the relationship between human development and fisheries footprint in the Mediterranean Sea and concluded that relatively low levels of human capital negatively impact fisheries sustainability, while relatively high levels positively impact fisheries. A comprehensive and cohesive approach, as well as coordinated and harmonized policies for blue economy sustainability, were proposed by [47, 48] and were also investigated in Indonesia. Creating a sustainable fisheries industry generates greater opportunities for long-term economic, social, and environmental growth. For instance, research by [49–51] on the relationship between sustainable fisheries production and economic growth in South Asian countries demonstrated that fisheries production directly influences sustainable development. The impacts of partial and complete replacement of fish production on growth performance, feed utilization, histological condition, and economic revenue status were also examined by [52]. According to [52], utilizing underutilized or discarded marine materials constitutes a sustainable approach for establishing a circular economy and creating high-value-added materials [53]. Examined the impact of Nile Tilapia (Oreochromis niloticus) initial stocking body sizes on water quality, growth efficiency, and economic output. Regarding growth performance and financial returns, the researchers concluded that the medium stocking fish size performed best among the three treatments.

Similarly, [54] investigated how Arctic fish skin crafts affected the sustainable development of coastal regions in Southwest Alaska. The study found that Indigenous fish skin practices can significantly reduce poverty, promote gender equality, improve health and well-being, ensure clean water and sanitation, foster decent work and economic growth, reduce social inequality, and address climate change, marine issues, responsible consumption, and responsible production. While earlier research primarily focused on coastal or marine pollution and tourism development, this study explores tourism development's diverse contributions to fisheries sector sustainability in achieving blue growth. Additionally, to achieve sustainable marine development, we examine tourism development's impact on fisheries production in EU seas. Most research employs time-series analysis, while this study utilizes static and panel regressions to examine how tourism and economic expansion affect fisheries production sustainability. The following constitute some of the potential novelties of this study: First and foremost, it examines the relationships among fisheries productivity, tourism development, economic expansion, energy consumption, governance, and population density in EU seas from 1990 to 2022. More crucially, to attain sustainable blue growth, we also investigate the effects of tourism development and economic growth factors on the fisheries sector in the EU27 seas. We divide the EU27 region into two alternative groups: the EU14 developed countries and the EU13 developing countries. Because our data spans 1990 to 2022 and exhibits non-stationary behavior, we employ unit root and cointegration analysis to evaluate stationarity. Using the method of moments quantile regression (MMQR), we examine the interactions among tourism development, governance, economic growth, energy consumption, and population density. Regarding factors affecting fisheries sector sustainability, we also conduct static and panel regression analyses. Finally, we project how future tourism development variables will affect fisheries and ocean resource sustainability.

6.3 Modeling Global Tourism Competitiveness and Fisheries

Two factors merit consideration in developing fisheries in developing countries: (1) the impact of multinational economic activities on cluster and tourism competitive advantage; and (2) the role of interconnectedness/networks among cluster participants. However, several studies on cluster development, tourism, and Porter's theory emphasized the international tourism competitiveness of EU member countries and regions in a global economy. A novel strategy termed "global tourism competitiveness of fisheries development" is proposed in light of the significance of these challenges for both fisheries sustainability and the tourism industry's competitiveness. The model below presents the proposed fisheries development paradigm for tourism competitiveness. The three additions to this revised model are the incorporation of governance, the emphasis on fossil fuel consumption, and the population density indicator. This revised model is based on Porter's diamond model and includes numerous variables identified by [55] and the [56] models.

First, governance should be recognized as one of the key factors in developing sustainable fisheries, as it significantly impacts fisheries sustainability in both the EU13 developing countries and the EU14 developed countries. The degree and nature of networks among each component affecting the cluster's tourism competitiveness should be documented and illustrated. The following model was developed based on recent investigations by [57–60]:

$$FP_{it} = f\left(TDit, ENit, EGit, GVit, PDit\right)$$

The dependent variable is fisheries production (FP), measured as total fisheries output in metric tons. Data for these variables were obtained from the World Development Indicators [61]. The independent variables include: (GV), which reflects government effectiveness and was obtained from the official European statistics website [62]. The analysis utilizes data for EU economies from 1990 to 2022, with each variable treated as per capita values. (TD) represents international tourism revenues, proxied by expenditures of incoming foreign tourists in the reporting economy for travel-related products, obtained from the European Commission database [63]. (EN) represents fossil fuel consumption in the fisheries and marine sectors, obtained from the official European data website [62]. (PD) represents population density, calculated as the annual average population divided by geographical area. (EG) represents gross domestic product, measured in constant US dollars and retrieved from the European Commission database [63]. To accomplish the objectives of this study, we collected annual data for the selected countries from the World Bank's World Development Indicators database (WDI), the European Commission's database (Eurostat), and the official portal for European data (Europa) for the years 1990 to 2022. Table 6.1 lists the data displays, measurement units, sources, and transformations.

In this part, we discuss the findings. To start, Table 6.2 displays the chosen components' statistical characteristics. (i.e., maximum, minimum, mean, and standard deviation). The values of FP, TD, EN, EG, and GV's respective means are 4.926, 55.750, 2.751, 4.379, and 1.867. The outstanding standard deviation level for each of the study's components is 0.825, 0.577, 0.761, 0.95, and 0.069, respectively. We can get closer to the unit root test thanks to the factors' descriptive properties.

We employ dynamic ordinary least squares (DOLS), fully modified ordinary least squares (FMOLS), and fixed-effects ordinary least squares (FEOLS) for comparative purposes. The FEOLS approach incorporates Driscoll and Kraay standard errors, which are robust to common forms of cross-sectional dependence and autocorrelation up to a specified lag. According to [65], the primary concerns in estimating dynamic cointegrated panels stem from heterogeneity issues related to variations

Table 6.1 Data, units, transformation, and source

Data	Measurement unit	transformation	Source
Fisheries production	Metric tons	Natural logarithm transformation	WDI
Tourism development	Current US Dollars	Natural logarithm transformation	WDI
Energy input	Thousand tonnes	Natural logarithm transformation	Eurostat
Economic growth	Constant US dollars	Natural logarithm transformation	WDI
Governance	% of confidence for governance	Natural logarithm transformation	Eurostat
Population density	People per sq.km	Natural logarithm transformation	Eurostat

Source: [64]

Table 6.2 Results of descriptive statistics

Variable	Observations	Mean	Std. Dev.	Min	Max
FP	864	4.926	0.825	3.211	6.310
TD	864	5.750	0.577	4.071	5.910
EN	864	2.751	0.761	0.214	4.600
EG	864	4.379	0.395	3.130	5.250
GV	864	1.867	0.069	1.480	1.980

Source: Author's calculation

in cross-sectional means and differential cross-sectional adjustment to the cointegrating equilibrium. To address these concerns, Pedroni's FMOLS model incorporates individual-specific intercepts and permits varying serial correlation characteristics of the error processes across different panel members. Monte Carlo simulation results demonstrated that the DOLS estimator was unbiased relative to both OLS and FMOLS estimators in finite samples. This prompted [66] to extend the DOLS estimator to panel data contexts. The DOLS estimator further mitigates endogenous feedback by incorporating lead and lag differences, thereby correcting for endogeneity. Given the limitations of previous estimation techniques, a panel quantile regression approach was employed to evaluate distributional and heterogeneous effects across quantiles [67]. The seminal work by [67] introduced the panel quantile regression methodology. Unlike conventional least-squares regressions, which yield estimates of the conditional mean of the dependent variable given specific values of the independent variables, quantile regressions are typically employed to estimate the conditional median or various other quantiles of the response variable. Quantile regressions demonstrate greater robustness to outliers in the estimation process. Additionally, they are particularly suitable when there is minimal or no relationship between the conditional means of two variables [68].

However, the present study utilized the [16] method of moments quantile regression (MMQR) with fixed effects. While quantile regression is robust to outliers, it does not account for potential unobserved heterogeneity among individuals in a panel. The MMQR technique permits individual-level effects to influence the entire distribution rather than merely shifting means, as demonstrated in studies by [17, 69], among others, thereby enabling the identification of conditional heterogeneous covariance effects of fisheries sustainability determinants. The MMQR estimation approach is particularly appropriate when the panel data model incorporates individual-level effects and contains endogenous explanatory variables. The MMQR technique is also straightforward to implement as it generates non-crossing estimates of the regression quantiles. For a location-scale model specification, the computation of the conditional quantiles QY(τ|X) takes the following form:

$$Y_{it} = \alpha_{i+} X_{it}{}' \beta' + \left(\delta_i + Z_{it} \gamma' \right) U_{it} \tag{6.1}$$

Where parameters that need to be evaluated and the probability, $P(\delta_i + Z_{it}\gamma' > 0) = 1(\alpha, \beta', \delta, \gamma')$. Z is a k-vector of identifiable components of X that are differentiable transformations with element l provided by $(\alpha_i, \delta_i), i = 1, \ldots, n$, represents the individual i fixed effects.

$$Z_l = Z_l(X), l = 1, \ldots, k \tag{6.2}$$

The explanatory variable X_{it} follows an independent and identically distributed distribution for each individual unit i and remains temporally independent across time t. Following Machado and Silva (2019), the error term U_{it} is orthogonal to X_{it} and identically distributed across both individuals and time periods. Moreover, the error term is normalized to satisfy the required moment conditions, which exempt the model from stringent homogeneity assumptions. The specification of Equation (1) is presented as follows;

$$Q_Y(\tau | X_{it}) = (\alpha_{i+} \delta_i q(\tau)) + X_{it} \beta' + Z_{it} \gamma' q(\tau) \tag{6.3}$$

This study defines the explanatory variable vector X_{it} to contain five logarithmically transformed indicators: the natural logarithm of economic growth (LEG), the natural logarithm of energy input (LEN), the natural logarithm of population density (LPD), the natural logarithm of tourism development (LTD), and the natural logarithm of governance (LGV). The scalar coefficient expression $X_{it}\beta_i(\tau) + \alpha_i(\tau)$ captures the quantile-based fixed effect for each individual unit i, while $QY(\tau|X_{it})$ represents the conditional quantile distribution of the dependent variable. In this case, the dependent variable is the natural logarithm of fish production (FP), which is conditioned on the explanatory vector X_{it}. Distinct from conventional least-squares fixed effects, the individual effect in this framework is not limited to a simple intercept shift. Instead, individual heterogeneous coefficients are permitted to fluctuate across the conditional quantiles of the endogenous dependent variable Y, and all heterogeneous parameters remain time-invariant. The τ-th sample quantile

$q(\tau)$ is eventually derived through the numerical solution of the quantile optimization function.

$$\min_q \sum_i \sum_t P_\tau \left(R_{it} - \left(\delta_i + Z_{it} \gamma' \right) q \right) \tag{6.4}$$

Where $P_\tau (A) = (\tau - 1)\, AI\, \{A \leq 0\} + \tau AI\, \{A > 0\}$ denotes the check function.

6.4 The Outcome of Tourism Global Competitiveness Impacts From Fisheries

Prior to parameter estimation, a set of preliminary diagnostic tests is conducted to examine the time-series properties of the analyzed variables. This study first assesses the presence of cross-sectional dependence (CD) across the panel dataset. Cross-sectional correlation is capable of distorting true coefficient values; if ignored, unobserved cross-sectional interdependencies substantially diminish the statistical efficiency of panel data estimation [70]. Accordingly, addressing cross-sectional dependence is essential to ensure accurate and reliable empirical outcomes. This research employs the [71] CD test to identify cross-sectional correlation among sampled economies. As presented in Table 6.3, all quantitative variables exhibit significant cross-sectional dependence across countries, with the exception of fisheries productivity. To mitigate potential size distortion caused by such interdependence, the current study adopts cross-sectionally robust techniques for unit root testing, cointegration analysis, and panel regression estimation.

This study employs the Levin–Lin–Chu (LLC) unit root test [72] and the Im–Pesaran–Shin (IPS) unit root test [73] to objectively evaluate the integration characteristics of the analyzed variables. Panel unit root tests enhance statistical precision by integrating cross-sectional and time-series information to verify data stationarity against the null hypothesis of a unit root. The LLC test imposes a homogeneous framework, assuming that all individual series within the panel revert to equilibrium at a consistent stationary rate. In contrast, the IPS test relaxes this homogeneity constraint and allows heterogeneous autoregressive parameters across individual panel units, with the alternative hypothesis specifying that at least one series in the panel is stationary. This study prioritizes the LLC framework for fundamental stationarity assessment, owing to its superior statistical power for moderately small panel datasets. Both tests are jointly applied to examine the influence of

Table 6.3 Cross-sectional dependence results

Variables	FP	TD	EN	EG	GV	PD
Pesaran (2004)	(18.82)***	(85.53)***	(65.38)***	(83.55)***	(31.30)***	(99.27)***

Remark: *** refer importance at the 1%, scale
Source: [64]

cross-sectional dependence on stationarity behavior. As demonstrated in Table 6.4, all variables are non-stationary at their level form but become stationary following first-order differentiation. Accordingly, every variable incorporated within the empirical model follows an identical integration order of I(1).

This study utilizes the [65] panel cointegration test and the bootstrapped [74] panel cointegration test, as summarized in Table 6.5, to examine the existence of

Table 6.4 Panel unit root test result

	Difference		First difference	
Variable	LLC	IPS	LLC	IPS
FP	3.067	−3.700	−24.13***	−22.737***
TD	−1.525	−2.717	−11.423***	−13.251***
EN	−8.657	−5.187	−14.730***	−16.930***
EG	−1.571	−0.210	−18.74***	−68.36***
GV	−1.09	−0.254	−23.47***	−23.47***
PD	−2.189	−5.018	−11.914***	−14.911***

Remark: *** refer importance at the 1%, scale
Levin, Lin & Chu test (LLC), and Im, Pesaran, and Shin W-stat test (IPS)
Source: [64]

Table 6.5 Panel cointegration test for EU countries

Test		Without trend	With trend	
A) Pedroni Residual Cointegration Test:				
Alternative hypothesis: common AR coefficients (within dimension):				
Panel v-Statistic		−2.174 (0.985)	−1.913 (0.972)	
Panel rho-Statistic		2.357 (0.990)	1.303 (0.903)	
Panel PP-Statistic		−1.692* (0.045)	−4.467*** (0.000)	
Panel ADF-Statistic		−1.571** (0.058)	−4.244*** (0.000)	
Alternative hypothesis: common AR coefficients (between dimensions)				
Group rho-Statistic		−3.060	(0.998)	
Group PP-Statistic		−5.457***	(0.000)	
Group ADF-Statistic		−5.400***	(0.000)	
B) Westerlund (2007) Bootstrap Panel Cointegration:				
Statistics	Value	Z-Value	p-value	Robust p-value
Gt	−2.409	−1.095	0.095	0.081*
Ga	−7.225	3.100	0.999	0.985
Pt	−11.677	−1.408	0.080	0.074*
Pa	−5.673	1.569	0.942	0.999

Remark: ***, **, and * refer importance at the 1%, 5%, and 10% scales respectively
Values in parentheses are *p*-values
Source: [64]

robust long-run relationships among the analyzed variables [65]. developed an all-encompassing panel cointegration framework grounded in the Engle–Granger two-step approach. In its initial computational stage, this technique accounts for panel heterogeneity by eliminating individual-specific deterministic trends and short-run dynamic components. The test produces seven residual-based statistical indicators, which are categorized into two distinct groups listed in Table 6.5: pooled within-dimension tests that assume homogeneous stochastic processes across panel units and grouped between-dimension tests that allow heterogeneous individual behavioral patterns. Meanwhile, the [74] method comprises four innovative statistical tests under the null hypothesis of no cointegration. Unlike conventional residual-based approaches, this technique focuses on structural dynamic characteristics rather than residual fluctuations, thereby overcoming restrictive common factor assumptions. As highlighted by [75], violations of common factor constraints substantially weaken the reliability of residual-based cointegration estimations, which justifies the necessity of adopting structural dynamic testing procedures.

The relaxation of such restrictive assumptions eliminates the requirement for consistent adjustment speeds across short-run and long-run dynamic paths. The bootstrap algorithm proposed by [74] is adopted to mitigate size distortion originating from cross-sectional dependence and generate robust critical values. As reported in Table 6.5, both the [65] conventional cointegration test and the [74] bootstrapped cointegration test provide statistically significant evidence confirming the presence of long-run cointegrated relationships among variables.

Table 6.6 presents the long-run estimation outcomes derived from the FEOLS, FMOLS, and DOLS regression techniques. Despite minor numerical discrepancies across estimators, the three methods yield broadly consistent coefficient magnitudes and significance levels in Model 1. Among all explanatory variables, tourism development exhibits the most stable statistical performance across the three specifications. The empirical results indicate that tourism expansion exerts a detrimental influence on fisheries sustainability. Specifically, a 1% increase in tourism development reduces fishery sustainability by 0.59% under the FEOLS framework and 0.94% according to the FMOLS estimation, while the corresponding DOLS coefficient shows weak statistical significance. Such empirical findings are consistent with [1–3] The negative tourism-induced pressure is predominantly attributed to

Table 6.6 Model 1 panel estimation for the EU-27 region from 1990 to 2022

Long-run coefficient	DOLS		FMOLS		FEOLS	
	Coefficient	Std. Error	Coefficient	Std. Error	Coefficient	Std. Error
TD	−0.426	(0.041)	−0.948	(0.016)***	−0.597	(0.064)***
EN	−0.405	(0.074)***	−2.047	(0.019)***	−0.105	(0.044)***
EG	0.311	(0.089)	2.674	(0.032)***	0.707	(0.097)***
GV	0.832	(0.252)***	0.037	(0.021)*	0.882	(0.527)*
PD	0.928	(0.317)***	1.110	(0.096)***	0.461	(0.071)***

Note: ***, **, and * indicate significance at the 1%, 5%, and 10% levels respectively
Values in parentheses are P-values
Source: [64]

cruise ship activities. As a prevalent international tourism carrier, cruise vessels discharge massive untreated waste into transboundary marine waters on an annual basis. Large volumes of passenger sewage are arbitrarily released in unregulated open sea areas, causing irreversible marine contamination.

Consistent with theoretical expectations, energy input presents a significant negative correlation with fisheries sustainability within the European Union. The inhibitory coefficients of energy input are recorded at 0.10%, 2.04%, and 0.40% for FEOLS, FMOLS, and DOLS, respectively. This empirical evidence aligns with [36]. Traditional wild fishing vessels predominantly rely on diesel and gasoline combustion, generating substantial carbon emissions. Approximately 70% of such anthropogenic emissions are absorbed by seawater, triggering water quality degradation and marine pollution. Notably, the EU Emissions Trading Scheme (ETS) currently exempts fishery industries from carbon regulatory constraints, granting the sector immunity from standard climate mitigation policies. Economic growth demonstrates significant positive impacts on EU fisheries sustainability. A 1% increase in economic growth improves fishery sustainability by 0.09% and 0.03% under FEOLS and FMOLS estimations, whereas the DOLS coefficient presents inconsistent fluctuation. This empirical outcome validates the growth hypothesis for EU fishery economies and echoes the conclusions of [49–51], who confirmed the positive and inelastic growth-driven effects on fishery expansion. Favorable economic conditions construct a competitive and viable market ecosystem to sustain long-term fishery development across European regions.

Similarly, governance quality exerts significant promoting effects on fisheries sustainability. A 1% improvement in governance enhances fishery sustainability by 0.88%, 0.03%, and 0.83% based on FEOLS, FMOLS, and DOLS results. These findings support the growth hypothesis and are empirically consistent with [76–78]. Benefiting from its status as the world's largest seafood consumption market, high seafood import dependency, and dominant regulatory influence, the European Union plays a decisive role in global fishery governance formulation. Population density displays statistically significant positive coefficients across all estimation techniques, with marginal effects of 0.46% (FEOLS), 1.11% (FMOLS), and 0.92% (DOLS). This evidence contradicts the modernization hypothesis proposed by [42] but is consistent with [40, 41], Collectively, FMOLS, FEOLS, and DOLS estimations strongly validate the growth hypothesis. The positive population density coefficients emphasized the necessity of intensified administrative supervision to mitigate overexploitation and excessive resource consumption in densely populated coastal regions. Table 6.7 summarizes the heterogeneous outcomes of panel quantile regression. Tourism expansion imposes significant negative pressure on fisheries sustainability across most quantiles, with only the lowest quantile recording a marginal and insignificant tourism impact. Such distributional heterogeneity verifies the substantial regulatory effectiveness of tourism management policies in high-yield fishery economies. The growth hypothesis is universally valid across all fisheries production quantiles within the EU27 sample, indicating that tourism development generally takes policy precedence over marine conservation regardless of national fishery output scales. Owing to high industrial transformation costs,

Table 6.7 Panel quantile estimation results for EU27 countries

Variables	Method of moments quantile regression with fixed effects										
	Location	Scale	Quantiles								
			0.1	0.2	0.3	0.4	0.5	0.6	0.7	0.8	0.9
TD	(−0.064)***	(−0.520)***	(−0.213)	(−0.845)**	(−0.941)***	(−0.867)***	(−0.784)***	(−0.547)***	(−0.399)***	(−0.290)***	(−0.279)***
EN	(−0.044)***	(−0.549)	(−0.490)	(−0.332)	(−0.100)	(−0.219)	(−0.485)	(−0.966)	(−0.362)	(−0.073)*	(−0.092)**
EG	(0.097)***	(0.455)	(0.490)	(0.332)	(0.048)	(0.050)	(0.194)	(0.044)	(0.036)	(0.097)	(0.227)
GV	(0.527)*	(6.266)***	(1.736)***	(1.241)	(1.449)***	(1.303)***	(1.627)***	(2.123)***	(2.388)***	(2.150)***	(2.373)***
PD	(0.071)***	(0.198)***	(0.296)	(0.700)***	(0.618)***	(0.652)***	(0.589)***	(0.220)*	(0.082)	(0.126)	(0.280)***

Notes: *, **, and *** indicate significance at 10%, 5%, and 1% levels, respectively
Source: [64]

economies at preliminary production quantiles face substantial obstacles in transitioning from traditional tourism modes to sustainable developmental frameworks.

Furthermore, the adverse influence of energy input is statistically significant solely within upper fishery-production quantiles, with its marginal magnitude gradually increasing at higher quantile levels. This empirical tendency corroborates [36], highlighting that renewable energy optimization facilitates sustainable fishery progression. Governance generates positive sustainability effects for economies at both the lowest and highest production quantiles. Such heterogeneity may stem from divergent fishing extraction technologies between developed EU14 economies and emerging EU13 nations. Consistent with [76–78], significant governance-driven growth effects are identified for major fishery producers, further verifying the robustness of the growth hypothesis. Population density presents significant positive correlations with fisheries productivity across middle and upper quantiles, while its marginal influence gradually weakens at higher quantile levels. This distributional characteristic supports the empirical arguments of [40, 41, 79], confirming that rational population density management promotes sustainable fishery objectives. Conversely, economic growth exhibits weak quantile-variant effects on fishery sustainability. This outcome validates the effectiveness of fishery-oriented economic policies in high-output nations. Additionally, the insignificant distributed trend of GDP contradicts [49–51] while aligning with [42]. The empirical evidence implies that economic expansion retains a higher policy priority than environmental protection across the EU27 region.

Table 6.8 presents the sub-sample long-run estimation outcomes for developed EU14 economies based on FEOLS, FMOLS, and DOLS techniques. Although the three estimators differ slightly in statistical significance, their coefficient magnitudes remain generally consistent. The empirical outcomes reveal that tourism development exerts a pronounced negative influence on fisheries sustainability within the EU14 panel. A 1% increase in tourism development reduces fishery sustainability by 0.58% (FEOLS) and 1.21% (FMOLS), while the DOLS estimation displays marginal statistical significance. These sub-sample findings are consistent with [4, 5, 18]. Coastal tourism imposes multiple environmental pressures on marine ecosystems, among which public health contamination constitutes a critical

Table 6.8 Model 2 panel estimation for the EU-14 region from 1990 to 2022

Long-run coefficient	DOLS		FMOLS		FEOLS	
	Coefficient	Std. error	Coefficient	Std. error	Coefficient	Std. error
TD	−0.042	(0.041)	−1.210	(0.023)***	−0.586	(0.064)***
EN	−0.405	(0.074)***	−1.812	(0.021)***	−0.104	(0.044)**
EG	0. 311	(0.089)	0.492	(0.017)***	0.702	(0.097)***
GV	0.832	(0.252)***	2.278	(0.027)***	0.879	(0.527)*
PD	0.928	(0.317)***	1.443	(0.097)***	0.458	(0.071)***

Note: ***, **, and * indicate significance at the 1%, 5%, and 10% levels respectively
Values in parentheses are P-values
Source: [64]

concern. Tourism-intensive seasons generate excessive marine wastewater, inevitably deteriorating oceanic habitats and threatening aquatic biota.

In line with theoretical expectations, energy input maintains a statistically significant negative association with fisheries sustainability across all three estimators for the EU14 developed economies. The sustainability reduction coefficients are recorded at 0.10%, 1.81%, and 0.40% for FEOLS, FMOLS, and DOLS, respectively. This evidence supports [38, 39], who emphasized that fossil fuel combustion exacerbates marine pollution and restricts fishery resource availability in industrialized economies. Economic growth yields significant positive impacts on EU14 fisheries sustainability. A one-percentage increase in economic growth improves fishery performance by 0.70% and 0.49% under FEOLS and FMOLS, whereas the DOLS coefficient lacks stability. This result validates the growth hypothesis for developed European nations and diverges from [45], who identified adverse economic impacts on resource security. Nevertheless, the finding is consistent with [52–54]. Fisheries productivity inherently interacts with diverse marine economic activities, including shipping, aquaculture, and recreational tourism. For instance, the annual maintenance expenditure for repairing abandoned fishing gear and damaged vessels generates persistent economic burdens for the fishery industry. Governance quality presents robust positive significance across two estimation specifications within the EU14 sample. A 1% improvement in governance promotes fisheries output by 0.87%, 2.27%, and 0.83% according to FEOLS, FMOLS, and DOLS estimations. Such empirical outcomes align with [32, 80]. EU institutional frameworks strongly advocate Sustainable Development Goal 14 (Life Below Water) and the 2030 United Nations Sustainable Development Agenda. The revised ocean governance agenda further strengthens the marine ecological commitments embedded within the European Green Deal. Population density also retains significant positive coefficients throughout all estimation techniques, with marginal effects of 0.45% (FEOLS), 1.44% (FMOLS), and 0.92% (DOLS). This empirical trend supports the modernization hypothesis proposed by [44, 45], while contradicting [4]. The growth hypothesis is strongly validated across all estimators, demonstrating that rational population regulation effectively maintains intergenerational fish resource availability and optimizes seafood consumption structures.

Table 6.9 reports the panel quantile regression results for the EU14 developed economies. Tourism development generates significant negative impacts on fisheries sustainability across most quantiles, excluding the lowest quantile. This distributional evidence confirms the robust efficiency of marine tourism regulatory policies within industrialized European nations. The universal validity of the growth hypothesis across all quantiles indicates that tourism development generally overrides marine environmental legislation in EU14 economies. Due to substantial industrial transformation costs, countries at lower fishery-production quantiles face constraints in transitioning from conventional tourism infrastructure to renewable and eco-friendly developmental systems. Notably, the adverse tourism influence weakens among high-yield fishery economies, which may be attributed to advanced fishing extraction technologies adopted in developed nations. Consistent with [4, 5, 18], the quantile estimations verify the growth hypothesis for highly industrialized

Table 6.9 Panel quantile estimation results for EU14 countries

Variables	Method of moments quantile regression with fixed effects										
	Location	Scale	Quantiles								
			0.1	0.2	0.3	0.4	0.5	0.6	0.7	0.8	0.9
TD	(−0.064)***	(−0.673)***	(−0.667)	(−0.788)**	(−0.915)***	(−0.789)***	(−0.673)***	(−0.455)***	(−0.356)***	(−0.277)***	(−0.336)***
EN	(−0.044)**	(−0.116)***	(−0.147)	(−0.115)**	(−0.014)	(−0.082)**	(−0.116)***	(−0.112)***	(−0.109)***	(−0.106)***	(−0.049)
EG	(0.097)***	(0.091)	(0.523)	(0.307)	(0.306)	(0.028)	(0.091)	(0.013)	(0.011)	(0.105)	(0.213)
GV	(0.527)*	(1.607)***	(1.379)	(1.200)	(1.472)*	(1.362)	(1.607)***	(2.388)***	(2.349)***	(2.088)***	(2.160)***
PD	(0.071)***	(0.604)***	(0.090)	(0.715)***	(0.625)***	(0.645)***	(0.604)***	(0.168)	(0.061)	(0.041)	(0.203)**

Notes: *, **, and *** indicate significance at 10%, 5%, and 1% levels, respectively
Source: [64]

fishery producers, providing critical policy implications for marine resource governance.

Furthermore, energy input exerts significant negative pressure on fisheries sustainability across medium and upper quantiles, while its inhibitory effect remains negligible at the 1st, 3rd, and 9th quantiles. The marginal magnitude of energy-related pollution intensifies alongside higher quantile levels. Governance presents insignificant coefficients at the 1st, 2nd, and 4th quantiles but generates strong positive promotional effects for medium and upper quantiles, with its influence progressively amplified across higher distribution levels. Population density positively correlates with fishery performance within low-to-medium quantiles, whereas its impact becomes statistically insignificant at the highest quantile; simultaneously, the marginal population density effect gradually diminishes across ascending quantiles. Additionally, economic growth displays trivial and insignificant influences on fisheries production throughout all quantiles. This reflects the high efficiency of specialized fishery economic policies in high-output developed economies and demonstrates that marine resource sustainability remains relatively independent of macroscopic economic expansion within the EU14 region.

Table 6.10 summarizes the sub-sample estimation outcomes for developing EU13 economies using the FEOLS, FMOLS, and DOLS panel regression techniques. The three estimators yield broadly comparable coefficient values with minor numerical deviations. Empirically, tourism development exhibits a significant negative correlation with fisheries sustainability under FEOLS and FMOLS, while the DOLS coefficient presents unstable statistical performance. Quantitatively, a 1% increase in tourism development reduces fisheries sustainability by 0.24% and 2.60% for FEOLS and FMOLS, respectively. This finding corroborates [18, 21, 22], who emphasized that unregulated tourism expansion undermines fishery resource security across developing economies.

Energy input maintains a consistent negative association with fisheries sustainability across all three estimation specifications within the EU13 sample. A one-percentage rise in energy input lowers fisheries sustainability by 0.29% (FEOLS), 0.74% (FMOLS), and 0.37% (DOLS). Such evidence validates the arguments of [37, 81], stating that fossil fuel consumption and associated carbon emissions hinder the sustainable development of fishery industries in developing regions.

Table 6.10 Model 3 panel estimation for the EU-13 region from 1990 to 2022

Long-run coefficient	DOLS		FMOLS		FEOLS	
	Coefficient	Std. error	Coefficient	Std. error	Coefficient	Std. error
TD	−0.036	(0.142)	−2.606	(0.034)***	−0.248	(0.096)**
EN	−0.376	(0.069)***	−0.744	(0.034)***	−0.294	(0.045)***
EG	0.073	(0.140)	1.228	(0.025)***	0.453	(0.095)***
GV	0.050	(0.475)	3.964	(0.039)***	0.685	(0.660)
PD	1.089	(0.750)	2.953	(0.013)***	0.653	(0.093)***

Note: ***, **, and * indicate significance at the 1%, 5%, and 10% levels respectively
Values in parentheses are P-values
Source: [64]

Population density also presents distinct empirical characteristics for the EU13 panel. Improved population density regulation generates positive fishery outcomes, with sustainability increasing by 0.65% (FEOLS) and 2.953% (FMOLS), whereas the corresponding DOLS coefficient lacks stability. This positive nexus is consistent with [46, 82], who confirmed the population-driven growth effect for developing European fishery economies. Economic growth serves as a critical driving force for EU13 fisheries sustainability. Under FEOLS and FMOLS frameworks, a 1% increment in economic growth enhances fisheries output by 0.45% and 1.22%, respectively, while the DOLS estimation shows inconsistent fluctuation. The significantly positive economic growth coefficients explicitly verify the growth hypothesis for developing EU economies, aligning with [52, 54]. Moreover, governance quality displays heterogeneous estimation results across different regression techniques. Governance exerts a prominent positive influence exclusively under the FMOLS specification, with a 1% governance improvement boosting fisheries sustainability by 3.96%. The statistically significant FMOLS coefficient further supports the growth hypothesis for the EU13 panel, whereas FEOLS and DOLS produce unstable governance coefficients. This empirical heterogeneity echoes [29, 33, 35], highlighting that improved ocean governance facilitates fishery expansion during the initial developmental stage of emerging economies.

Table 6.11 presents the heterogeneous distributional results derived from the method of moments quantile regression (MMQR) for the EU13 economies. Tourism development imposes significant adverse impacts on fisheries sustainability across most quantiles, excluding the eighth and ninth quantiles. The inhibitory magnitude of tourism gradually weakens from −0.700 to −0.092 alongside ascending quantiles, with statistical insignificance observed within the two highest production quantiles. Energy input maintains robust negative correlations with fishery sustainability across nearly all quantiles, except for the second quantile, and its detrimental marginal effect intensifies at higher quantile levels.

Similarly, economic growth significantly promotes fisheries sustainability from the first to the seventh quantile, with its promotional magnitude declining steadily from 0.647 to 0.213 as quantiles increase; nevertheless, such growth-driven effects become insignificant for the eighth and ninth quantiles. Governance exhibits mixed distributional characteristics, generating significant positive influences in the early-stage and high-order quantiles while showing insignificant impacts within the medium quantiles (4th–7th). Notably, the promotional strength of governance continuously strengthens across higher quantiles for the EU13 sample. Population density positively contributes to fisheries sustainability from the second to the ninth quantile, with only the first quantile displaying statistical insignificance. Consistent with other variables, the marginal contribution of population density expands progressively at elevated quantile levels.

Panel FMOLS and DOLS estimations confirm the results obtained from the panel FEOLS. The panel FEOLS coefficients exhibit the same signs and levels of statistical significance as those of the panel FMOLS and DOLS. This consistency indicates that the FMOLS and DOLS results are reliable and can be used for drawing conclusions. Although the coefficients from FEOLS share the same signs as

Table 6.11 Panel quantile estimation results for EU13 countries

Variables	Method of moments quantile regression with fixed effects										
	Location	Scale	Quantiles								
			0.1	0.2	0.3	0.4	0.5	0.6	0.7	0.8	0.9
TD	(−0.096)**	(−0.246)**	(−0.700)***	(−0.645)***	(−0.290)***	(−0.293)***	(−0.246)**	(−0.204)**	(−0.264)***	(−0.092)	(−0.044)
EN	(−0.045)***	(−0.169)***	(−0.351)**	(−0.108)	(−0.199)***	(−0.227)***	(−0.169)***	(−0.116)**	(−0.331)***	(−0.345)***	(−0.277)***
EG	(0.095)***	(0.431)***	(0.647)***	(0.513)***	(0.494)***	(0.505)***	(0.431)***	(0.335)***	(0.213)**	(0.080)	(0.015)
GV	(0.660)	(0.342)	(3.299)***	(2.518)***	(0.987)**	(0.255)	(0.342)	(0.088)	(0.730)	(1.395)***	(1.849)***
PD	(0.093)***	(0.701)***	(0.072)	(0.545)**	(0.669)***	(0.721)***	(0.701)***	(0.773)***	(0.686)***	(0.767)***	(0.801)***

Notes: *, **, and *** indicate significance at 10%, 5%, and 1% levels, respectively
Source: [64]

those from FMOLS and DOLS, their significance levels differ slightly. Nevertheless, the panel DOLS estimates can be considered reliable and free from serial correlation and endogeneity problems.

To examine the impact of tourism growth on fisheries sustainability across countries with different levels of economic development, the EU-27 nations were divided into two groups: 14 developed countries (EU14) and 13 developing countries (EU13). The estimated effects of tourism growth on fisheries sustainability in the EU14 developed nations are presented in Tables 6.8 and 6.9. Similarly, the estimated effects in the EU13 developing nations are shown in Tables 6.10 and 6.11.

The findings in Tables 6.8 and 6.10 reveal that both energy use and tourism growth exert significant negative effects on fisheries sustainability. Moreover, the results indicate that the negative impact of tourism expansion on fisheries sustainability is stronger in the EU13 developing countries than in the EU14 developed countries. The precise magnitudes of this effect are –0.248 for the EU13 and – 0.586 for the EU14. This suggests that implementing sustainable tourism development strategies in the fisheries sector would enhance fisheries sustainability more effectively in the EU13 developing countries compared to the EU14 developed nations. This difference may stem from tax exemptions on seafood consumption in coastal tourist areas and rising demand linked to ongoing tax law reforms in the EU14 developed countries.

The findings further show that energy use has a more substantial negative impact on fisheries sustainability in the EU14 developed countries than in the EU13 developing countries, with magnitudes of –0.104 and – 0.294, respectively. This implies that the adoption of renewable energy policies and marine environmental taxation measures targeting energy consumption in the fisheries sector could lead to greater improvements in fisheries sustainability in the EU14 developed countries relative to the EU13 developing nations.

However, the results also indicate that governance has a significant positive effect on fisheries sustainability in the EU14 developed nations, but not in the EU13 developing countries. The magnitudes of this impact are 0.879 for the EU14 and 0.685 for the EU13. This suggests that the application of ocean governance and marine ecology policy frameworks in the fisheries sector could generate substantially higher gains in fisheries sustainability in the EU14 developed countries than in the EU13 developing nations.

Furthermore, population density demonstrates a stronger positive influence on fisheries sustainability in the EU13 developing countries compared to the EU14 developed nations, with impact magnitudes of 0.653 and 0.458, respectively. This implies that establishing specialized institutions and effective regulations for managing population density in coastal areas could yield greater improvements in fisheries sustainability in the EU13 developing nations than in the EU14 developed countries.

Finally, the results show that investment in fisheries infrastructure and fleets has a stronger positive and significant effect on fisheries sustainability in the EU14 developed nations than in the EU13 developing countries. The magnitudes of this impact are 0.702 for the EU14 and 0.453 for the EU13. This indicates that wealthier

EU14 nations are better positioned to enhance fisheries sustainability through such investments compared to the EU13 developing nations.

6.5 Conclusion and Solution for Sustainable Fisheries Tourism

This study examines the empirical relationships between fisheries production, tourism development, energy input, governance, economic growth, and population density, taking into account the growth hypothesis. It contributes to the literature on marine and ocean resources that investigates blue growth within the context of fisheries and sustainable development. Using panel data extending over a period of 31 years across 27 EU nations, the study applied three essential estimation approaches: panel cointegration, panel unit root testing, and panel estimation techniques. The variable governance (measured as the percentage of confidence for governance) was incorporated into the model as a quantitative control for fisheries production and was also used to evaluate the level of sustainable development. The panel cointegration testing approach and panel unit root analysis demonstrated that after the variables were processed to I(1), they maintained a non-spurious long-run relationship. Although the coefficient sizes derived from the various estimators are similar, traditional panel cointegration estimation methodologies indicated some differences in the statistical significance of the coefficients. For a more thorough examination of the empirical relationships, the MMQR approach was employed. This method enables the investigation of the heterogeneous effects of the exogenous factors across multiple quantiles of the conditional distribution of fisheries sustainability. The coefficient estimations from FEOLS, FMOLS, and DOLS supported the growth hypothesis for the fisheries-producing economies of the EU27 member nations.

In Table 6.7, the MMQR results showed that the conditional distribution of fisheries sustainability is adversely and significantly affected across all quantiles from the second to the ninth quantiles, particularly at the first quantile of tourism development. This suggests that tourist growth has a significant detrimental impact on the development of the fisheries sector in both the early and late stages of development in the EU27 nations. Similarly, energy input is significantly negative at the eighth and ninth quantiles, indicating that the recent (higher) quantiles of the conditional distribution of fisheries output are influenced by the negative link between energy input and fisheries sustainability. This implies that energy input has a relatively small negative impact on the growth of the fisheries sector during the early and middle stages of transition in the EU27 nations.

Fisheries sustainability rose statistically significantly across all quantiles from the first to the ninth, except at the second quantile. It is important to emphasize that nations with effective governance and better-established fisheries are more economically diversified than emerging economies, which rely significantly more on fish imports. This suggests that the fisheries sector may account for a substantially larger part of the marine resources profile in industrialized nations.

The MMQR estimate of GDP growth is insignificant across all quantiles, as shown by a visual review of Table 6.7. This suggests that economic expansion has no significant effects on the sustainability of fisheries at any quantile, just as it has no effects on the quantiles of fisheries productivity. Similarly, the impacts of population density on fisheries production are minimal at the quantiles where fisheries sustainability was strongest. This means that nations that rely less on conventional fisheries methods in the fisheries business tend to have denser populations. In this instance, dependence does not automatically indicate abundance. This provides further proof of the resource curse brought on by deindustrialization.

Tables 6.9 and 6.11 show that the rise of tourism has a negative and significant impact on the sustainability of fisheries, except at the earlier quantiles in the EU14 and the more recent quantiles in the EU13 nations, respectively. However, tourist growth's impact on fisheries sustainability was shown to be most significant at the most recent quantile and lowest at the oldest quantile. This suggests a further decline in the sustainability of fisheries from the lowest quantile to the highest quantile through a further upward increase in the development of tourism from the lowest to the highest quantile. This deterioration can be observed in the pollution of the shoreline and the decline of marine life.

From a different perspective, the negligible effect of conventional energy consumption in the fisheries sector might be linked to the insignificance of the negative consequences of energy input in the previous quantiles. Meanwhile, the government's fisheries policies aimed at preventing the ongoing spread of coastal pollution and unethical fisheries methods may be associated with the significance of the adverse effects of energy input. Compared to the lower fisheries sustainability in the EU13 developing nations that employ conventional energy, fisheries sustainability is considerably more critical to the inhabitants of the EU14 developed countries. Therefore, to gain voter support, renewable energy input aligned with popular demand would implement more robust measures against coastal pollution and the destruction of marine ecosystems.

On the other hand, the current government fisheries measures to prevent the continual development of overfishing and unsustainable fisheries practices might be linked to the insignificance of the beneficial effects of governance in the medium quantiles. In contrast to the lower production levels in the EU13 developing nations, inhabitants of the EU14 developed countries place greater emphasis on fisheries policy and regulation. Thus, to win voter support, an effective enabling environment that complies with widespread public desire would implement additional rules against overfishing and unethical practices.

The operations of the fisheries sector, which may depend on economic growth and sustainable development, may be given higher significance by both the electorate and the elected representatives, as observed in Tables 6.9 and 6.11. Due to the dominance of other competitiveness determinants and a stronger presence of the resource curse, the influence of economic growth may be observed at the latest quantiles in the EU13 emerging and EU14 developed nations. The improvement in sustainable fisheries development in the EU13 developing nations (but not in the EU14 developed countries) suggests that economic growth has a positive and significant impact in the early and medium quantiles.

However, population density exhibited a positive but insignificant link with fisheries sustainability at the medium quantiles (from the 6th to the eighth quantiles) in the EU14 developed nations, as shown in Table 6.9. Population density is positive and significant throughout most quantiles in Tables 6.9 and 6.11. This finding supports the growth hypothesis in the body of previous research.

The data may suggest several explanations for the various quantiles in the context of fisheries-producing countries in the EU13 and EU14. One possible reason for the middle quantiles in the EU14 nations might be the result of ineffective population density restrictions in coastal areas. This scenario arises when inefficient population density input crowds out all other fisheries practices that rely on tradable economic sectors. As a result, these countries become increasingly reliant on imports, meaning that a significant portion of their economic development will consist of fish importing and exporting processes.

The migration of the export-oriented fisheries industry, which is labor-intensive, from EU13 developing market economies to EU14 developed market economies in a bid to boost profit margins might shift the economic growth dynamics in favor of imports and lead to a significant reduction in the levels of fisheries production in the EU13 countries at higher quantiles. The strictness of fisheries policy on companies relying on marine resources and export-oriented activities may also directly affect the growth hypothesis in the EU14 developed nations. This situation may force the fisheries sector to reduce its production and marine waste. Many businesses may relocate their operations to other less developed countries with less onerous taxation, environmental policies, and abundant marine resources in order to capture more profit, due to the high expense of employing skilled experts and new technologies in the EU13 developing countries.

The fact that the energy input and tourism development factors for the fisheries sector remain notably negative may indicate that fisheries operations in both the EU14 developed and EU13 developing nations still follow the conventional approach. This finding implies that more effort should be made to shift tourism development and energy input away from conventional practices and traditional resources toward sustainable and environmentally friendly practices, in order to increase sustainable development in the fisheries industry in both EU14 and EU13 countries. In other words, promoting fish production from sustainable and regenerative sources is essential. Additionally, the method of developing tourism sustainably should be broadly adopted by all EU members in conjunction with the sustainability of blue growth.

Where the EU13 developing nations lack the means to adopt sustainable and renewable sources, they could introduce modest environmental tax laws for the fisheries sector. These policies would involve mild taxation on the fisheries sector, which relies on marine resources. The revenue raised from these tariffs should be reinvested into long-term infrastructure development for sustainable fisheries.

Finally, this analysis could be made more comprehensive by including other factors such as competition, globalization, and financial development, all of which influence the sustainability of the EU region's fisheries business and marine environment. Additionally, this research could be conducted at the individual country level, such as in the Republic of China, or in other geographic areas such as ASEAN, BRICS, the Group of Seven, or the Next 11 nations, while applying cutting-edge econometric techniques.

This kind of modeling undoubtedly has limitations; this work is no exception. Due to data limitations, the modeling could not evaluate the potential advantages of fisheries and coastline tourism, which still requires further study. Moreover, based on the scope of the present study, the model may be examined in the context of other nations to gain a broader contextual perspective. Although the research employed modern econometric methodologies to demonstrate these relationships, it still faced several limitations. For instance, the present findings were evaluated only through econometric methods and included only a restricted set of factors covering the period from 1990 to 2022. Future research may examine the impact of public-private investments in sustainable tourism and renewable resources on fisheries productivity using the existing dataset and additional parameters. Second, the current framework is limited to EU nations only, which could be addressed by integrating several other nations with substantial emissions. Thirdly, researchers should also investigate the economic complexity of tourist growth and fisheries sustainability over multiple periods to account for structural breaks in the data.

References

1. Ferter K et al (2023) Integrating complementary survey methods to estimate catches in Norway's complex marine recreational hook-and-line fishery. ICES J Mar Sci 80(1):107–121. https://doi.org/10.1093/icesjms/fsac216
2. Rochgiyanti R, Arisanty D, Hastuti KP (2023) Fishing tourism of conflict area in South Kalimantan. TSSJ 39:516–528. https://doi.org/10.47577/tssj.v39i1.7960
3. Deely J, Hynes S, Cawley M, Hogan S (2023) Modelling domestic marine and coastal tourism demand using logit and travel cost count models. Economic Analysis and Policy 77:123–136. https://doi.org/10.1016/j.eap.2022.11.003
4. Komppula R, Suni J, Turunen L, Gartner WC, Vainikka A (2022) Determinants of the value of fishing tourism experience among anglers. IJSMM 22(1/2):19. https://doi.org/10.1504/IJSMM.2022.121258
5. Lankia T, Venesjärvi R, Pouta E (2022) Importance-performance analysis of the fishing tourism service structure: recreational anglers' preferences on the remote salmon river of Teno in Finland. Fish Res 254:106425. https://doi.org/10.1016/j.fishres.2022.106425
6. Chudik A, Pesaran H (2015) Large panel data models with cross-sectional dependence: a survey. Online. https://academic.oup.com/edited-volume/41981/chapter/355335206?searchresult=1&utm_source=TrendMD&utm_medium=cpc&utm_campaign=Oxford_Academic_Books_TrendMD_0. Accessed 20 May 2026
7. Flaten O, Lien G, Koesling M, Løes A-K (2010) Norwegian farmers ceasing certified organic production: characteristics and reasons. J Environ Manag 91(12):2717–2726. https://doi.org/10.1016/j.jenvman.2010.07.026

8. Breitung J, Pesaran MH (2008) Unit roots and cointegration in panels. In: Mátyás L, Sevestre P (eds) The econometrics of panel data, In Advanced studies in theoretical and applied econometrics, vol. 46. Springer Berlin Heidelberg, Berlin, Heidelberg, pp 279–322. https://doi.org/10.1007/978-3-540-75892-1_9
9. Andrews DWK (2005) Cross-Section regression with common shocks. Econometrica 73(5):1551–1585. https://doi.org/10.1111/j.1468-0262.2005.00629.x
10. Cerrato M, Sarantis N (2002) The cross sectional dependence puzzle. London Guildhall University. Online. http://fmwww.bc.edu/RePEc/res2002/Cerrato.pdf. Accessed 20 May 2026
11. Balsalobre-Lorente D, Shahbaz M, Roubaud D, Farhani S (2018) How economic growth, renewable electricity and natural resources contribute to CO2 emissions? Energy Policy 113:356–367. https://doi.org/10.1016/j.enpol.2017.10.050
12. Bekun FV, Alola AA, Sarkodie SA (2019) Toward a sustainable environment: nexus between CO2 emissions, resource rent, renewable and nonrenewable energy in 16-EU countries. Sci Total Environ 657:1023–1029. https://doi.org/10.1016/j.scitotenv.2018.12.104
13. Isham J (2005) The varieties of resource experience: natural resource export structures and the political economy of economic growth. World Bank Econ Rev 19(2):141–174. https://doi.org/10.1093/wber/lhi010
14. Mehlum H, Moene K, Torvik R (2006) Institutions and the resource curse. Econ J 116(508):1–20. https://doi.org/10.1111/j.1468-0297.2006.01045.x
15. Leite MC, Weidmann J (1999) Does mother nature corrupt? Natural resources, corruption, and economic growth. International Monetary Fund. Online. https://books.google.com/books?hl=en&lr=&id=UFYYEAAAQBAJ&oi=fnd&pg=PA3&dq=Leite,C.,+Weidmann,+J.+(1999).+Does+Mother+Nature+Corrupt:+Natural+Resources,+Corruption,+and+Economic+Growth+IMF+Working+Papers.+International+Monetary+Fund.&ots=o29De_iChY&sig=s0RvFto7iCj2jrPO3Qpuken_QN4. Accessed 20 May 2026
16. Machado JAF, Santos Silva JMC (2019) Quantiles via moments. J Econ 213(1):145–173. https://doi.org/10.1016/j.jeconom.2019.04.009
17. Koenker R (2004) Quantile regression for longitudinal data. J Multivar Anal 91(1):74–89. https://doi.org/10.1016/j.jmva.2004.05.006
18. Gideran CN (2023) Indonesia tourism pandemic coastal plastic waste: a comprehensive review: Indonesia tourism pandemic coastal plastic waste: a comprehensive review. Journal of Oceanography and Aquatic Science 1(1):14–19
19. Das GK (2022) Coastal environments of India: a coastal west bengal perspective. In: In Springer Water. Springer International Publishing, Cham. https://doi.org/10.1007/978-3-031-18846-6
20. Pushan ZA, Rahman E, Islam N, Aich N (2022) A critical review of the emerging research on the detection and assessment of microplastics pollution in the coastal, marine, and urban Bangladesh. Front Environ Sci Eng 16(10):128. https://doi.org/10.1007/s11783-022-1563-2
21. Smith TF, Elrick-Barr CE, Thomsen DC, Celliers L, Le Tissier M (2023) Impacts of tourism on coastal areas. Camb Prisms Coast Futures 1:e5. https://doi.org/10.1017/cft.2022.5
22. Downs CA, Cruz OT, Remengesau TE (2022) Sunscreen pollution and tourism governance: science and innovation are necessary for biodiversity conservation and sustainable tourism. Aquat Conserv 32(5):896–906. https://doi.org/10.1002/aqc.3791
23. Chen X, Xu Q, Li L (2023) Illegal, unreported, and unregulated fishing governance in disputed maritime areas: reflections on the international legal obligations of states. Fishes 8(1):p. 36. https://doi.org/10.3390/fishes8010036
24. Spijkers J, Mackay M, Turner J, McNeill A, Travaille K, Wilcox C (2023) Diversity of global fisheries governance: types and contexts. Fish Fish 24(1):111–125. https://doi.org/10.1111/faf.12713
25. Jarvis RM, Young T (2023) Pressing questions for science, policy, and governance in the high seas. Environ Sci Pol 139:177–184. https://doi.org/10.1016/j.envsci.2022.11.001
26. Guggisberg S, Jaeckel A, Stephens T (2022) Transparency in fisheries governance: achievements to date and challenges ahead. Mar Policy 136:104639. https://doi.org/10.1016/j.marpol.2021.104639

27. Okafor-Yarwood I, Kadagi NI, Belhabib D, Allison EH (2022) Survival of the richest, not the fittest: how attempts to improve governance impact African small-scale marine fisheries. Mar Policy 135:104847. https://doi.org/10.1016/j.marpol.2021.104847
28. Jentoft S, Chuenpagdee R (2022) Interactive learning and governance transformation for securing blue justice for small-scale fisheries. Administration & Society 54(7):1255–1282. https://doi.org/10.1177/00953997211073947
29. Dudayev R, Hakim LL, Rufiati I (2023) Participatory fisheries governance in Indonesia: are octopus fisheries leading the way? Mar Policy 147:105338. https://doi.org/10.1016/j.marpol.2022.105338
30. Kelkar N, Arthur RI (2022) A review of governance and tenure in inland capture fisheries and aquaculture systems of India. Online. https://books.google.com/books?hl=en&lr=&id=wL1qEAAAQBAJ&oi=fnd&pg=PP1&dq=Kelkar,+N.,+%26+Arthur,+R.+I.+(2022).+A+review+of+governance+and+tenure+in+inland+capture+fisheries+and+aquaculture+systems+of+India.+doi:10.4060/cb8615en&ots=43o6zSOa5g&sig=lUJVzKdVu4wfxr1-PtME1opbuaU. Accessed 20 May 2026
31. Acuña-Alonso C, Varandas S, Álvarez X, Martinho A (2023) Analysis of the evolution of a fisheries management plan based on environmental governance: living laboratory in the Olo River, Portugal. Fish Res 260:106595. https://doi.org/10.1016/j.fishres.2022.106595
32. Linke S, Gillette MB, Jentoft S (2022) Institutionalizing injustice? aligning governance orders in swedish small-scale fisheries. In: Jentoft S, Chuenpagdee R, Bugeja Said A, Isaacs M (eds) Blue justice, In MARE Publication Series, vol. 26. Springer International Publishing, Cham, pp 529–547. https://doi.org/10.1007/978-3-030-89624-9_28
33. Ynacay-Nye A, Hisano S, Suryawan AS (2023) Alternatives to sustainable seafood certifications: transitions of small-scale fisheries governance in northeastern Japan. J Rural Stud 97:269–280. https://doi.org/10.1016/j.jrurstud.2022.12.002
34. Day S (2026) Pacific Salmon fisheries management: an (unusual) example of polycentric governance involving indigenous participation at multiple scales. The Center for Growth and Opportunity. Online. https://www.thecgo.org/books/the-environmental-optimism-of-elinor-ostrom/chapter-3-chapter-3-pacific-salmon-fisheries-management-an-unusual-example-of-polycentric-governance-involving-indigenous-participation-at-multiple-scales/. Accessed 20 May 2026
35. Sorensen J, Kincl L, Weil R, Dzugan J, Christel D (2022) Fisheries governance and associated health implications: current perspectives from US commercial fishermen. Mar Policy 141:105119. https://doi.org/10.1016/j.marpol.2022.105119
36. Marshall Z, Brockway PE (2020) A net energy analysis of the global agriculture, aquaculture, fishing and forestry system. Biophys Econ Sust 5(2):9. https://doi.org/10.1007/s41247-020-00074-3
37. Firouzi S, Kosari-Moghaddam A, Radgoudarzi M (2022) Resource use efficiency of WARM-WATER fish culture upon different pond sizes. Env Prog and Sustain Energy 41(2):e13778. https://doi.org/10.1002/ep.13778
38. Sala A et al (2022) Energy audit and carbon footprint in trawl fisheries. Sci Data 9(1):428. https://doi.org/10.1038/s41597-022-01478-0
39. Van Neste SL, Couture-Guillet A (2022) The urban politicization of fossil fuel infrastructure: mediatization and resistance in energy landscapes. Environment and Planning C: Politics and Space 40(8):1801–1818. https://doi.org/10.1177/23996544221111657
40. Mukherjee J, Saha NC, Karan S (2022) Bioaccumulation pattern of heavy metals in fish tissues and associated health hazards in human population. Environ Sci Pollut Res 29(15):21365–21379. https://doi.org/10.1007/s11356-021-17297-6
41. Siddhnath, Ranjan A, Mohanty BP, Saklani P, Dora KC, Chowdhury S (2022) Dry fish and its contribution towards food and nutritional security. Food Rev Intl 38(4):508–536. https://doi.org/10.1080/87559129.2020.1737708
42. Bilal M, Qadir A, Yaqub A, Hassan HU, Irfan M, Aslam M (2023) Microplastics in water, sediments, and fish at Alpine River, originating from the Hindu Kush Mountain, Pakistan:

implications for conservation. Environ Sci Pollut Res 30(1):727–738. https://doi.org/10.1007/s11356-022-22212-8
43. Wang L, Su M, Kong H, Ma Y (2021) The impact of marine technological innovation on the upgrade of China's marine industrial structure. Ocean & Coastal Management 211:105792. https://doi.org/10.1016/j.ocecoaman.2021.105792
44. Albasri H, Sammut J (2022) A comparative study of sustainability profiles between small-scale mariculture, capture fisheries and tourism communities within the Anambas Archipelago Small Island MPA, Indonesia. Aquaculture 551:737906. https://doi.org/10.1016/j.aquaculture.2022.737906
45. Ramos J, Drakeford B (2021) Blue economy and tourism: is there a potential to create sustainable jobs? In handbook of research on human capital and people management in the tourism industry. IGI Global Scientific Publishing, pp 243–264. Online. https://www.igi-global.com/chapter/blue-economy-and-tourism/263358. Accessed 20 May 2026
46. Yıldırım DÇ, Yıldırım S, Bostancı SH, Turan T (2022) The nexus between human development and fishing footprint among mediterranean countries. Mar Pollut Bull 176:113426. https://doi.org/10.1016/j.marpolbul.2022.113426
47. Nurlita L, Yanuar T, Syah R, Indradewa R, Hamdi E (2021) Human capital implementation strategy in the Marina company. American International Journal of Business Management (AIJBM) 4(8):152–156
48. Najib H, Nawangsari LC (2021) Effect of intellectual capital on organizational sustainability with employee innovative behavior as intervening variables in Pt. Jaya maritime services. EJBMR 6(1):158–163. https://doi.org/10.24018/ejbmr.2021.6.1.714
49. Rahmizal M (2022) Analysis of Indonesia marine fisheries with economic growth, population and effort effectiveness. European Journal of Formal Sciences and Engineering 5(1):53–61. https://doi.org/10.26417/ejef.v1i1.p17-22
50. Setiyowati H, Nugroho M, Halik A (2022) Developing a blue economy in depok west java, Indonesia: opportunities and challenges of Neon Tetra fish cultivation. Sustainability 14(20):13028. https://doi.org/10.3390/su142013028
51. Coppola D, Lauritano C, Palma Esposito F, Riccio G, Rizzo C, De Pascale D (2021) Fish waste: from problem to valuable resource. Mar Drugs 19(2):116. https://doi.org/10.3390/md19020116
52. Abo-Taleb HA et al (2021) Effect of a new feed Daphnia magna (Straus, 1820), as a fish meal substitute on growth, feed utilization, histological status, and economic revenue of grey mullet, mugil cephalus (Linnaeus 1758). Sustainability 13(13):7093. https://doi.org/10.3390/su13137093
53. Shamsuddin M et al (2022) Effects of stocking larger-sized fish on water quality, growth performance, and the economic yield of Nile Tilapia (Oreochromis niloticus L.) in floating cages. Agriculture 12(7):942. https://doi.org/10.3390/agriculture12070942
54. Palomino E, Pardue J (2021) Alutiiq fish skin traditions: connecting communities in the COVID-19 era. Heritage 4(4):4249–4263. https://doi.org/10.3390/heritage4040234
55. Crouch GI, Ritchie JB (1999) Tourism, competitiveness, and societal prosperity. J Bus Res 44(3):137–152
56. Dwyer L, Kim C (2003) Destination competitiveness: determinants and indicators. Curr Issue Tour 6(5):369–414. https://doi.org/10.1080/13683500308667962
57. Anwar A, Siddique M, Dogan E, Sharif A (2021) The moderating role of renewable and non-renewable energy in environment-income nexus for ASEAN countries: evidence from method of moments quantile regression. Renew Energy 164:956–967. https://doi.org/10.1016/j.renene.2020.09.128
58. Ehigiamusoe KU, Lean HH, Smyth R (2020) The moderating role of energy consumption in the carbon emissions-income nexus in middle-income countries. Appl Energy 261:114215. https://doi.org/10.1016/j.apenergy.2019.114215

59. Dogan E, Seker F (2016) Determinants of CO2 emissions in the European Union: the role of renewable and non-renewable energy. Renew Energy 94:429–439. https://doi.org/10.1016/j.renene.2016.03.078
60. Apergis N, Payne JE (2010) The emissions, energy consumption, and growth nexus: evidence from the commonwealth of independent states. Energy Policy 38(1):650–655. https://doi.org/10.1016/j.enpol.2009.08.029
61. World Bank (n.d.) Fish to 2030: prospects for fisheries and aquaculture
62. European Commission (2022) Joint research centre. Clean energy technology observatory, ocean energy in the european union: status report on technology development, trends, value chains and markets: 2022. Publications Office, LU. Online. https://data.europa.eu/doi/10.2760/162254. Accessed 30 Dec 2025
63. European Commission (n.d.) The European green deal—European commission. Online. https://commission.europa.eu/strategy-and-policy/priorities-2019-2024/european-green-deal_en. Accessed 8 June 2025
64. Alsaleh M, Wang X, Nan Z, Liu R, Sun Q (2023) Retracted: impact of coastal tourism demand on fisheries industry sustainability: a suggested framework for blue growth. Nat Res Forum 50(2):998–1024. https://doi.org/10.1111/1477-8947.12332
65. Pedroni P (2004) Panel cointegration: asymptotic and finite sample properties of pooled time series tests with an application to the PPP hypothesis. Economic Theory 20(03):597–625. https://doi.org/10.1017/S0266466604203073
66. Kao C, Chiang M-H (2000) On the estimation and inference of a cointegrated regression in panel data. In: Advances in econometrics, vol 15. Emerald (MCB UP), Bingley, pp 179–222. https://doi.org/10.1016/S0731-9053(00)15007-8
67. Koenker R, Bassett G (1978) Regression quantiles. Econometrica 46(1):33. https://doi.org/10.2307/1913643
68. Binder M, Coad A (2011) From average Joe's happiness to miserable Jane and cheerful John: using quantile regressions to analyze the full subjective well-being distribution. J Econ Behav Organ 79(3):275–290. https://doi.org/10.1016/j.jebo.2011.02.005
69. Canay IA (2011) A simple approach to quantile regression for panel data. Econ J 14(3):368–386
70. Phillips PCB, Sul D (2003) Dynamic panel estimation and homogeneity testing under cross section dependence. Econ J 6(1):217–259. https://doi.org/10.1111/1368-423X.00108
71. Pesaran MH (2021) General diagnostic tests for cross-sectional dependence in panels. Empir Econ 60(1):13–50. https://doi.org/10.1007/s00181-020-01875-7
72. Levin A, Lin C-F, James Chu C-S (2002) Unit root tests in panel data: asymptotic and finite-sample properties. J Econ 108(1):1–24. https://doi.org/10.1016/S0304-4076(01)00098-7
73. Im KS, Pesaran MH, Shin Y (2003) Testing for unit roots in heterogeneous panels. J Econ 115(1):53–74. https://doi.org/10.1016/S0304-4076(03)00092-7
74. Westerlund J (2007) Testing for error correction in panel data*. Oxf Bull Econ Stat 69(6):709–748. https://doi.org/10.1111/j.1468-0084.2007.00477.x
75. Kremers JJM, Ericsson NR, Dolado JJ (1992) The power of cointegration tests. Oxf Bull Econ Stat 54(3):325–348. https://doi.org/10.1111/j.1468-0084.1992.tb00005.x
76. Climent-Pérez P, Galán-Cuenca A, García-d'Urso NE, Saval-Calvo M, Azorin-Lopez J, Fuster-Guillo A (2023) Automatic Fish size estimation from uncalibrated fish market images using computer vision and deep learning. In: García Bringas P, Pérez García H, Martinez-de-Pison FJ, Villar Flecha JR, Troncoso Lora A, De La Cal EA, Herrero Á, Martínez Álvarez F, Psaila G, Quintián H, Corchado Rodriguez ES (eds) 17th International Conference on Soft Computing Models in Industrial and Environmental Applications (SOCO 2022), In Lecture Notes in Networks and Systems, vol 531. Springer Nature Switzerland, Cham, pp 319–329. https://doi.org/10.1007/978-3-031-18050-7_31
77. Schrobback P, Hoshino E, Pascoe S, Curtotti R (2022) Market integration of domestic and imported seafood: insights from the sydney fish market*. *Aus J Agri & amp; Res Econ* 66(1):216–236. https://doi.org/10.1111/1467-8489.12452

78. Pincinato RBM, Gasalla MA, Garlock T, Anderson JL (2022) Market incentives for shark fisheries. Mar Policy 139:105031. https://doi.org/10.1016/j.marpol.2022.105031
79. Hernández-Barrero S, Valderrama Barco M, Barreto Reyes CG, Paramo J, Sierra Sierra L, Stotz W (2022) Is traditional fisheries management correctly addressing the possible causes of fish production decline? The relationship between environmental degradation and artisanal river fisheries in the Magdalena River basin, Colombia. Mar Freshw Res 73(12):1475–1488. https://doi.org/10.1071/MF22025
80. Aguión A et al (2022) Establishing a governance threshold in small-scale fisheries to achieve sustainability. Ambio 51(3):652–665. https://doi.org/10.1007/s13280-021-01606-x
81. Kumar SS, Bibin C, Naresh K, Parthiban S, Karthikeyan K, Akash S (2022) Experimental investigation of diesel engine using fish wastage oil. In: Presented at the recent trends in science and engineering. AIP Publishing, Krishnagiri, India, p 020182. https://doi.org/10.1063/5.0074309
82. Seisay M (2022) Human capital investment and financing for Africa's blue economy: conference on unlocking the full potential of the fisheries sector within the framework of a sustainable blue economy. Online. http://repository.au-ibar.org/handle/123456789/1175. Accessed 20 May 2026

Chapter 7
Do Global Competitiveness Factors Shape Fisheries Sustainability Practices

Acronyms

EU	European Union
CFP	Common Fisheries Policy
AGM	The Annual General Meeting
CO_2	Carbon Dioxide Emissions
MMQR	Method of Moments Quantile Regression
EUR	Euro
CD	Cross-sectional Dependency
FP	Fisheries Production
WDI	World Bank Development Indicators
GDP	Gross Domestic Product
PD	Population Density
DOLS	Dynamic Ordinary Least Squares
FMOLS	Fully Modified Ordinary Least Squares
FEOLS	Fixed-Effects Ordinary Least Squares
Europa	European Data Portal
Eurostat	European Commission's Database
LLC	Lin & Chu Test
IPPS	Im, Pesaran, and Shin Test
EU27	27 European Union Member States
EU13	13 European Union Member States
BRICS	Brazil, Russia, India, China, South Africa
ASEAN	Association of Southeast Asian Nations
EU14	14 European Union Member States
OLS	Ordinary Least Squares
LGDP	Natural Logarithm of Economic Growth
LINN	Natural Logarithm of the Innovation Ecosystem
LHCI	Natural Logarithm of Human Capital Development
LMRK	Natural Logarithm of Market Size
STECF	Scientific, Technical and Economic Committee for Fisheries
EAPO	European Association of Fish Producer Organisations
INN	Innovation Ecosystem and Intellectual Property
MRK	Represents Market Size

M. Alsaleh, *Fisheries Sustainability in the EU*, Urban Sustainability,
https://doi.org/10.1007/978-981-92-4496-6_7

HCI	Represents Human Capital
PPP	Purchasing Power Parity
LENB	Natural Logarithm of the Enabling Environment

According to research on fisheries, there are two principal trade-offs among these sustainability pillars. First, a fisheries industry's improving ecological health diminishes economic returns for fishermen. Second, individual fishermen's financial success undermines fisheries communities' social objectives. This paper aims to investigate the dynamic impact of global competitiveness on the sustainability of the fisheries industry in 27 European countries by examining fisheries production, enabling environments, innovation ecosystems, market size, human capital, and economic growth between 1990 and 2022. At higher quantiles, the findings demonstrated a strong positive association between enabling environment and fisheries sustainability by adopting the novel Method of Moments Quantile Regression (MMQR) with fixed effects. Additionally, in EU27 countries, the impact of human capital was positive and substantial from the first to ninth quantiles. Notably, the findings reveal that the EU14 developed countries exhibit a greater level of development than the EU13 developing countries, which considerably benefits fisheries sustainability. Market size and economic growth were also found to constrain fisheries sustainability in EU14 developed and EU13 developing countries, supporting the growth hypothesis for fisheries-producing countries. The innovation ecosystem was found to enhance fisheries sustainability across all quantiles. Notably, the findings demonstrate that the significant positive impact of the innovation ecosystem, market size, and economic development on fisheries sustainability is more pronounced in the emerging countries of the EU13 than in the developed countries of the EU14. With effective and environmentally sustainable technology, policymakers can enhance market competitiveness in the EU14 and EU13 countries and advance sustainable development objectives.

7.1 Fisheries Global Competitiveness Market

The health of our seas, fisheries, communities, and economies is impacted by the environmental effects of commercial fisheries and fishing practices in the EU and globally [1]. The fisheries industry is incredibly diverse, and the environmental concerns discussed here constitute the main obstacles to sustainable fisheries development: (1) Increasing demand, improved technology, and evolving management have enabled fish populations to be subjected to exceptionally high fishing pressure, leading to depletion or even complete collapse. (2) Heavy or intensive fishing equipment may disrupt ecosystems during fish capture and result in long-term damage. (3) Numerous marine creatures accidentally captured by fisheries but possessing little economic value are separated from the profitable harvest and returned to the ocean, often dead or injured [1].

Many of the assessed European fish populations continue to be overfished or outside acceptable biological limits, according to the Scientific, Technical and Economic Committee for Fisheries (STECF) [2]. In fact, following a decade of recovery, the North-East Atlantic's proportion of overfished stocks increased from 38% to 43%, while conditions in the Mediterranean and Black Seas remained dire, with 83% of assessed populations overfished. Multiple fisheries managed separately but targeting the same stock, the potential for ecological or socioeconomic linkages within an ecosystem, and multispecies capture constitute the causes of EU market failures [1]. First, although fishing mortality has decreased, it appears unlikely that all Common Fisheries Policy (CFP) targets will have been achieved by 2020. This represents one of the significant challenges facing fisheries sector sustainability. Although the landing obligation constitutes a cornerstone of fisheries management, it has yet to meet its objectives. Second, alleged "unfair" competition between EU and non-EU producers may require specific trade measures, such as consumer education, that would assist in differentiating EU seafood products. Third, certain EU markets prioritize imported species, and vessels and producers are constrained by access and trade agreements. Fourth, while the EU is a logical leader in fisheries management, such initiatives encounter challenges in regions where third-country fleets are subject to lax regulations (capacity-building, cooperation agreement terms, unregulated marine zones, among others) [2].

The Annual General Meeting (AGM) of the European Association of Fish Producer Organisations (EAPO) was held in Gothenburg, Sweden, in 2022 [2]. The implications of Brexit, the COVID-19 pandemic, the Russia–Ukraine conflict, and increasing competition for space were identified by EAPO as the primary concerns facing the fisheries sector at the event [2]. As one of the sectors the EU targets to receive the most substantial funding from this fund, fisheries is the subject of particular concern. The fund was established to assist businesses headquartered in the EU in adjusting to increased trade friction with the UK. France alone will receive approximately €130 million of the roughly €650 million specifically designated to support European fisheries activities in UK waters [2].

Europeans continued to consume more fish and seafood products at home in 2021. According to [3], EU household expenditure on fish and seafood increased by 7% from 2020, continuing the upward trend observed between 2019 and 2020. The COVID-19 pandemic's lingering effects, which increased demand for at-home consumption, likely contributed to the rise in domestic consumption [3] However, Euromonitor estimates indicate that out-of-home consumption has also increased; this growth is anticipated to continue through 2024 before stabilizing in 2026 [3]. A 10% trade balance deficit—or €1.80 billion more than in 2020—resulted from higher imports of fish and seafood products and lower exports to third countries in 2021. The total value of imports increased more substantially than volumes due to higher prices [3]. This rise may be partially attributed to the euro's depreciation against the Norwegian krone, which increased import prices from Norway, one of the leading EU suppliers. Additionally, following the COVID-19 pandemic, countries such as Germany, France, Italy, the Netherlands, and Spain, among others, gradually began reopening their hotels, restaurants, and coffee shops, resulting in substantial increases in demand for high-value species [3].

Imports and fisheries-related production declined between 2019 and 2022. According to [4] the total supply of fish and seafood products fell to one of its lowest recorded levels between 2011 and 2020. The decline in the industry's sustainability may be attributed to the effects of Brexit, the COVID-19 pandemic, and geopolitical turmoil, which most adversely affected logistics and production operations. According to estimates by [4] apparent EU consumption was 10.41 million t of live weight equivalent, or 23.28 kg per person. Additionally, this figure represented a decade low. The most significant increase in apparent consumption was observed in Bulgaria at +6%, which contrasted with the overall negative trend for the EU [4]. This may be related to the influence of the geopolitical crisis, the COVID-19 pandemic, and Brexit challenges, which led to decreased fisheries industry efforts to promote sustainable development, or to fishing quotas, which were typically lower in 2022 than in 2020. The increasing fish production in the EU zone, which peaked in 2020 at 5.6 million t, is supporting the European fish market. EU fish production is anticipated to grow at a 3% rate during the projected period 2023–2028, driven by escalating demand from the food sector [5].

The research questions addressed are: (1) How does global competition influence the sustainability of fisheries production and associated marine ecology in European seas between 1990 and 2022? (2) Are EU member states more susceptible to the effects on fisheries sustainability caused by global competitiveness variables? (3) Can the fisheries sector become more competitive to meet EU objectives for sustainable ocean and marine development by 2030? The primary objective of this research is to examine the sustainability of fisheries production in relation to global competitiveness determinants in European Union seas between 1990 and 2022. (2) To comprehensively analyze the effects of global competition between EU13 developing and EU14 developed countries from 1990 to 2022 on fisheries production and marine ecosystem sustainability. (3) To examine the magnitude of global competitiveness factors' influence on fisheries production sustainability between developing and developed countries from 1990 to 2022.

Many contend that recent developments have made competition a vital determinant of sustainable development prospects and dynamics. Trade barriers have been significantly reduced, and sometimes even eliminated, due to crucial policy changes such as the World Trade Organization, improvements to the common agricultural and fisheries policy, and the most recent EU expansion [5]. Consequently, the macro-marketing environment is undergoing substantial transformation, significantly increasing export-country rivalry. The competitive environment in which the fisheries sector operates presents new risks and challenges. Additionally, challenges related to sustainable food security, marine environment sustainability, and ocean resource sustainability are of paramount importance globally [5]. Since a food product must be competitive while satisfying domestic and international organizations' concerns about ocean resource sustainability, marine ecology sustainability, and a sustainable blue food system, competitiveness is becoming an increasingly complex issue [5]. The highly competitive global market presents no exception for the EU's sustainable fisheries sector. The fisheries sector accounts for a significant portion of the EU food market, and over the past decade, imports have grown dramatically [5]. Since a dearth of pertinent literature fails to adequately explain the observed changes in the fisheries industry market, research examining competitiveness and sustainability factors affecting the fisheries industry may help develop future marketing strategies for sustainable fisheries.

The assumption that disturbance factors are independent and that no cross-sectional dependence (CD) exists among the stated variables across countries constitutes a critical limitation in earlier research on EU states. According to [6–10], these factors might represent common or idiosyncratic shocks, spillovers, or a combination thereof. Incorporating CD into the analysis is crucial since a cross-sectional group of EU economies will likely exhibit mutual dependence due to further economic integration. EU countries are particularly susceptible to global shocks that influence market size, economic development, economic growth, and global competitiveness. According to [9], if CD is not addressed in research, the findings regarding the relationship between economic growth and the enabling environment will be biased and inconsistent.

The current research will assist in determining whether the global competitiveness profile of fisheries-producing countries will be similarly affected by the growth-mitigating influence of natural resources. Previously published research employed more aggregate exogenous variables that did not distinguish between point-source and diffuse-source resources [11, 12]. In contrast, this research utilizes a highly specialized natural resource indicator (fisheries production sustainability). We focused on fisheries in EU economies because point-source resources typically generate more adverse consequences than diffuse sources [13, 14]. Because sound policy has been demonstrated to mitigate or even reverse the resource dependency-induced growth-mitigating effect [15], our sample spans developed and developing economies. We also incorporated regulatory control variables, such as the enabling environment, for more robust inferences. The EU27 fisheries-producing countries include developed nations such as Germany, Italy, and Sweden, as well as emerging countries such as Estonia, Bulgaria, and Malta. The study's timeframe spans from 1990 to 2022.

Because the selected countries exhibit different levels of economic development, our study employs the cutting-edge method of moments quantile regression of [16] to examine the validity of the economic growth-induced resource sustainability nexus in the presence of the fisheries industry and other control variables across different quantiles of the conditional distribution of fisheries sustainability. Through the inclusion of fixed effects, this method provides empirical insights into distributional heterogeneity relationships. By allowing for heterogeneous economic growth-fisheries production correlations at different blue resource sustainability conditional quantiles, the technique captures patterns that conventional mean regression may fail to detect. Consequently, the growth hypothesis is thoroughly investigated in fisheries-producing countries.

To our knowledge, this study is the first to consider distributional heterogeneity when examining resource sustainability induced by economic growth in the EU27 countries, taking into account fisheries production, enabling environment, human capital, market size, innovation ecosystem, and economic growth. Evaluating the growth hypothesis at different quantiles of the fisheries sustainability conditional distribution is necessary for several reasons. Conditional quantile estimates, in contrast to conditional mean estimations, demonstrate greater robustness to outliers' effects emanating from the outcome variable [17].

Furthermore, conditional mean estimates fail to adequately capture the distributional effects of global competition on fisheries sustainability. Quantile regression, on the other hand, offers more intuitive appeal, particularly in panel regressions, as it disaggregates the distributional influence of independent factors on the dependent variable across multiple quantile ranges. Consequently, it facilitates easier categorization of the heterogeneous impacts across different cross-sectional groups. As a result, conditional quantile estimates provide information that conditional mean estimations do not.

The remaining sections of this chapter are structured as follows. Section 7.2 briefly reviews the literature related to the growth hypothesis. Section 7.3 outlines the data and methodologies. Section 7.4 presents and discusses the estimation results, and Sect. 7.5 provides conclusions and policy recommendations based on the empirical findings.

7.2 Fisheries Global Competitiveness Historical Development

Sustainable competitiveness refers to the capacity to create and maintain inclusive wealth in the fisheries industry without compromising its ability to sustain or enhance its current wealth levels. For instance, [1, 2, 18, 19] examined the competitiveness of the Indonesian fisheries processing industry and found that Depok City ranked first with exceptionally high competitiveness, possessing advantages in three areas: facilities and infrastructure, production, and the utilization of science and technology [2]. In Mexico, [3] examined the impact of market competition on the fisheries sector, asserting that market competition can substantially enhance fisheries economic growth [20]. assessed stakeholders' perceptions of competitiveness variables in the fisheries industry and their effects on implementing economic policies in Ecuador and concluded that for coastal populations that depend on fisheries activities for economic survival, greater competitiveness is essential for developing appropriate policies for the fisheries industry's sustainable development. [21] examined the competitiveness of fisheries exports in Malaysia and found that the country experiences a significant fish trade imbalance and that its fisheries industry may lack sufficient international competitiveness.

According to [22], Shandong Province is the most competitive province in the fisheries industry among China's coastal provinces and possesses high potential for a sustainable fisheries development approach. They employed the Analytic Hierarchy Process (AHP) and the Delphi method to establish the evaluation index system and analyze the competitiveness of the fisheries industry across 11 coastal provinces in China.

Numerous studies, including [4, 23–27] have examined how poor governance and enforcement encourage overfishing, the significance of sound policy for fisheries management, and various strategies for effective and sustained enforcement [5, 28] among others, investigated the legal and policy status of the fisheries industry in Indonesia and India, respectively, and asserted that community partnership and co-management constitute the most suitable approaches to achieve an enabling environment for fisheries industry sustainability in those countries. According to evaluations of fisheries management plans by [29–31] based on environmental governance, there has been notable improvement in fish abundance in Europe, rising from 1229 (in 2008) to 1142 (in 2013) and 1781 (in 2018), with the decline in 2013 being attributed to a nearby wildfire.

While local knowledge structures are theoretically acknowledged, [32] investigated alternatives to sustainable seafood certifications in Japan and noted that although influential political and economic actors advocate for sustainable seafood certification to address external shocks, they do not draw upon local knowledge. In the USA, [33, 34] investigated the fundamental design of fisheries management regulations to conserve and sustain marine resources and found that the enabling environment constitutes a significant factor outside fishermen's direct control and that when combined with unpredictable weather, access to reliable crew, and rapid market price shifts, it exerts severe consequences on their general well-being.

Many studies, including those by [35–39] stated that technology must work in conjunction with equitable and resilient policy innovations to transform aquatic food systems and render them socially, economically, and environmentally sustainable over the long term [40–42]. investigated future developments in marine fisheries in Europe by integrating management, technological advancement, and public policy, and referred to the Fisheries-Energy-Environment Nexus resource management approach. This approach would better utilize systems thinking to optimize trade-offs and synergies among competing domains to achieve fisheries, conservation, and other environmental objectives [43, 44]. Investigated the relationship between technological innovation and marine fisheries growth in ten coastal regions of China and concluded that technological innovation is essential for fostering marine fisheries growth.

From another perspective, several studies, including those by [45–47], confirmed that increased investment in improving human capital could potentially offset, at least partially, the effects of decommissioning in the fisheries sector. In the Mediterranean Sea, [48, 49] examined the relationship between human development and fisheries footprint and concluded that relatively low levels of human capital adversely affect fisheries sustainability, while relatively high levels exert positive impacts. A comprehensive and cohesive approach, along with coordinated and harmonized policies for blue economy sustainability, was proposed by [50, 51] and was also investigated in Indonesia.

According to [52–54] sustainable fisheries refer to the marketing and distribution of fish products through aquaculture or wild catch in ways that would not compromise the long-term sustainability of species while also securing the future of marine ecosystems. The influence of market development on the expansion of the Asian fisheries sector was examined by [55–57], who concluded that efficient fisheries management plans are crucial to fisheries market sustainability [58–60] iInvestigated fish market development for sustainable seafood product development in Europe, demonstrating how integrating domestic and international markets can promote sustainable fisheries development and fisheries market expansion.

Studies by [61–65], among others, have estimated that unsustainable fisheries result in losses exceeding 20% of the global fisheries industry's value. By creating a sustainable fisheries sector, there exists greater potential for long-term economic, social, and environmental development [66–68] investigated optimal levels of fish production growth in China to support maximum growth performance with economic efficiency, demonstrating that sustainable fisheries development can enhance economic performance and economic growth. In East Asia, [69–72], among others, studied the economic impact of fish production and trade and concluded that developing countries could substantially boost their economies by focusing on production and high-quality fisheries exports.

While earlier research primarily focused on the fisheries sector, this study explores the effects of global competitiveness variables on fisheries industry sustainability to achieve blue growth. Additionally, to promote sustainable development, we also examine how global competitiveness factors affect fisheries production in EU seas. Most research employs time-series analysis, while this study utilizes static and panel regressions to examine how economic development and global competition affect fisheries production sustainability. The following constitute some of this paper's potential innovations: First and foremost, it examines the relationships among fisheries production, enabling environment, innovation ecosystem, human capital, market size, and economic development in EU seas from 1990 to 2022. More crucially, to attain sustainable blue growth, we also investigate the diverse effects of economic development and global competitiveness factors on the fisheries sector in the EU27 seas. We divide the EU27 region into two alternative groups: the EU14 developed countries and the EU13 developing countries. Because our data spans 1990 to 2022 and exhibits non-stationary behavior, we employ unit root and cointegration analysis to evaluate stationarity. Additionally, by adopting the MMQR approach, the Method of Moments Quantile Regression, we examine the interactions among market size, enabling environment, human capital, and innovation growth. We also conduct static and panel regression analysis on the variables affecting economic development and global competitiveness. Finally, we project how future changes in global competitiveness will affect fisheries and ocean resource sustainability.

7.3 Fisheries Competitiveness Market Modeling

Although the international competitiveness of EU member countries and the region in the global economy was emphasized in several cluster development, competitiveness, and Porter's theory studies, two factors merit consideration in fisheries development in developing countries: (1) the influence of multinational economic activities on the cluster and competitive advantage and (2) the function of interconnectedness/networks among cluster participants. Given the significance of these concerns for the long-term competitiveness of fisheries, a novel strategy termed "global competitiveness of fisheries development" is proposed. The model below presents the proposed fisheries development paradigm for enhanced global competitiveness. The three contributions to this new model are the incorporation of governance, the focus on innovation, and the human capital indicator. This revised model is based on Porter's diamond model and incorporates numerous characteristics identified [73, 74] models.

First, governance should be recognized as one of the key factors in sustainable fisheries development, as it significantly impacts the competitiveness of the fisheries industry in the 13 developing countries comprising the EU13. Since they are interconnected, the degree and nature of networks among each competitiveness component within the cluster should be identified and illustrated. The following model was developed based on recent investigations by [75–78] (2010):

$$FP_{it} = f\left(ENBit, INNit, MRKit, HCIit, GDP_{it}\right)$$

The dependent variable is fisheries production (FP), measured as total fisheries output in metric tonnes. Data for these variables were obtained from the [79] World Development Indicators. The independent variables include ENB, which represents government effectiveness and was obtained from the official European data website [80]. The analysis utilizes data for EU economies from 1990 to 2022, with each variable treated as per capita values. INN represents the innovation ecosystem and intellectual property activities, derived from the European Commission's database [81]. MRK represents market size, measured as GDP scaled by purchasing power parity (PPP) in current international USD. HCI represents human capital, identified by the average duration of education and learning in society. GDP represents gross domestic product, measured in constant US dollars and obtained from the European Commission's database [81].

To accomplish the objectives of this study, we collected annual data for the selected countries from the World Bank's World Development Indicators database (WDI), the European Commission's database (Eurostat), and the official portal for European data (Europa) for the years 1990 to 2022. The data displays, measurement units, sources, and transformations are presented in Table 7.1 is part, we'll go through the findings. Table 7.2 displays the statistical characteristics of each chosen element, including its maximum, minimum, mean, and standard deviation. The values of FP, ENB, INN, MRK, and HCI's respective means are 4.931, 1.867, 3.646,

Table 7.1 Data, units, transformation, and source

Data	Measurement unit	Transformation	Source
Fisheries production	Metric tons	Natural logarithm transformation	WDI
Enabling environment	% of confidence in governance	Natural logarithm transformation	WDI
Innovation ecosystem	Trademark applications, direct nonresident	Natural logarithm transformation	Eurostat
Market size	GDP per PPP (current international $)	Natural logarithm transformation	WDI
Human capital	The average number of years of education (years)	Natural logarithm transformation	Eurostat
Economic growth	Constant US dollars	Natural logarithm transformation	Eurostat

Source: [82, 83]

Table 7.2 Results of descriptive statistics

Variable	Observations	Mean	Std. dev.	Min	Max
FP	864	4.931	0.827	3.211	6.310
ENB	864	1.867	0.069	1.482	1.979
INN	864	3.646	0.331	2.223	4.354
MRK	864	4.408	0.229	3.744	5.044
HCI	864	1.888	0.019	1.837	1.921
EG	864	4.371	0.398	3.130	5.248

Source: [82, 83]

4.408, 1.888, and 4.371. The outstanding amount of standard deviation for each of the study's components is 0.827, 0.069, 0.331, 0.229, 0.019, and 0.398, respectively. We can get closer to the unit root test thanks to the factors' descriptive properties.

We employ fixed-effects ordinary least squares (FEOLS), dynamic ordinary least squares (DOLS), and fully modified ordinary least squares (FMOLS) for comparative purposes. The FEOLS approach incorporates Driscoll and Kraay standard errors, which are robust to common forms of cross-sectional dependence and autocorrelation up to a specified lag. According to [84], the primary concerns in estimating dynamic cointegrated panels stem from heterogeneity issues related to variations in cross-sectional means and differential cross-sectional adjustment to the cointegrating equilibrium. To address these concerns, Pedroni's FMOLS model incorporates individual-specific intercepts and permits varying serial correlation characteristics of the error processes across different panel members. Monte Carlo simulation results demonstrated that the DOLS estimator was unbiased relative to both OLS and FMOLS estimators in finite samples. This motivated [85] to extend the DOLS estimator to panel data contexts. The DOLS estimator further mitigates endogenous feedback by incorporating lead and lag differences, thereby correcting for endogeneity.

A panel quantile regression approach was employed to evaluate the distributional and heterogeneous effects across quantiles due to the limitations of previous estimation techniques [86] In their seminal study, the panel quantile regression methodology was introduced by [87]. Unlike conventional least-squares regressions, which yield estimates of the conditional mean of the dependent variable given specific values of the independent variables, quantile regressions are typically employed to estimate the conditional median or various other quantiles of the response variable. Quantile regressions demonstrate greater robustness to outliers in the estimation process. Additionally, they are particularly suitable when there is minimal or no relationship between the conditional means of two variables [88].

However, the researchers adopted the Method of Moments Quantile Regression (MMQR) of [16] with fixed effects in this current investigation. While quantile regression is robust to outliers, it does not account for potential unobserved heterogeneity among individuals in a panel. The MMQR technique permits individual-level effects to influence the entire distribution rather than merely shifting means, as demonstrated in studies by [17, 89], among other authors, thereby enabling the identification of conditional heterogeneous covariance effects of fisheries sustainability determinants. In situations where the panel data model incorporates individual-level effects and contains endogenous explanatory variables, the MMQR estimation approach is particularly appropriate. The MMQR technique is also straightforward to implement as it generates non-crossing estimates of the regression quantiles. For a location-scale model specification, the estimation of the conditional quantiles QY(τ|X) takes the following form:

$$Y_{it} = \alpha_{i+} X_{it}{}' \beta' + \left(\delta_i + Z_{it} \gamma' \right) U_{it} \tag{7.1}$$

Where P(δi + Zitγ >0) = 1. (α, β, δ, γ) are parameters requiring estimation. Z is a k-vector of identifiable components of X that are differentiable transformations with element l given by: (αi, δi), i = 1,…, n, represents the individual i fixed effects.

$$Z_l = Z_l\left(X\right), l = 1, \ldots, k \tag{7.2}$$

For each given i, X_{it} is independently and identically distributed and is also independent over time (t). According to Machado and Silva (2019), U_{it} is orthogonal to X_{it}, independently and identically distributed across individuals (i) and across time (t), and is normalized to satisfy the moment criteria, which, among other factors, did not necessitate strict exogeneity. Eq. (1) implies the following:

$$Q_Y\left(\tau | X_{it}\right) = \left(\alpha_{i+} \delta_i q\left(\tau\right)\right) + X_{it} \beta' + Z_{it} \gamma' q\left(\tau\right) \tag{7.3}$$

The independent variables in the current study include the natural logarithm of economic growth (LGDP), the natural logarithm of the innovation ecosystem (LINN), the natural logarithm of human capital development (LHCI), the natural logarithm of market size (LMRK), and the natural logarithm of the enabling environment (LENB). The scalar coefficient X_{it}—αi(τ) ≡ αi + δiq(τ) represents the quantile-specific fixed effect for individual i, and QY(τ|X_{it}) denotes the conditional quantile distribution of the dependent variable Yit (natural logarithm of fisheries production), which is contingent upon the location of the independent variable X_{it}. The individual-level effects, in contrast to conventional least-squares fixed effects, do not represent a simple intercept shift. Rather, these time-invariant parameters exhibit heterogeneous effects that can vary across the quantiles of the conditional distribution of the dependent variable Y. The τ-th sample quantile, denoted by q(τ), is obtained by solving the following optimization problem:

$$\min_q \sum_i \sum_t P_\tau \left(R_{it} - \left(\delta_i + Z_{it} \gamma' \right) q \right) \tag{7.4}$$

Where the check function P_τ (A) = (τ—1) AI {A ≤ 0} + TAI {A > 0} is shown.

7.4 Fisheries Global Competitiveness Outcomes

Several standard preliminary tests are conducted to examine the time-series properties of the variables before estimating the unknown parameters. We first assess whether cross-sectional dependence (CD) exists within the panel. Cross-sectional dependence may distort the true parameter values of coefficient estimates. If ignored, cross-sectional dependence arising from common unobserved variables could significantly diminish panel data efficiency gains [90]. Consequently, this issue must be addressed to obtain reliable coefficient estimates. We employ the [91] CD test to examine cross-sectional dependence within the panel. The remaining quantitative variables in Table 7.3 demonstrate substantial cross-sectional dependence across countries, with the exception of fisheries production. Therefore, to minimize the likelihood of size distortions, the researchers who adopted unit root and cointegration testing analyses, as well as panel estimation procedures, must employ methodologies that are robust to the effects of cross-sectional dependence.

We employ the [92], LLC, and [93] IPS unit root tests to assess the integration properties of the studied variables in an unbiased manner. Panel unit root tests aim

Table 7.3 Cross-sectional dependence results

Variables	FP	ENB	INN	MRK	HCI	GDP
Pesaran (2004)	(18.82)***	(31.30)***	(43.28)***	(98.67)***	(99.27)***	(83.55)***

Remark: *** refer importance at the 1%, scale
Source: [82, 83]

to enhance test power by combining cross-sectional and time-series information. Various panel unit root tests have been developed to test the unit root null hypothesis against alternative hypotheses [92]., test the unit root null hypothesis against a homogeneous alternative, wherein each series in the panel is stationary with an identical rate of mean reversion [93], test the unit root null hypothesis against the alternative that at least one series in the panel is stationary. The novel test introduced in this study focuses on panel stationarity, which naturally directs attention toward the LLC framework. This assumption is relaxed in the Im-Pesaran-Shin (IPS) panel test, which allows each individual to possess a distinct autoregressive parameter. For moderately small panel datasets, such as the one employed in the present study, the (LLC) panel unit root test has been demonstrated to exhibit superior power relative to other comparable unit root tests. We utilize these two tests to examine how cross-sectional dependence affects the panel unit root tests. Table 7.4 demonstrates that all variables are non-stationary at levels but become stationary at first differences across all unit root test specifications. This indicates that the variables utilized for estimation are integrated of order one, I(1).

We employ the [84] panel cointegration test and the [94] bootstrapped panel cointegration test in Table 7.5 to determine whether genuine long-run relationships exist among the variables under consideration [84]. proposes a comprehensive framework for panel cointegration testing based on the principles of the Engle and Granger two-step methodology. Pedroni's approach accounts for heterogeneity by removing individual-specific deterministic trends and short-run dynamics in the initial processing stage. From the estimated residuals, Pedroni generates seven alternative test statistics, which can be categorized as either tests assuming a common process, generally referred to as "pooled" or "within-dimension" tests, or tests assuming individual processes, commonly referred to as "grouped" or "between-dimension" tests, as presented in Table 7.5. The [94] method introduces four novel tests with the null hypothesis of no cointegration. Because the dynamics are structural rather than residual-based, the test relaxes the common factor restrictions imposed on residual-based tests. The violation of common factor restrictions can substantially reduce the power of residual-based cointegration tests, thereby necessitating structural dynamics.

Table 7.4 Panel unit root test results

	Difference		First difference	
Variable	LLC	IPS	LLC	IPS
FP	3.067	−3.700	−24.13***	−22.737***
ENB	−1.090	−0.254	−23.47***	−23.47***
INN	−1.209	−3.942	−7.818***	−11.92***
MRK	−3.775	−3.465	−6.894***	−10.42***
HCI	−2.189	−5.018	−11.91***	−14.91***
GDP	−1.571	−0.210	−18.74***	−68.36***

Remark: *** refer importance at the 1%, scale
Levin, Lin & Chu test (LLC), and Im, Pesaran, and Shin W-stat test (IPS)
Source: [82, 83]

Table 7.5 Panel cointegration test for EU countries

Test	Without trend	With trend
A) Pedroni Residual Cointegration Test:		
Alternative hypothesis: common AR coefficients (within dimension):		
Panel v-Statistic	−1.711 (0.956)	−2.356 (0.990)
Panel rho-Statistic	2.600 (0.995)	0.565 (0.714)
Panel PP-Statistic	−1.621* (0.052)	−7.871*** (0.000)
Panel ADF-Statistic	−2.011** (0.022)	−8.361*** (0.000)
Alternative hypothesis: common AR coefficients (between dimensions)		
Group rho-Statistic	−3.161	(0.999)
Group PP-Statistic	−7.829***	(0.000)
Group ADF-Statistic	−8.284***	(0.002)

B) Westerlund (2007) Bootstrap Panel Cointegration:				
Statistics	Value	Z-value	p-value	Robust p-value
Gt	−2.409	−1.095	0.095	0.081*
Ga	−7.225	3.100	0.999	0.985
Pt	−11.677	−1.408	0.080	0.074*
Pa	−5.673	1.569	0.942	0.999

Remark: ***, **, and * refer importance at the 1%, 5%, and 10% scales, respectively
Values in parentheses are *p*-values
Source: [82, 83]

With this constraint relaxed, it is no longer necessary for adjustment procedures in the short run and the long run to assume identical values. Westerlund's (2007) bootstrap methodology enables us to mitigate the distortionary effects of cross-sectional dependence and obtain robust critical values. According to the results presented in Table 7.5, cointegration is strongly supported by both the Pedroni (2004) test and the Westerlund (2007) bootstrapped cointegration test.

Table 7.6 presents the results of the FMOLS, DOLS, and FEOLS estimation procedures. Although there are variations in their statistical significance levels, Model 1 in Table 7.6 demonstrates that the coefficient estimates obtained from all three specifications are, on average, relatively similar. Regarding statistical significance and coefficient magnitude, fisheries production in the EU region exhibits the greatest stability across all three estimators. According to the FMOLS estimator, a one percent increase in the enabling environment exerts a positive influence on fisheries sector sustainability of 0.19% and 0.88%, with weak though statistically significant support in the FEOLS estimation specification. This finding is consistent with the principal conclusions of [23–25]. Due to its position as the world's largest seafood market, its capacity to import more than half of its requirements, and its status as a significant fisheries power, the EU has emerged as a crucial actor in regulating international fisheries.

Table 7.6 Model 1 Panel estimation for the EU27 region from 1990 to 2022

Long-run coefficient	DOLS		FMOLS		FEOLS	
	Coefficient	Std. error	Coefficient	Std. error	Coefficient	Std. error
ENB	0.887	(0.498)*	0.192	(0.016)***	0.032	(0.184)
INN	0.099	(0.084)	0.042	(0.011)***	0.039	(0.013)***
MRK	1.417	(0.377)***	0.088	(0.057)***	0.470	(0.121)***
HCI	9.816	(4.630)***	1.792	(0.089)***	7.318	(1.55)***
GDP	0.037	(0.219)	0.065	(0.016)***	0.234	(0.053)***
Long-run coefficient	DOLS		FMOLS		FEOLS	
	Coefficient	Std. error	Coefficient	Std. error	Coefficient	Std. error
ENB	0.887	(0.498)*	0.192	(0.016)***	0.032	(0.184)
INN	0.099	(0.084)	0.042	(0.011)***	0.039	(0.013)***
MRK	1.417	(0.377)***	0.088	(0.057)***	0.470	(0.121)***
HCI	9.816	(4.630)***	1.792	(0.089)***	7.318	(1.55)***
GDP	0.037	(0.219)	0.065	(0.016)***	0.234	(0.053)***

Note: ***, **, and * indicate significance at the 1%, 5%, and 10% levels respectively
Values in parentheses are P-values
Source: [82, 83]

Furthermore, as anticipated, the innovation ecosystem in the EU region demonstrates a positive and statistically significant relationship with fisheries industry sustainability. However, its coefficient values appear inconsistent across DOLS specifications, increasing fisheries industry sustainability by 0.03% to 0.04% for a one percent increase in the innovation ecosystem according to the FEOLS and FMOLS estimators, respectively. This finding is consistent with earlier results by [35–37]. Research in fisheries and marine innovation is crucial for understanding the marine environment, developing the blue economy of the future, and realizing Europe's goal of becoming the first climate-neutral continent by 2050.

Market size also substantially affects fisheries sector sustainability in the EU region. A one percent increase in market size enhances sustainability by 0.47%, 0.08%, and 1.41% according to the FEOLS, FMOLS, and DOLS estimators, respectively. This supports the growth hypothesis for the panel of fisheries-producing countries. This outcome corroborates the principal findings of [52–54] who documented a positive and inelastic influence of market size on fisheries industry growth. The European fish industry is thriving for several reasons, chief among them Europeans' increasing expenditure on fish and shellfish despite higher prices.

Similarly, all estimated specifications indicate that human capital in the EU region is substantial and exerts a positive impact. A one percent increase in human capital development enhances sustainability by 7.31%, 1.79%, and 9.8% according to the FEOLS, FMOLS, and DOLS estimators, respectively. This result aligns with the modernization hypothesis, though it conflicts with findings [46, 47], The FMOLS, FEOLS, and DOLS specifications provide strong and statistically significant evidence supporting the growth hypothesis. "Human capital input" refers to the labor required for fish capture, while immaterial capital refers to the investment needed for intangible assets, such as fisheries rights ownership necessary for participating in rights-regulated fisheries.

Economic growth exerts a highly positive effect on fisheries sector sustainability in the EU region. Specifically, a one percent increase in economic growth enhances sustainability by 0.23% and 0.06% according to the FEOLS and FMOLS estimators, although its coefficient appears inconsistent across DOLS specifications. This supports the growth hypothesis for the panel of fisheries-producing countries. This result confirms the critical findings of [61–63] who documented a positive and inelastic influence of economic expansion on fisheries sector development. The EU region's fisheries sector can achieve sustained growth by creating appropriate conditions for an economically viable and competitive market.

Table 7.7 demonstrates that the impact of the enabling environment on the fisheries sector is statistically significant at all quantiles according to the panel quantile regression estimates, except for the median quantiles, where the impact on fisheries production is negligible at the median quantiles (4th and 5th quantiles). This outcome supports the effectiveness of marine regulations relevant to the fisheries sector in higher marine-producing countries. The findings demonstrate that the growth hypothesis is valid in the EU27 for fisheries production only at the lowest and highest quantiles. This suggests that development takes precedence over marine policy in countries at median income levels. Additionally, due to associated investment costs and lower production levels, countries at median quantiles may encounter difficulties transitioning their fisheries industries from nonrenewable to renewable resource sources. The fisheries sector benefits from an enabling environment in the highest and lowest-producing countries. This may be attributable to the differing extraction technologies employed in the established EU14 countries and the emerging EU13 economies. According to [23–25], who verified the growth hypothesis for several countries producing exceptionally high levels of fisheries output, these findings portend negative environmental implications. Additionally, our results align with recent empirical research by [1], who supported the growth hypothesis, another critical framework for fishery resources. The findings also align with [2] who demonstrated the validity of the growth hypothesis for leading fisheries resource producers.

Additionally, across all quantiles, there exists a substantial and positive correlation between innovation ecosystem outcomes and fisheries production. The magnitude of the innovation ecosystem's impact increases at higher quantiles. Consequently, this is consistent with the conclusions reached by [35–37] that the innovation ecosystem propels the fisheries sector toward more sustainable growth. Additionally, there exists a strong and positive correlation between market size outcomes and fisheries production at median and above-median quantiles. Conversely, the impact of market size becomes more substantial at lower quantiles. This supports the findings [52–54], which demonstrated that market efficiency drives the fisheries sector to attain sustainable competitiveness.

Additionally, there exists a substantial and positive correlation between human capital outcomes and fisheries production at median and upper quantiles. Similarly, the magnitude of human capital's impact increases at lower quantiles. This supports the findings of [45–47] who concluded that human capital development propels the fisheries sector toward sustainable objectives. Additionally, the impact of economic

Table 7.7 Panel quantile estimation results for EU27 countries

Variables	Method of moments quantile regression with fixed effects										
	Location	Scale	Quantiles								
			0.1	0.2	0.3	0.4	0.5	0.6	0.7	0.8	0.9
ENB	(0.184)	(0.520)	(1.734)***	(1.083)***	(0.546)	(0.060)	(0.498)*	(0.937)***	(1.206)***	(1.521)***	(1.721)***
INN	(0.013)***	(0.549)***	(0.310)**	(1.006)***	(1.160)***	(1.128)***	(1.061)***	(0.952)***	(0.753)***	(0.563)***	(0.313)**
MRK	(0.121)***	(0.455)***	(0.081)	(0.041)	(0.166)***	(0.185)**	(0.354)***	(0.287)***	(0.219)***	(0.177)***	(0.487)
HCI	(1.55)***	(6.266)***	(0.075)	(0.060)	(0.035)***	(0.028)***	(0.038)***	(0.025)***	(0.030)***	(0.030)***	(0.280)*
GDP	(0.053)***	(0.198)	(0.862)***	(0.827)***	(0.886)***	(0.733)***	(0.611)***	(0.477)***	(0.271)***	(0.063)	(0.048)

Notes: *,**, and *** indicate significance at 10%, 5%, and 1% levels, respectively
Source: [82, 83]

expansion as measured by GDP on fisheries production is substantial at median and upper quantiles but negligible at lower quantiles. This finding supports the effectiveness of fisheries policy in promoting economic expansion in higher-producing countries. According to [62, 63] economic performance efficiency drives the fisheries sector to pursue sustainable growth.

Table 7.8 presents the results of the FMOLS, DOLS, and FEOLS estimation procedures. Model 2 in Table 7.8 demonstrates that although the coefficient estimates from the three specifications differ in terms of statistical significance, they are, on average, relatively similar. Regarding statistical significance and coefficient magnitude, the enabling environment in the EU14 developed countries is consistent across two specifications. A one percent increase in the enabling environment positively impacts fisheries production by 0.38% according to the FEOLS estimator and 0.59% according to the FMOLS estimator; however, the coefficient appears inconsistent across DOLS specifications. This finding is consistent with the principal conclusions of [18, 19] (2021, 2]. The EU communication is essential to the EU's implementation of Sustainable Development Goal 14 on life below water and the United Nations' 2030 Agenda for Sustainable Development. Identifying accessible interpretations of previous commitments demonstrates the EU's strong dedication to ocean conservation. The revised agenda plays a significant role in implementing the blue component of the European Green Deal.

According to all specifications, the market size in the EU14 developed countries demonstrates a positive and statistically significant relationship with the fisheries industry. A one percent increase in market size enhances fisheries production by 0.39%, 0.43%, and 0.60% according to the FEOLS, FMOLS, and DOLS estimators, respectively. This finding supports earlier results from [55–57]. Enhanced market intelligence, transparency, and efficiency in fisheries industry sustainability may be achieved through direct monitoring of the quantities, values, and prices of fisheries products from the point of first sale to retail, including imports and exports.

A one percent increase in human capital enhances fisheries production by 9.34%, 10.30%, and 11.61% according to the FEOLS, FMOLS, and DOLS estimators, respectively, demonstrating that human capital in the EU14 developed countries exerts a positive and significant effect. This result conflicts with [49] but supports the modernization theories proposed by [48]. The growth hypothesis receives strong

Table 7.8 Model 2 panel estimation for the EU14 region from 1990 to 2022

Long-run coefficient	DOLS		FMOLS		FEOLS	
	Coefficient	Std. error	Coefficient	Std. error	Coefficient	Std. error
ENB	0.504	(0.365)	0.598	(0.236)**	0.388	(0.156)**
INN	0.056	(0.021)	0.026	(0.013)	0.016	(0.018)
MRK	0. 607	(0.227)***	0.431	(0.137)***	0.391	(0.088)***
HCI	11.616	(3.238)***	10.301	(1.858)***	9.348	(1.251)***
GDP	1.039	(0.322)***	0.136	(0.080)*	0.091	(0.053)*

Note: ***, **, and * indicate significance at the 1%, 5%, and 10% levels respectively
Values in parentheses are P-values
Source: [82, 83]

statistical support across all specifications. Human fishing practices influence both the fish population size and its evolution over generations. According to research, when fisheries predominantly target large fish, they exert evolutionary pressure favoring smaller species' survival. Beyond certain thresholds, excessive fishing pressure can create food supply challenges.

Additionally, economic development exerts a highly positive effect on the fisheries sector development in the EU14. A one percent increase in economic growth enhances fisheries production by 0.09%, 0.13%, and 1.03% according to the FEOLS, FMOLS, and DOLS estimators, respectively. The growth hypothesis is supported for the panel of developed EU14 countries. This finding contrasts with [64], who documented an adverse and inelastic influence of economic expansion on fisheries security, while it corroborates the key conclusions of [64–66]. Other economic activities, including shipping, fisheries, aquaculture, and recreation, are impacted by fisheries production. For instance, the expenditure associated with repairing vessels and other equipment damaged by abandoned fishing gear represents a substantial annual cost for the fisheries sector.

According to the panel quantile regression estimates in Table 7.9, the enabling environment's influence on fisheries sustainability is statistically significant at both above-median and upper quantiles. This outcome confirms the effectiveness of marine policy relevant to fisheries production in the developed EU14 countries. The findings demonstrate that the growth hypothesis holds for the fisheries industries in the developed EU14 countries at above-median and upper quantiles. This suggests that marine policy takes precedence over fisheries production expansion among the EU14 developed countries. The investment costs associated with transitioning from conventional to renewable resource sources may present greater challenges for EU14 developed countries at lower quantiles, which are at less advanced developmental stages in terms of fisheries production. In the EU14 countries with the highest fisheries production levels, the enabling environment's positive effects on the fisheries sector are less pronounced. This may be attributable to the diverse extraction technologies employed in the EU14 developed countries' fisheries industries. These findings carry significant policy implications and align with [18] findings, which supported the growth hypothesis for several developed countries and documented exceptionally high levels of fisheries production. Our results are also consistent with previous empirical research by [19] which supported the growth hypothesis, another critical framework for fisheries resources. The findings concur with those of [2], who demonstrated the validity of the growth hypothesis for leading fisheries resource producers.

Additionally, innovation ecosystem outcomes are substantially and positively correlated with the fisheries sector at median and above-median quantiles. However, the innovation ecosystem's influence is minimal at lower and upper quantiles. Similarly, the magnitude of the innovation ecosystem's influence increases with higher quantiles. Market size outcomes are also strongly and positively correlated with the fisheries industry across all quantiles; however, market size's influence is negligible at the second quantile. Conversely, the magnitude of market size's impact decreases at higher quantiles. Additionally, there exists a significant and positive

Table 7.9 Panel quantile estimation results for EU14 countries

Variables	Method of moments quantile regression with fixed effects										
	Location	Scale	Quantiles								
			0.1	0.2	0.3	0.4	0.5	0.6	0.7	0.8	0.9
ENB	(0.156)**	(0.143)*	(0.774)	(0.434)	(0.320)	(0.319)	(0.404)	(1.135)***	(1.627)***	(1.861)***	(2.131)***
INN	(0.018)	(0.301)*	(0.011)**	(0.017)	(0.014)**	(0.015)**	(0.016)***	(0.015)***	(0.020)**	(0.025)	(0.035)
MRK	(0.088)***	(0.199)***	(0.653)*	(0.021)	(0.112)**	(0.094)**	(0.229)***	(0.263)***	(0.195)***	(0.183)***	(0.213)***
HCI	(1.251)***	(0.303)	(3.170)	(0.703)	(1.371)*	(0.125)	(1.054)***	(1.386)***	(1.438)***	(1.322)***	(1.573)***
GDP	(0.053)*	(0.189)***	(2.599)*	(0.213)	(0.406)*	(0.525)	(0.424)***	(0.523)***	(0.845)**	(0.711)**	(0.777)***

Notes: *,**, and *** indicate significance at 10%, 5%, and 1% levels, respectively
Source: [82, 83]

correlation between human capital outcomes and the fisheries sector at median and upper quantiles. However, no correlation exists at lower quantiles. Similarly, the magnitude of market size's impact diminishes with higher quantiles. The majority of quantiles demonstrate a substantial impact of economic expansion on fisheries production as measured by GDP. At the second and fourth quantiles, the effects of economic expansion are statistically insignificant. This finding supports the effectiveness of fisheries policy in promoting economic expansion in higher-producing countries. This suggests that in the EU14 developed countries, economic growth takes precedence over marine-resource sustainability.

The outcomes of the various panel estimation techniques employed in Model 3—namely, FMOLS, DOLS, and FEOLS—are presented in Table 7.10. These techniques yield nearly identical coefficient values. The results demonstrate that enabling environments exert a positive and substantial influence on fisheries production in the EU13 developing countries. A one percent increase in the enabling environment enhances fisheries production by 0.85% according to the DOLS estimator. The positive and statistically significant coefficient of the enabling environment in the DOLS methodology confirms the existence of the growth hypothesis in the EU13 developing countries. However, the coefficient appears inconsistent across FEOLS and FMOLS specifications. This conclusion is consistent with those of [29–31], who assert that during the early stages of development, fisheries production increases as the degree of ocean governance and the availability of enabling environments rise.

Innovation ecosystem outcomes are closely associated with fisheries production across FEOLS, FMOLS, and DOLS methodologies. A one percent expansion in the innovation ecosystem increases fisheries production by 0.07%, 0.08%, and 0.37% according to the FEOLS, FMOLS, and DOLS estimators, respectively. This result is consistent with assertions made by [41, 42, 44, 95] that the innovation ecosystem enhances fisheries production in developing countries.

The empirical findings demonstrate a positive relationship between market size and fisheries production in the EU13 developing countries. A one percent increase in market size results in fisheries production gains of 0.42%, 0.52%, and 0.69% according to the FEOLS, FMOLS, and DOLS estimators, respectively. This finding,

Table 7.10 Model 3. Panel estimation for the EU13 region from 1990 to 2022

Long-run coefficient	DOLS		FMOLS		FEOLS	
	Coefficient	Std. error	coefficient	Std. error	coefficient	Std. error
ENB	0.855	(0.766)**	0.330	(0.472)	0.097	(0.327)
INN	0.376	(0.166)**	0.083	(0.050)*	0.078	(0.026)***
MRK	0.692	(0.436)***	0.527	(0.304)*	0.429	(0.195)**
HCI	0.772	(0.958)*	6.776	(4.036)*	6.209	(2.708)**
GDP	0.550	(0.701)**	0.300	(0.139)**	0.272	(0.081)***

Note: ***, **, and * indicate significance at the 1%, 5%, and 10% levels respectively
Values in parentheses are P-values
Source: [82, 83]

which demonstrates that market efficiency enhances fisheries production, is supported by [58, 59] for European countries and [44, 95] EU developing countries.

The empirical findings also demonstrate a positive relationship between human capital and fisheries production in the EU13 developing countries. A one percent increase in human capital enhances fisheries production by 6.20%, 6.77%, and 0.77% according to the FEOLS, FMOLS, and DOLS estimators, respectively. This finding, which demonstrates that human capital development enhances fisheries production, is supported by [50, 51, 95] for developing countries and [72] for developing EU countries.

More importantly, the results demonstrate that economic development exerts a positive and substantial impact on fisheries sustainability in the EU13 developing countries. A one percent increase in GDP enhances fisheries production by 0.27%, 0.30%, and 0.55% according to the FEOLS, FMOLS, and DOLS estimators, respectively. The growth hypothesis in the EU13 developing countries is confirmed by the positive and statistically significant GDP coefficients across all three methodologies. This result is consistent with research from developing countries by [69, 70].

Table 7.11 presents the results of the Method of Moments Quantile regression. First, it can be confirmed that enabling environments positively impact fisheries sustainability, not at below-median quantiles but at median and upper quantiles. According to the table, significant variation in fisheries production attributable to the enabling environment increases from 0.806 to 0.380 as the quantile rises (from the 1st to the 3rd and from the 5th onward). The influence of the enabling environment is statistically negligible at quantiles from the second to the fourth.

Furthermore, all quantiles except the median quantiles (5th, 6th, and 7th quantiles) demonstrate a strong and positive correlation between innovation ecosystem outcomes and fisheries sustainability. Similarly, the magnitude of the innovation ecosystem's influence increases with higher quantiles. Additionally, all quantiles except the uppermost (8th and 9th) demonstrate a strong and positive correlation between market size outcomes and fisheries sustainability. Conversely, the impact of market size at higher quantiles for the EU13 developing countries is diminishing.

Similarly, human capital input exerts a positive impact on fisheries sustainability across all quantiles, and this correlation is statistically significant from the first to the ninth quantile. In contrast to other variables, higher quantiles result in a lesser magnitude of human capital input's impact. Furthermore, it is clearly verifiable that economic expansion positively impacts fisheries sustainability. Table 7.11 demonstrates that the improvement in fisheries sustainability due to economic development is significant and increases from 0.483 to 0.513 as the quantile rises (from the 5th to the 9th). This holds true for both median and upper quantiles.

Panel FMOLS and DOLS estimations were employed to validate the results obtained from the FEOLS panel model. The findings indicate that the FEOLS coefficients exhibit the same directional signs and comparable levels of statistical significance as those estimated through FMOLS and DOLS. This consistency confirms the robustness and reliability of the FMOLS and DOLS estimators for drawing valid conclusions. Although the significance levels differ slightly across the FEOLS, FMOLS, and DOLS estimations, the coefficients remain consistent

Table 7.11 Panel quantile estimation results for EU13 countries

Variables	Method of moments quantile regression with fixed effects										
	Location	Scale	Quantiles								
			0.1	0.2	0.3	0.4	0.5	0.6	0.7	0.8	0.9
ENB	(0.327)	(0.252)	(0.806)**	(0.221)	(0.156)	(0.166)	(0.904)***	(0.973)***	(1.204)***	(1.501)***	(1.380)***
INN	(0.026)***	(0.435)**	(0.675)***	(0.614)***	(0.583)***	(0.525)***	(0.210)	(0.129)	(0.027)	(0.566)***	(0.885)***
MRK	(0.195)**	(0.903)**	(2.060)***	(2.440)***	(1.844)***	(1.792)***	(1.611)***	(1.488)***	(0.979)***	(0.237)	(0.477)
HCI	(2.708)**	(0.170)***	(0.426)***	(0.443)***	(0.399)***	(0.424)***	(0.400)***	(0.367)***	(0.287)***	(0.125)***	(0.104)***
GDP	(0.081)***	(0.120)	(0.249)	(0.251)	(0.275)	(0.638)	(0.483)**	(0.429)**	(0.398)**	(0.334)*	(0.513)*

Notes: *, **, and *** indicate significance at 10%, 5%, and 1% levels, respectively
Source: [82, 83]

in sign. Moreover, the DOLS estimator is considered highly reliable because it effectively addresses issues related to serial correlation and endogeneity. To further investigate the influence of global competitiveness on fisheries sustainability, the EU27 countries were divided into two categories: 14 developed economies (EU14) and 13 developing economies (EU13). This classification enables a comparative assessment based on differences in economic development levels. The estimated effects of global market competitiveness on fisheries sustainability in the EU14 developed countries are presented in Tables 7.8 and 7.9, whereas Tables 7.10 and 7.11 report the corresponding estimates for the EU13 developing countries.

The results reported in Tables 7.8 and 7.10 demonstrate that global market competition exerts a significant positive effect on fisheries sustainability. Furthermore, environmental factors exhibit a stronger positive influence on fisheries sustainability in the EU14 developed countries than in the EU13 developing countries. The estimated magnitudes of this effect are 0.097 for EU14 and 0.388 for EU13. These findings imply that the adoption of ocean governance frameworks and marine ecological taxation policies within the fisheries sector may enhance fisheries sustainability more effectively in developed EU14 countries than in developing EU13 nations. In contrast, the innovation ecosystem has a more substantial positive impact on fisheries sustainability in EU13 developing countries than in EU14 developed nations. The estimated coefficients for EU13 and EU14 are 0.078 and 0.016, respectively. This suggests that environmental taxation strategies and green technology policies targeting innovation ecosystems in the fisheries sector could significantly strengthen fisheries sustainability in EU13 economies. The findings further indicate that market size contributes more strongly to fisheries sustainability in EU13 developing nations compared to EU14 developed countries. The corresponding effect sizes are 0.429 and 0.391, respectively. This implies that adopting market expansion strategies may lead to greater improvements in fisheries sustainability in EU13 economies relative to EU14 nations. One possible explanation is the existence of tax exemptions on seafood demand and consumption under revised taxation systems in wealthier EU14 countries.

Additionally, the results reveal that human capital contributes more substantially to fisheries sustainability in EU14 developed countries than in EU13 developing nations. The estimated coefficients are 9.348 for EU14 and 6.209 for EU13. This finding suggests that investments in specialized institutions, technical expertise, and skill development within the fisheries sector can substantially improve fisheries sustainability, particularly in developed economies.

Finally, economic development is found to have a stronger positive influence on fisheries sustainability in EU13 developing nations than in EU14 developed countries. The estimated impact values are 0.272 for EU13 and 0.091 for EU14. This indicates that investments in fisheries infrastructure, modernization of fleets, and broader fisheries sector development initiatives may generate greater sustainability gains in developing EU13 countries compared with their EU14 counterparts.

7.5 Conclusion and Solutions for Fisheries Competitiveness

This study investigates the empirical relationships among fisheries production, enabling environments, innovation ecosystems, market size, human capital, and economic growth within the framework of the growth hypothesis. The study contributes to the growing literature on marine resources, blue economy development, and fisheries sustainability. Using panel data covering 27EU countries over 31 years, the study employed panel unit root tests, panel cointegration techniques, and several panel estimation approaches. Governance quality, represented by the enabling environment variable (% confidence in governance effectiveness), was incorporated to evaluate fisheries governance and sustainable development outcomes. The results indicate that all variables are integrated of order one, I(1), and exhibit a stable long-run relationship based on panel cointegration analysis. Although the coefficient magnitudes remain broadly similar across estimation methods, variations in statistical significance were observed among traditional panel cointegration estimators. To provide a more comprehensive empirical assessment, the study adopted the Method of Moments Quantile Regression (MMQR), which allows the analysis of heterogeneous effects of explanatory variables across multiple quantiles of fisheries sustainability.

The FMOLS, DOLS, and FEOLS results support the growth hypothesis for fisheries-producing EU27 economies. The MMQR findings reported in Table 7.7 reveal that the enabling environment positively and significantly influences fisheries sustainability across all quantiles, particularly around the middle quantiles. Similarly, the innovation ecosystem demonstrates a positive and statistically significant relationship with fisheries sustainability across all quantiles, indicating that innovation contributes to fisheries sector development during both early and advanced stages of economic transition. Market size also positively affects fisheries sustainability, especially between the third and eighth quantiles. The results further suggest that developing countries with larger fisheries output tend to be more economically diversified compared to developed countries that rely heavily on seafood imports. Consequently, fisheries resources may contribute a larger share to economic activity in emerging economies. The MMQR estimates further show that economic growth positively affects fisheries sustainability from the first to the seventh quantiles, with the largest coefficients observed at the lower quantiles. The declining coefficients across higher quantiles imply that the impact of economic growth weakens as fisheries sustainability improves. Likewise, human capital contributes more strongly to fisheries sustainability at higher quantiles, suggesting that countries relying less on conventional fisheries practices tend to invest more effectively in human capital development. The findings also provide additional support for the resource curse hypothesis associated with deindustrialization. Existing fisheries regulations and environmental policies aimed at controlling overfishing and marine degradation may explain the weaker effects of enabling environments at lower quantiles in Tables 7.9 and 7.11. Citizens in developed EU14 countries generally place greater importance on fisheries governance and environmental regulation than those in EU13 developing countries. Consequently, policymakers in developed economies are more likely to introduce stricter sustainability-oriented fisheries policies to meet public expectations.

The innovation ecosystem becomes particularly important at medium fisheries production quantiles, especially within EU13 economies, where fisheries expansion creates opportunities for technological advancement and industrial development. However, stringent environmental regulations in higher quantiles may reduce the perceived benefits of innovation due to concerns regarding environmental degradation and compliance costs. The study further finds that the influence of economic growth varies across quantiles because of the interaction between competitiveness factors and the persistence of resource-dependence dynamics. Strong positive effects of economic growth in higher quantiles indicate improved sustainable fisheries development in both EU14 and EU13 countries. Similar findings from Mexico, Ecuador, Malaysia, China, and Vietnam reported by [20, 21, 96] support the conclusion that stronger economic performance can enhance fisheries sustainability. Therefore, the study recommends policies aimed at improving economic growth and competitiveness in fisheries-producing nations, particularly in the developed EU14 economies with high fisheries output.

Human capital is also found to be positively significant across nearly all quantiles. However, in developed EU14 countries, its impact becomes statistically insignificant at lower quantiles, possibly due to inefficient allocation of skilled labor and technological resources in fisheries production. This situation may encourage dependence on seafood imports and reduce domestic fisheries production. Furthermore, the migration of labor-intensive fisheries industries from EU13 to EU14 economies in search of higher profits may shift economic growth dynamics toward import dependence while reducing fisheries production in developing economies. Strict fisheries policies and environmental regulations in developed EU14 countries may additionally constrain fisheries production and increase operational costs. As a result, fisheries firms may relocate operations to less regulated developing countries with lower taxation levels and more abundant marine resources. Market size remains positively significant across medium quantiles, with the strongest effects observed at lower quantiles. This suggests that improvements in domestic fisheries markets, international seafood trade, and broader market competitiveness contribute positively to fisheries sustainability. Nevertheless, the continued reliance on conventional fisheries practices in both EU14 and EU13 countries highlights the need for a transition toward environmentally sustainable and regenerative fisheries systems. Accordingly, the study recommends promoting sustainable fish production methods and renewable marine-resource practices. For EU13 developing nations lacking sufficient financial capacity to adopt sustainable technologies, moderate environmental taxation policies may provide an alternative mechanism. Revenue generated from such taxes could be reinvested into long-term fisheries infrastructure and sustainability initiatives.

Finally, future research may extend this analysis by incorporating additional variables such as tourism, globalization, and financial development, all of which may influence fisheries sustainability and marine environmental quality. The framework could also be applied to other regions such as ASEAN, BRICS, the G7, or the Next 11 economies using advanced econometric techniques. Despite its contributions, the study has limitations. Data constraints prevented the examination of the broader benefits of fisheries competitiveness, and the analysis was limited to EU countries between 1990 and 2022. Future studies may explore the role of public-private investment in fisheries and ocean energy, expand the sample to include other regions, and account for structural breaks when analyzing the economic complexity of fisheries competitiveness over time.

References

1. Yusuf M (2022) The competitiveness of Indonesian shrimp export in Malaysia and Singapore markets. Online. http://repository.unimus.ac.id/5474/1/6.%20THE%20COMPETITIVENESS%20OF%20INDONESIAN%20SHRIMP%20EXPORT%20IN%20MALAYSIA%20AND%20SINGAPORE%20MARKETS.pdf. Accessed 20 May 2026
2. Handaka AA, Lestari RP, Rizal A (2022) Analysis of Competitiveness of the Fisheries Processing Industry in West Java, Indonesia. AJFAR 16:8–19. https://doi.org/10.9734/ajfar/2022/v16i330372
3. Peñaloza-Talavera MF, Martínez-Arroyo JA, Valenzo-Jiménez MA (2022) Formation of a fishing and aquaculture cluster as a tool for regional competitiveness. In: Terzioğlu MK (ed) *Advances in econometrics, operational research, data science and actuarial studies, Contributions to Economics*. Springer International Publishing, Cham, pp 267–283. https://doi.org/10.1007/978-3-030-85254-2_16
4. Okafor-Yarwood I, Kadagi NI, Belhabib D, Allison EH (2022) Survival of the Richest, not the Fittest: How attempts to improve governance impact African small-scale marine fisheries. Mar Policy 135:104847. https://doi.org/10.1016/j.marpol.2021.104847
5. Kelkar N, Arthur RI (2022) A review of governance and tenure in inland capture fisheries and aquaculture systems of India. Online. https://books.google.com/books?hl=en&lr=&id=wL1qEAAAQBAJ&oi=fnd&pg=PP1&dq=Kelkar,+N.,+%26+Arthur,+R.+I.+(2022).+A+review+of+governance+and+tenure+in+inland+capture+fisheries+and+aquaculture+systems+of+India.+https://doi.org/10.4060/cb8615en&ots=43o6zSOa5g&sig=lUJVzKdVu4wfxr1-PtME1opbuaU. Accessed 20 May 2026
6. Chudik A, Pesaran H (2015) Large panel data models with cross-sectional dependence: a survey. Online. https://academic.oup.com/edited-volume/41981/chapter/355335206?searchresult=1&utm_source=TrendMD&utm_medium=cpc&utm_campaign=Oxford_Academic_Books_TrendMD_0. Accessed 20 May 2026
7. Flaten O, Lien G, Koesling M, Løes A-K (2010) Norwegian farmers ceasing certified organic production: characteristics and reasons. J Environ Manag 91(12):2717–2726. https://doi.org/10.1016/j.jenvman.2010.07.026
8. Breitung J, Pesaran MH (2008) Unit roots and cointegration in panels. In: Mátyás L, Sevestre P (eds) *The econometrics of panel data, Advanced Studies in Theoretical and Applied Econometrics*, vol 46. Springer Berlin Heidelberg, Berlin, Heidelberg, pp 279–322. https://doi.org/10.1007/978-3-540-75892-1_9
9. Andrews DWK (2005) Cross-section regression with common shocks. Econometrica 73(5):1551–1585. https://doi.org/10.1111/j.1468-0262.2005.00629.x
10. Cerrato M, Sarantis N (2002) The cross sectional dependence puzzle. London Guildhall University. Online. http://fmwww.bc.edu/RePEc/res2002/Cerrato.pdf. Accessed 20 May 2026
11. Balsalobre-Lorente D, Shahbaz M, Roubaud D, Farhani S (2018) How economic growth, renewable electricity and natural resources contribute to CO2 emissions? Energy Policy 113:356–367. https://doi.org/10.1016/j.enpol.2017.10.050

12. Bekun FV, Alola AA, Sarkodie SA (2019) Toward a sustainable environment: nexus between CO2 emissions, resource rent, renewable and nonrenewable energy in 16-EU countries. Sci Total Environ 657:1023–1029. https://doi.org/10.1016/j.scitotenv.2018.12.104
13. Isham J (2005) The varieties of resource experience: natural resource export structures and the political economy of economic growth. World Bank Econ Rev 19(2):141–174. https://doi.org/10.1093/wber/lhi010
14. Mehlum H, Moene K, Torvik R (2006) Institutions and the resource curse. Econ J 116(508):1–20. https://doi.org/10.1111/j.1468-0297.2006.01045.x
15. Leite MC, Weidmann J (1999) Does mother nature corrupt? Natural resources, corruption, and economic growth. International Monetary Fund. Online. https://books.google.com/books?hl=en&lr=&id=UFYYEAAAQBAJ&oi=fnd&pg=PA3&dq=Leite,C.,+Weidmann,+J.+(1999).+Does+Mother+Nature+Corrupt:+Natural+Resources,+Corruption,+and+Economic+Growth+IMF+Working+Papers.+International+Monetary+Fund.&ots=o29De-kCdS&sig=NDcr5kVNSfELzupOiiN-WefMC1Y. Accessed 20 May 2026
16. Machado JAF, Santos Silva JMC (2019) Quantiles via moments. J Econ 213(1):145–173. https://doi.org/10.1016/j.jeconom.2019.04.009
17. Koenker R (2004) Quantile regression for longitudinal data. J Multivar Anal 91(1):74–89. https://doi.org/10.1016/j.jmva.2004.05.006
18. Suryana AAH, Amalia DH (2021) Analysis of competitiveness fisheries processing industry in Indonesia. AJFAR 14:30–42. https://doi.org/10.9734/ajfar/2021/v15i530343
19. Suryana AAH, Perdiansyah (2021) Analysis of human resources competitiveness of capture fisheries in Indonesia. AJFAR 15:40–46. https://doi.org/10.9734/ajfar/2021/v15i230326
20. Menéndez Delgado ER, Rodríguez Ríos EB, Bernal Casasola D, Cerbán Jiménez MDM (2021) Artisanal fishery in Ecuador. A case study of Manta city and its economic policies to improve competitiveness of the sector. Mar Policy 124:104313. https://doi.org/10.1016/j.marpol.2020.104313
21. Soh BH, Lim G-T, Chua SY (2021) Competitiveness of malaysian fisheries exports: a constant market share analysis. *MJES* 58(2):175–198. https://doi.org/10.22452/MJES.vol58no2.1
22. Kang H, Yang Z, Zhang Z (2023) The competitiveness of China's seaweed products in the international market from 2002 to 2017. Aquacult Fish 8(5):579–586. https://doi.org/10.1016/j.aaf.2021.10.003
23. Chen X, Xu Q, Li L (2023) Illegal, unreported, and unregulated fishing governance in disputed maritime areas: reflections on the international legal obligations of states. Fishes 8(1):36. https://doi.org/10.3390/fishes8010036
24. Spijkers J, Mackay M, Turner J, McNeill A, Travaille K, Wilcox C (2023) Diversity of global fisheries governance: Types and contexts. Fish Fish 24(1):111–125. https://doi.org/10.1111/faf.12713
25. Jarvis RM, Young T (2023) Pressing questions for science, policy, and governance in the high seas. Environ Sci Pol 139:177–184. https://doi.org/10.1016/j.envsci.2022.11.001
26. Guggisberg S, Jaeckel A, Stephens T (2022) Transparency in fisheries governance: Achievements to date and challenges ahead. Mar Policy 136:104639. https://doi.org/10.1016/j.marpol.2021.104639
27. Jentoft S, Chuenpagdee R (2022) Interactive learning and governance transformation for securing blue justice for small-scale fisheries. Administration & Society 54(7):1255–1282. https://doi.org/10.1177/00953997211073947
28. Dudayev R, Hakim LL, Rufiati I (2023) Participatory fisheries governance in Indonesia: are octopus fisheries leading the way? Mar Policy 147:105338. https://doi.org/10.1016/j.marpol.2022.105338
29. Acuña-Alonso C, Varandas S, Álvarez X, Martinho A (2023) Analysis of the evolution of a fisheries management plan based on environmental governance: living laboratory in the Olo River, Portugal. Fish Res 260:106595. https://doi.org/10.1016/j.fishres.2022.106595
30. Aguión A et al (2022) Establishing a governance threshold in small-scale fisheries to achieve sustainability. Ambio 51(3):652–665. https://doi.org/10.1007/s13280-021-01606-x

31. Linke S, Gillette MB, Jentoft S (2022) Institutionalizing injustice? aligning governance orders in swedish small-scale fisheries. In: Jentoft S, Chuenpagdee R, Bugeja Said A, Isaacs M (eds) *Blue Justice, MARE Publication Series*, vol 26. Springer International Publishing, Cham, pp 529–547. https://doi.org/10.1007/978-3-030-89624-9_28
32. Ynacay-Nye A, Hisano S, Suryawan AS (2023) Alternatives to sustainable seafood certifications: transitions of small-scale fisheries governance in northeastern Japan. J Rural Stud 97:269–280. https://doi.org/10.1016/j.jrurstud.2022.12.002
33. Day S (2026) Pacific Salmon fisheries management: An (unusual) example of polycentric governance involving indigenous participation at multiple scales. The Center for Growth and Opportunity. Online. https://www.thecgo.org/books/the-environmental-optimism-of-elinor-ostrom/chapter-3-chapter-3-pacific-salmon-fisheries-management-an-unusual-example-of-polycentric-governance-involving-indigenous-participation-at-multiple-scales/. Accessed 20 May 2026
34. Sorensen J, Kincl L, Weil R, Dzugan J, Christel D (2022) Fisheries governance and associated health implications: current perspectives from US commercial fishermen. Mar Policy 141:105119. https://doi.org/10.1016/j.marpol.2022.105119
35. Schwing FB (2023) Modern technologies and integrated observing systems are 'instrumental' to fisheries oceanography: a brief history of ocean data collection. Fish Oceanogr 32(1):28–69. https://doi.org/10.1111/fog.12619
36. Scarnecchia DL, Schooley JD (2022) Trophies, technology, and timescape in fisheries management, as exemplified through Oklahoma's world record paddlefish *Polyodon Spathula*. Fisheries 47(9):381–394. https://doi.org/10.1002/fsh.10764
37. Ziegler R et al (2022) Social innovation for biodiversity: a literature review and research challenges. Ecol Econ 193:107336. https://doi.org/10.1016/j.ecolecon.2021.107336
38. Torres Cañete F, Oyanedel R, Gelcich S (2022) Adoption and impacts of fishing gear innovations: insights from a small-scale fishery in Chile. Fish Res 248:106200. https://doi.org/10.1016/j.fishres.2021.106200
39. Pathak S, Lavudya N, Kumar RM, Dash RK (2022) Use of innovative technologies for fisheries resource management. Just Agriculture 2(5):1–8
40. Kemp PS et al (2023) Future advances in UK marine fisheries policy: integrated nexus management, technological advance, and shifting public opinion. Mar Policy 147:105335. https://doi.org/10.1016/j.marpol.2022.105335
41. Karagiannis D, Metaxas T (2023) 'Fishing' for winemaking perfection in cloudy waters…underwater wineries: innovation genius or another marketing trick? implications on wine & food tourism. In: Sigala M, Haller C (eds) Technology advances and innovation in wine tourism. Springer Nature Singapore, Singapore, pp 231–250. https://doi.org/10.1007/978-981-19-8277-4_14
42. Nogueira LA, Lindeløv B, Olsen J (2023) From waste to market: exploring markets, institutions, and innovation ecosystems for waste valorization. Bus Strat Env 32(4):2261–2274. https://doi.org/10.1002/bse.3247
43. Li Y, Kong J, Ji J (2022) Environmental regulation, technological innovation and development of marine fisheries—evidence from ten coastal regions in China. Fishes 7(1):20. https://doi.org/10.3390/fishes7010020
44. Fu X-M et al (2022) Evaluation of the innovation ability of China's marine fisheries from the perspective of static and dynamic. Mar Policy 139:105032. https://doi.org/10.1016/j.marpol.2022.105032
45. Wang M, Janssen ABG, Bazin J, Strokal M, Ma L, Kroeze C (2022) Accounting for interactions between sustainable development goals is essential for water pollution control in China. Nat Commun 13(1):730. https://doi.org/10.1038/s41467-022-28351-3
46. Albasri H, Sammut J (2022) A comparative study of sustainability profiles between small-scale mariculture, capture fisheries and tourism communities within the Anambas Archipelago Small Island MPA, Indonesia. Aquaculture 551:737906. https://doi.org/10.1016/j.aquaculture.2022.737906

47. Ramos J, Drakeford B (2021) Blue economy and tourism: is there a potential to create sustainable jobs? In: Handbook of research on human capital and people management in the tourism industry. IGI Global Scientific Publishing, pp 243–264. Online. https://www.igi-global.com/chapter/blue-economy-and-tourism/263358. Accessed 20 May 2026
48. Yıldırım DÇ, Yıldırım S, Bostancı SH, Turan T (2022) The nexus between human development and fishing footprint among mediterranean countries. Mar Pollut Bull 176:113426. https://doi.org/10.1016/j.marpolbul.2022.113426
49. Seisay M (2022) Human capital investment and financing for Africa's blue economy: conference on unlocking the full potential of the fisheries sector within the framework of a sustainable blue economy. Online. http://repository.au-ibar.org/handle/123456789/1175. Accessed 20 May 2026
50. Nurlita L, Yanuar T, Syah R, Indradewa R, Hamdi E (2021) Human capital implementation strategy in the Marina company. American International Journal of Business Management (AIJBM) 4(8):152–156
51. Najib H, Nawangsari LC (2021) Effect of intellectual capital on organizational sustainability with employee innovative behavior as intervening variables in Pt. Jaya Maritime services. EJBMR 6(1):158–163. https://doi.org/10.24018/ejbmr.2021.6.1.714
52. Climent-Pérez P, Galán-Cuenca A, García-d'Urso NE, Saval-Calvo M, Azorin-Lopez J, Fuster-Guillo A (2023) Automatic fish size estimation from uncalibrated fish market images using computer vision and deep learning. In: García Bringas P, Pérez García H, Martinez-de-Pison FJ, Villar Flecha JR, Troncoso Lora A, De La Cal EA, Herrero Á, Martínez Álvarez F, Psaila G, Quintián H, Corchado Rodriguez ES (eds) *17th International Conference on Soft Computing Models in Industrial and Environmental Applications (SOCO 2022), Lecture Notes in Networks and Systems*, vol 531. Springer Nature Switzerland, Cham, pp 319–329. https://doi.org/10.1007/978-3-031-18050-7_31
53. Schrobback P, Hoshino E, Pascoe S, Curtotti R (2022) Market integration of domestic and imported seafood: insights from the Sydney Fish Market*. *Aus J Agri & amp; Res Econ* 66(1):216–236. https://doi.org/10.1111/1467-8489.12452
54. Pincinato RBM, Gasalla MA, Garlock T, Anderson JL (2022) Market incentives for shark fisheries. Mar Policy 139:105031. https://doi.org/10.1016/j.marpol.2022.105031
55. Malto MAD, Dumilag RV (2023) Market fish resources in Magallanes, Sorsogon, Philippines. *Philippine Journal of* Science 152(1) Online. https://www.researchgate.net/profile/Richard-Dumilag/publication/365890886_Market_Fish_Resources_in_Magallanes_Sorsogon_Philippines/links/63883b6abbdef30dc9887097/Market-Fish-Resources-in-Magallanes-Sorsogon-Philippines.pdf. Accessed: May 20, 2026
56. Shimose T, Kanaiwa M (2022) Influence of the body color and size on the market value of wild captured coralgroupers (Serranidae, Plectropomus): implications for fisheries management. Fish Res 248:106223. https://doi.org/10.1016/j.fishres.2021.106223
57. Macusi ED, Morales IDG, Macusi ES, Pancho A, Digal LN (2022) Impact of closed fishing season on supply, catch, price and the fisheries market chain. Mar Policy 138:105008. https://doi.org/10.1016/j.marpol.2022.105008
58. Ewald C-O, Haugom E, Kanthan L, Lien G, Salehi P, Størdal S (2022) Salmon futures and the Fish Pool market in the context of the CAPM and a three-factor model. Aquac Econ Manag 26(2):171–191. https://doi.org/10.1080/13657305.2021.1958105
59. Nielsen M et al (2022) Can small-scale fisheries survive market-based management? Nordic evidence. Fish Fish 23(1):256–272. https://doi.org/10.1111/faf.12614
60. J. Łuczyńska, M. J. Łuczyński, J. Nowosad, M. Kowalska-Góralska, and M. Senze, "Total mercury and fatty acids in selected fish species on the polish market: a risk to human health," IJERPH, vol. 19, no. 16, p. 10092, Aug. 2022, doi:https://doi.org/10.3390/ijerph191610092
61. McKinley E (2022) A blue economy for whom?: Linking marine social sciences with blue economy discourse. In: *Blue economy*. Routledge, pp 13–25. Online. https://www.taylorfrancis.com/chapters/edit/10.4324/9781003280248-4/blue-economy-emma-mckinley. Accessed 20 May 2026

62. Nunes AJP, Masagounder K (2022) Optimal levels of fish meal and methionine in diets for juvenile litopenaeus vannamei to support maximum growth performance with economic efficiency. Animals 13(1):20. https://doi.org/10.3390/ani13010020
63. Starkweather KE, Keith MH, Zohora FT, Alam N (2023) Economic impacts and nutritional outcomes of the 2017 floods in B angladeshi S hodagor fishing families. American J Hum Biol 35(1):e23826. https://doi.org/10.1002/ajhb.23826
64. Karimi MS, Khezri M, Khan YA, Razzaghi S (2022) Exploring the influence of economic freedom index on fishing grounds footprint in environmental Kuznets curve framework through spatial econometrics technique: evidence from Asia-Pacific countries. Environ Sci Pollut Res 29(4):6251–6266. https://doi.org/10.1007/s11356-021-16110-8
65. Alharthi M, Hanif I (2020) Impact of blue economy factors on economic growth in the SAARC countries. Maritime Business Review 5(3):253–269
66. Wang Q, Wang S (2022) Carbon emission and economic output of China's marine fishery—a decoupling efforts analysis. Mar Policy 135:104831. https://doi.org/10.1016/j.marpol.2021.104831
67. Gou L, Yang S, Ren L (2022) Coupling analysis of the quality and quantity of marine economic growth from the perspective of high-quality development. J Coast Res 38(1):218–228
68. Peng D, Yang Q, Yang H-J, Liu H, Zhu Y, Mu Y (2020) Analysis on the relationship between fisheries economic growth and marine environmental pollution in China's coastal regions. Sci Total Environ 713:136641. https://doi.org/10.1016/j.scitotenv.2020.136641
69. Ayilu RK, Fabinyi M, Barclay K (2022) Small-scale fisheries in the blue economy: review of scholarly papers and multilateral documents. Ocean & Coastal Management 216:105982. https://doi.org/10.1016/j.ocecoaman.2021.105982
70. Suh D, Pomeroy R (2020) Projected economic impact of climate change on marine capture fisheries in The Philippines. Front Mar Sci 7:232. https://doi.org/10.3389/fmars.2020.00232
71. Nunan F (2020) The political economy of fisheries co-management: challenging the potential for success on Lake Victoria. Glob Environ Chang 63:102101. https://doi.org/10.1016/j.gloenvcha.2020.102101
72. Fisu AA, Didiharyono D, Bakhtiar (2020) Economic & financial feasibility analysis of tarakan fishery industrial estate masterplan. IOP Conf Ser: Earth Environ Sci 469(1):012002. https://doi.org/10.1088/1755-1315/469/1/012002
73. Crouch GI, Ritchie JB (1999) Tourism, competitiveness, and societal prosperity. J Bus Res 44(3):137–152
74. Dwyer L, Kim C (2003) Destination competitiveness: determinants and indicators. Curr Issue Tour 6(5):369–414. https://doi.org/10.1080/13683500308667962
75. Anwar A, Siddique M, Dogan E, Sharif A (2021) The moderating role of renewable and non-renewable energy in environment-income nexus for ASEAN countries: evidence from method of moments quantile regression. Renew Energy 164:956–967. https://doi.org/10.1016/j.renene.2020.09.128
76. Ehigiamusoe KU, Lean HH, Smyth R (2020) The moderating role of energy consumption in the carbon emissions-income nexus in middle-income countries. Appl Energy 261:114215. https://doi.org/10.1016/j.apenergy.2019.114215
77. Dogan E, Seker F (2016) Determinants of CO2 emissions in the European Union: the role of renewable and non-renewable energy. Renew Energy 94:429–439. https://doi.org/10.1016/j.renene.2016.03.078
78. Apergis N, Payne JE (2010) The emissions, energy consumption, and growth nexus: evidence from the commonwealth of independent states. Energy Policy 38(1):650–655. https://doi.org/10.1016/j.enpol.2009.08.029
79. World Bank (n.d.) Fish to 2030: prospects for fisheries and aquaculture
80. European Commission (n.d.) The European green deal—European commission. Online. https://commission.europa.eu/strategy-and-policy/priorities-2019-2024/european-green-deal_en. Accessed 08 June 2025

81. European Commission. Joint Research Centre (2022) *Clean energy technology observatory, Ocean energy in the European Union: status report on technology development, trends, value chains and markets: 2022*. Publications Office, LU. Online. https://data.europa.eu/doi/10.2760/162254. Accessed 30 Dec 2025
82. Alsaleh M, Yuan Y, Longqi S (2024) Do global competitiveness factors impact the marine sustainability practices? An empirical evidence from fisheries sector. Bus Strat Env 33(7):6671–6688. https://doi.org/10.1002/bse.3839
83. Wang X, Alsaleh M, Nan Z (2024) Measuring the impact of innovation ecosystem and enabling environment on fisheries system performance. J Knowl Econ 16(4):13924–13958. https://doi.org/10.1007/s13132-024-02330-z
84. Pedroni P (2004) Panel cointegration: asymptotic and finite sample properties of pooled time series tests with an application to the PPP hypothesis. Economic Theory 20(03). https://doi.org/10.1017/S0266466604203073
85. Kao C, Chiang M-H (2000) On the estimation and inference of a cointegrated regression in panel data. In: Advances in econometrics, vol 15. Emerald (MCB UP), Bingley, pp 179–222. https://doi.org/10.1016/S0731-9053(00)15007-8
86. Sarkodie SA, Strezov V (2019) Effect of foreign direct investments, economic development and energy consumption on greenhouse gas emissions in developing countries. Sci Total Environ 646:862–871. https://doi.org/10.1016/j.scitotenv.2018.07.365
87. Koenker R, Bassett G (1978) Regression quantiles. Econometrica 46(1):33. https://doi.org/10.2307/1913643
88. Binder M, Coad A (2011) From Average Joe's happiness to Miserable Jane and Cheerful John: using quantile regressions to analyze the full subjective well-being distribution. J Econ Behav Organ 79(3):275–290. https://doi.org/10.1016/j.jebo.2011.02.005
89. Canay IA (2011) A simple approach to quantile regression for panel data. Econ J 14(3):368–386
90. Phillips PCB, Sul D (2003) Dynamic panel estimation and homogeneity testing under cross section dependence. Econ J 6(1):217–259. https://doi.org/10.1111/1368-423X.00108
91. Pesaran MH (2021) General diagnostic tests for cross-sectional dependence in panels. Empir Econ 60(1):13–50. https://doi.org/10.1007/s00181-020-01875-7
92. Levin A, Lin C-F, James Chu C-S (2002) Unit root tests in panel data: asymptotic and finite-sample properties. J Econ 108(1):1–24. https://doi.org/10.1016/S0304-4076(01)00098-7
93. Im KS, Pesaran MH, Shin Y (2003) Testing for unit roots in heterogeneous panels. J Econ 115(1):53–74
94. Westerlund J (2007) Testing for error correction in panel data*. Oxf Bull Econ Stat 69(6):709–748. https://doi.org/10.1111/j.1468-0084.2007.00477.x
95. Li C-J, Kim H-H, Yang J-W (2021) A Comparative study of fishery industry competitiveness in China's coastal provinces. International journal of advanced smart convergence 10(2):158–167. https://doi.org/10.7236/IJASC.2021.10.2.158
96. Kang B et al (2021) Climate change impacts on China's marine ecosystems. Rev Fish Biol Fisheries 31(3):599–629. https://doi.org/10.1007/s11160-021-09668-6

Chapter 8
European Fisheries Information Communication and Technology (ICT)

Acronyms

EU	European Union
ICTs	Information and Communications Technologies
IUU	Illegal, Unreported, and Unregulated
CO_2	Carbon Dioxide Emissions
MMQR	Method of Moments Quantile Regression
EUR	Euro
CD	Cross-sectional Dependency
FP	Fisheries Production
WDI	World Bank Development Indicators
GDP	Gross Domestic Product
PD	Population Density
DOLS	Dynamic Ordinary Least Squares
FMOLS	Fully Modified Ordinary Least Squares
FEOLS	Fixed-Effects Ordinary Least Squares
Europa	European Data Portal
Eurostat	European Commission's Database
LLC	Lin & Chu Test
IPPS	Im, Pesaran, and Shin Test
EU27	27 European Union Member States
EU13	13 European Union Member States
BRICS	Brazil, Russia, India, China, South Africa
ASEAN	Association of Southeast Asian Nations
EU14	14 European Union Member States
OLS	Ordinary Least Square
LEG	Natural Logarithm of Economic Growth
LCO_2	Natural Logarithm of Carbon Dioxide Emission
LHC	Natural Logarithm of Human Capital Development
LICT	Natural Logarithm of Information and Communication Technology
LGVR	Natural Logarithm of Governance
CFP	Common Fisheries Policy
ICT	Information and Communications Technology

M. Alsaleh, *Fisheries Sustainability in the EU*, Urban Sustainability,
https://doi.org/10.1007/978-981-92-4496-6_8

Here is a reason for concern, given the history of technology in the seas. Too often, the depletion and contamination of marine species have been exacerbated by technical advancements and more potent fishing equipment. By examining the roles of fisheries production, information and communication technology, human capital, governance, carbon dioxide emissions, and economic growth from 1990–2022, this paper seeks to investigate the dynamic effect of ICT on the sustainability of the fisheries industry in 27 European countries. The findings demonstrated a substantial positive relationship between information and communication technology and the fisheries sector at higher quantiles using the novel method of moments quantile regression (MMQR) with fixed effects. Additionally, in the EU27 countries, the impact of economic growth was positively significant across most quantiles. The findings reveal that the EU14 developed countries exhibit greater ICT and economic development than the EU13 developing countries, which significantly benefits fisheries sustainability. At lower quantiles, the data revealed a significant positive association between human capital and the fisheries sector. Results demonstrate that developing countries in the EU13 possess greater human capital than developed countries in the EU14, which considerably benefits fisheries sustainability. Conversely, across all quantiles in the EU27 region, the findings revealed a substantial positive relationship between carbon dioxide emissions and the fisheries sector. The results indicate that the EU14 developed countries exhibit a larger significant positive magnitude between carbon dioxide emissions and fisheries production than the EU13 developing countries. To accomplish sustainable development objectives, these results provide policymakers with insights into promoting technological diffusion factors in the EU14 and EU13 countries through effective and environmentally sustainable technology in the fisheries sector.

8.1 Fisheries and ICT in the EU

Digitalization is changing the economy, cultures, and the life of individuals in society. This transformation is significantly influenced by modern and traditional ICT (information and communication technology). ICT has influenced Information and communications activities, which have always played an indispensable role in the sustainable development of societies [1]. Information and communications technologies (ICTs) refer to a range of technologies which transmit information and communications services using a number of techniques, including computers and radio broadcasting. The term includes computers, but not those components that form part of a computer (such as a display screen or keyboard), and not communication facilities that transmit data between computers or any technology such as those that permit electronic mail. The adoption of new forms of ICT has become widespread throughout the sector, including: resource estimation, the catching and culturing of fish, and the processing and sale of fish catch [1]. Specialized forms of ICT may exist for specific fisheries applications, such as fish-finding sonars; other ICT forms include GPS for navigation and location systems; mobile phones for personal communication, and for commerce and, for a few, access to the Internet and even radio programs for the fisheries community; Web-based information and networking resources for Fisheries; [1] Various forms of these technologies can be introduced even into relatively isolated communities, and when taken up by users they can offer a positive benefit to their lives.

Fisheries are becoming less labor-intensive due to the influence of artificial intelligence and can thus expand and generate four times as much output in a much shorter timeframe. The application of artificial intelligence spans the position of any labor, for example, feeders, monitors of the quality of the water, fish gatherers, etc. AI is used to protect endangered fish from becoming extinct. The adoption of AI in the fisheries industry globally supports a more sustainable approach toward open-sea fishing while at the same time keeping an eye on the entire world [2]. AI in the fight against IUU (illegal, unregulated, and unreported) fisheries has gained enormous attention and contributed significantly to curbing this challenge. With this application of artificial intelligence, it is possible to save as much as 30% of expenses by minimizing input waste in aquaculture, thus rendering fish-producing systems fully in control and with significantly reduced maintenance needs and input costs [2]. Reducing costs will be facilitated by the implementation of technologies that will ensure enhanced operational performance, automation of processes, safety and security measures, as well as minimized adverse environmental effects [2]. Enhanced marine connectivity will transform how marine stakeholders share and manage information [2].

European governments are dominated by the thought of the future in the short term, as there is a new regulation being signed which prevents overfishing [3]. Some industry sectors were given carte blanche by EU member states and fish freely at the cost of the sustainable exploitation of marine resources in order to get short-term profits and benefits for a narrow range of persons. Europe itself is already paying the price by fishing excessively in order to produce more and thus weaken its sector, in the face of decreasing marine stocks, environmental damage to waters Europe needs to exploit, resulting in a loss of revenue, decreasing fishing employment, and fish for consumption [3].

Global sustainability and biodiversity objectives adopted at the United Nations Biodiversity Conference in Montreal, Canada, in December 2022 further boost a critically important topic—marine protection worldwide [4]. On the one hand, the sea serves as a vital food source, while the fishing industry represents an important source of revenue for many coastal communities in Europe and globally [4]. On the other hand, as a consequence of severe resource over-exploitation, a real threat exists of specific species facing extinction or an irreversible collapse of entire ecosystems [4]. Despite substantial reform of the EU fisheries policy since 2013, overfishing still occurs in EU member states, particularly in the Mediterranean Sea and Black Sea. These problems are often compounded by concerns about pollution and the overall quality of the marine environment. Excessive use of fertilizers and pesticides, particularly on farmland, leads to an increase in nutrient concentration in water, which contributes to an accelerated growth of algae, which may in turn suffocate marine life and reduce the water's oxygen levels [5].

For developing a healthy marine environment, we must first have a clear understanding of each of the factors contributing to it. The latter, in any case, is an issue complicated by complexity [6]. Scientific expertise in this context plays a vital role in enabling us to get a better picture and find the most appropriate ways of tackling the issue of restoring the ocean's health [6]. As a response to these needs and in order to allow for improvement of quantities and qualities of information that is made available to marine experts and policymakers, the EU has a broad research agenda in this field [6].

The main research questions are: (1) How do Information and Communication Technology factors affect Fisheries Production Sustainability and Marine Ecology in the European Seas during 1990–2022? (2) Are EU Member States more vulnerable to the impact of ICT factors on fisheries production sustainability? (3) Can the Fisheries Sector be more environmentally friendly to meet the EU goals for Sustainable Ocean and Marine Development by 2030? (1) Examine the sustainability of fisheries production and its relationship with Information and Communication Technology in the European seas from 1990 to 2022. (2) Conduct a comprehensive analysis of the impacts of ICT factors on Fisheries production sustainability and Marine Ecology between EU13 developed countries and EU14 developing countries in 1990–2022. (3) Investigate how ICT factors influence Fisheries Production Sustainability in developed and developing countries from 1990 to 2022.

It has been argued that modern-day information and communication technologies have enabled them to be a significant enabler of sustainable development and dynamics [7]. A significant part of the reduction or removal of trade barriers has been achieved due to certain important policy changes such as the reform of the Common Fisheries Policy (CFP) in the EU and changes made to Common Agricultural and Fisheries policies and the recent reforms within the European Union which leads to significant shift in the macro-marketing environment with substantial increased rivalry between countries of origin of fish [7]. Additionally, the increasing complexities in the issues of sustainable food security, sustainability of the marine environment, sustainability of ocean resources, and challenges to sustainable development have placed a heavy emphasis on ICTs on the global stage [7]. Hence, the question of Information and Communication Technology has become an Increasingly Complex Issue [7]. The sustainability of the EU Fisheries sector is no exception, as it is operating in the Highly Competitive Global Market [7]. EU Fisheries Sector, which holds significant shares in the EU market for food products, has experienced a sharp increase in its share of total EU fish imports over the last 10 years [7].

Since the relevant literature did not provide any good explanations for the changes to ICT in the fisheries industry, we propose that ICT affecting factors and the fishery industry sustainability literature could aid in formulating effective marketing strategies for sustainable fisheries. The assumption that disturbance terms are independent of one another and there is no cross-sectional dependence (CD) amongst the specified variables across the countries could be regarded as a major shortcoming of earlier EU states' research [8–12]. argued that such factors could either reflect common or idiosyncratic shocks, spill-over effects, or both. Inclusion of CD is essential as a cross-sectional group of the economies of the EU will likely experience the same levels of dependence on the increased integration of global economies.

The economies in the EU are vulnerable to common shocks affecting both economic and technology development, such as widespread diffusion of the digital economy and economic development [11]. explained that if CD is neglected by research work, research regarding the relationship between economic growth and technology transfer will lead to biased results and inconsistent estimations.

The present research aims at determining whether the information and communication technology profiles of the fisheries-producing countries are equally influenced by the growth-diminishing influence of natural resources. Past research [13, 14] employed aggregate exogenous variables, ignoring whether the resource was a point or diffuse source in origin. On the contrary, this study uses the point source version of an extremely specific natural resource, such as sustainability in the fisheries industry. Our choice to focus on fisheries in the EU economies is because in the fisheries sector point sources usually create a much more adverse impact than diffuse sources.

[15, 16] Since sound environmental policies have proved to be capable of neutralizing or even reversing reliance on the resources that constrain growth effects [17], the sample comprises developed and developing economies.

We have included the regulatory control variable of governance to give us the best inferences. The EU27 countries for the fisheries sector comprise countries of developed economies like Germany, Italy, and Sweden and countries with emerging economies such as Estonia, Bulgaria, and Malta. This study covers the period from 1990 to 2022.

Because the economies selected have different levels of economic development status, the novel method of moments quantile regression (MMQR) of [18] is used to explore the robustness of the growth-induced resource sustainability nexus for the fisheries industry together with other controls across quantiles of the conditional distribution of fisheries sustainability. MMQR approach incorporates fixed effects into the estimation of the conditional distribution, and this method can capture how these parameters vary across different parts of the distributional spectrum. Henceforth, we rigorously analyze the growth hypothesis in Fisheries-producing economies.

This study is the first, to the researchers' knowledge, which examines growth-induced resource sustainability in EU27 countries with inclusion of human capital, governance, information and communication technology, carbon dioxide (CO_2) emissions, and fisheries production.

There are at least two reasons to evaluate the growth hypothesis at different quantiles of the conditional distribution of fisheries sustainability: 1. As compared to conditional mean estimation, conditional quantile estimation is more resilient to extreme observations that might be on the outcome variable (dependent variable) [19]. 2.

Conditional mean estimations fail to appropriately portray the distributional role of ICT on fisheries sustainability, while in the case of quantile regressions, a more straightforward interpretation arises, particularly in panel regressions where it permits the deconstruction of the distributional effect of independent variables on the dependent variable in different quantile groups and thus makes the comparison of their different role on different segments of the cross-sectional dimension.

As a consequence, the findings obtained by the estimation of conditional quantiles may provide more insight than what can be gained from the estimated conditional means. In the rest of the paper, Sect. 8.2 provides a short literature review of growth hypotheses. Data and methodology are provided in Sect. 8.3.

Section 8.4 gives the estimation results, while Sect. 8.5 provides concluding remarks and policy recommendations based on our findings.

8.2 Fisheries and ICT Historical Development

Numerous studies, [1–5], stated that the combination of technology and appropriate policy innovations should be applied to both improve the social, economic, and environmental sustainability of the aquatic food systems in the long term [6, 20, 21]. have examined prospects of the European marine fisheries in the future by modeling management, technology development, and policy together. The Fisheries-Energy-Environment Nexus resource management concept will take advantage of systems thinking better, enabling optimization of trade-offs and synergies among various domains so as to reach fisheries, conservation, and other environmental goals [22, 23]. explored the impact of technological innovation on marine fisheries growth in 10 Chinese coastal regions and confirmed that technological innovation is essential to developing marine fisheries.

Some studies including [24–29] have focused on the importance of appropriate policy in fisheries management, various means of strengthening sustainable fisheries enforcement, and the significance of proper enforcement for tackling overfishing; [7, 30] identified that co-management between local community and government has been the most suitable mechanism for ensuring enabling conditions in Indonesia and India marine fisheries respectively; the fisheries management plan analysis based on environmental governance in European region by [31–33] found that the fish stocks of abundance has greatly improved from 1229 in 2008 to 1142 in 2013 and 1781 in 2018 with a decrease in 2013 mainly due to the wildfires near by, while [34] pointed that local knowledge systems is theoretically respected by all, however, politicians and business entrepreneurs do not put it into practice and they argue for sustainable seafood certification as a remedy to external shocks. Alternative solutions toward sustainable seafood certification in Japan have also been investigated by [34]. In the USA, [35, 36] studied the core structure of fisheries regulation aimed at conserving marine resources and argued that the enabling environment, which is beyond the control of fishermen and therefore severely impacts overall fishermen welfare when combined with fluctuating weather conditions, access to sufficient and capable crew members, and quick changes in market price.

Meanwhile, several studies [37–39] stated that further investment in human capital can partially offset, if not fully, the consequences of fisheries sector decommissioning [40, 41]. examined the relationship between human development and fisheries footprint in the Mediterranean region, and concluded that a high level of human capital favors fisheries sustainability; meanwhile, a low level of human capital in the fisheries industry shows a negative influence upon fisheries sustainability. [42, 43] analyzed issues regarding human capital investment in the blue economy in Indonesia and presented a coherent and comprehensive investment plan together with coordinated policies and harmonized regulations to ensure blue economy sustainability.

Achieving sustainability in the fisheries sector would mean more opportunities for sustained economic, social and environmental development; For example, [44–46] study shows that the relationship between sustainable fisheries production and economic development in South Asia has indicated a direct link between fisheries production and sustainable development, and it also examines impact of full/partial substituting fish products on growth performance, feed efficiency, histology condition and economic income status [47] [48]. explored the initial stocking body weight of Nile Tilapia (Oreochromis niloticus) on water quality, growth efficiency, and economic profit, and it suggests that medium stocking fish weight has the best result on both growth performance and financial returns in three treatments; similar findings have been observed on the impact of Arctic fish skin craft on sustainable development in coastal regions of southwest Alaska by [49]. The study demonstrates that Indigenous fish skin crafting is an influential method that has a positive impact on poverty reduction, gender equality, and health/well-being, clean water & sanitation, decent work & economic growth, reduced inequality, climate change, marine resources, responsible consumption & production.

A sustainable fishery would mean that the catch rate would not endanger the existing fish populations and habitats as well as marine life. Only true sustainable fisheries can maintain ecological resources as well as jobs, food, and revenue for human beings [50]. analyzes fisheries and environmental pollution and find out that the fisheries processing waste has a great potential to substitute fuel used for Indonesian bioenergy generation; fishes, which is the major vertebrates, is still least known about its contribution toward global carbon export flux at region and global level [51] suggests that it's the variability estimates for fish contributions to total carbon flux from previous field studies, that has leads to huge variations in carbon flux contributions to total export carbon flux and fish contributions for total carbon export flux; similar research has been done by [52] examining the impact of fish stocks on the carbon fluxes on benthic and pelagic environment and points out benthic is much influential on aquatic CO_2 and CH net emission.

By using double logit model [53, 54] revealed that the impact of environmental regulation on China marine fisheries growth showed a "U-shape" relationship with China marine fisheries environment; however the links between fisheries sector and its environment has yet to be fully analyzed or synthesized though the studies on environment impact from fisheries sector has been rapidly expanding and research have focused on putative mechanisms underlying the changing environment and marine ecology in reviews [55–57]; compared with previous studies mostly concerning only about fisheries sector, this research investigates role of various ICT variables on sustainable development of fisheries industry for purpose of blue growth and we study the impact of ICT on fisheries production determining factors with sustainable development for EU seas, compared with previous research using only time-series analysis; this paper uses both static and panel regression to study the effect of economic development and technology diffusion on fisheries production sustainability. Our potential novelties are: First and foremost, the investigation of relations between fisheries production, ICT, governance, human capital, CO, and economic development in the EU seas for 1990–2022. And importantly, to achieve sustainable blue growth, the impact of technology diffusion factors and economic development on the EU27 marine fisheries sector is also studied. The EU27 region is divided into two regions: EU14 developed countries and EU13 developing countries. As we analyze the time period from 1990 to 2022 and have non-stationary data, we utilize unit root and cointegration tests for stationarity analysis. Then, we investigate the interaction effects of ICT, governance, human capital, CO, and economic development in the fisheries industry by using the MMQR method (method of moments quantile regression); further, we carry out static and panel regression tests on factors that affect economic development and technology diffusion. Ultimately, we forecast the future impact of technology diffusion factors on the fisheries industry and marine resources.

8.3 Porter's Theory and Fisheries Application

Several studies on cluster development, technology diffusion, and Porter's theory emphasize the international competitiveness of EU member states and regions in the global economy, but in fisheries development in developing countries, we would consider two considerations: (1) the contribution of MNCs economic activities to the cluster and technology diffusion advantage and (2) the relevance of interconnectedness/networks between clusters. A new approach, "information and communication technology of fisheries development," is suggested based on the relevance of these issues for sustainable fisheries development. Below figure is the proposed model of fisheries development paradigms for technology transfer, modified based on Porter's diamond model and several indicators given in studies by [59, 60]. Three additions are made, which are the importance of digitalization, technology, telecommunication, and innovation, as these are assets in any industry.

First, Governance plays a major role in fisheries technology transfer throughout all EU 27 member states as one of the core components for the long-term sustainability of the industry. And, the interaction of those elements implies that the nature and extent of the networks between each of the technology diffusion elements (information and communication technology) inside the cluster must be identified and portrayed. Based on new researches by [61, 62, 63, 64] we have formulated:

$$FP_{it} = f\left(ICTit, GVRit, HCit, ECit, CO_2it\right)$$

Where Yt is fisheries production (FP) expressed as annual fish harvest, total output, in tons. Data for these variables were obtained from the World Development Indicators database [58]. X: GVR: Government effectiveness, as calculated by the EU Official Statistics portal [59]. Used data in terms of economies of EU Countries during the period 1990–2022, using per capita variables. ICT: Information and communication technology activities obtained from the European Commission data (Eurostat, 2022). CO_2: Carbon dioxide emissions (t CO_2). HC: Human Capital, which is defined as the average duration of education and training in the society. EC: economic growth measured by gross domestic product in current US dollars obtained from European Commission data [60] (Table 8.1).

Now, let us look at the results of our estimation. In Table 8.2, the statistical features of each of the components selected are presented, including the maximum value, minimum value, average, and standard deviation. The average value of each component selected for study is FP = 4.891, ICT = 3.630, GVR = 1.866, HC = 1.888, EG = 4.364, and CO_2 = 0.918. The corresponding values for standard deviation are 0.819, 0.326, 0.070, 0.019, 0.402, and 0.242. Based on the above factors with descriptive properties, we may proceed to the unit root test.

Table 8.1 Data, units, transformation, and source

Data	Measurement unit	Transformation	Source
Fisheries production	Metric tons	Natural logarithm transformation	WDI
Governance	% of confidence in governance	Natural logarithm transformation	WDI
ICT	Technological and information application	Natural logarithm transformation	Eurostat
Carbon dioxide emissions	Million metric tons	Natural logarithm transformation	WDI
Human capital	The average number of years of education	Natural logarithm transformation	Eurostat
Economic growth	Constant US dollars	Natural logarithm transformation	Eurostat

Table 8.2 Results of descriptive statistics

Variable	Observations	Mean	Std. Dev.	Min	Max
FP	864	4.891	0.819	3.211	6.310
ICT	864	3.630	0.326	2.223	4.354
GVR	864	1.866	0.070	1.482	1.979
HC	864	1.888	0.019	1.837	1.921
EG	864	4.364	0.402	3.310	1.921
CO_2	864	0.918	0.242	0.429	1.817

Source: [61]

In the analysis of the results, we utilized FEOLS, DOLS, and FMOLS, all for comparison. FEOLS with Driscoll and Kraay errors, a set of robust standard errors that are resistant to common forms of serial correlation and cross-sectional dependence, are appropriate under these conditions. Regarding estimation methods in dynamic cointegrated panels, [62] state that the main sources of problems arise from heterogeneity problems with regard to different individual cross-section means and differential cross-section response toward the cointegrating equilibrium. This model specification is addressed by Pedroni's FMOLS estimator, which includes common specific intercepts and different individual heterogeneous correlations of the error process. An MC simulation reported in [62] suggested that the DOLS estimator is unbiased in finite samples with respect to both OLS and FMOLS estimators, and it is these characteristics that motivated [63] to extend the DOLS estimator in a panel data framework. Also, the DOLS estimator corrects for endogeneity by adding lead and lag differences, thus eliminating feedback endogenous problems.

The analysis used panel quantile regression to analyze how the distributional and heterogeneous impacts vary across different quantiles because previous estimators fail to resolve issues like these [64, 65]. introduced the panel quantile regression technique in their path-breaking work. Instead of estimates of the conditional mean of the response variable Y, conditioned on given values of the independent variables, quantile regression is generally used to estimate either the conditional median or any other specific quantiles of the response variable. Furthermore, quantile regression estimation is resistant to the influence of outliers and is particularly useful in situations where there is no or only little correlation between conditional means of variables [66].

However, for the current study, we used [18]. MMQR method of moments quantile regression estimation under fixed-effects specification. Quantile regression is indeed resistant to outliers, but it did not properly deal with unobserved heterogeneity between individual members of the panel. In the case of individual-level effects, the MMQR method can be considered as providing information about the entire distribution instead of only the conditional mean and is documented as being able to detect condition-heterogeneous effects between fisheries sustainability components in studies such as [19, 67], among others. The MMQR method of moments quantile regression is applicable when the panel data model has a fixed effect and an endogenous independent variable. It has an added advantage of simplicity since non-crossing estimates for regression quantiles are derived. In a location-scale model framework, estimation of the conditional quantiles QY(|X) follows this equation:

$$Y_{it} = \alpha_{i+} X_{it}{}' \beta' + \left(\delta_i + Z_{it} \gamma' \right) U_{it} \tag{8.1}$$

Where P (δi + Zitγ >0) = 1. (α, β, δ, γ) are parameters requiring estimation. Z is a k-vector of identifiable components of X that are differentiable transformations, with element l given by: (αi, δi), i = 1,…, n, represents the individual i fixed effects.

$$Z_l = Z_l\left(X\right), l = 1,\ldots,k \tag{8.2}$$

For each given i, Xit is independently and identically distributed and is also independent over time (t). According to Machado and Silva's (2019) moment criteria, which among other factors do not necessitate strict exogeneity, Uit is orthogonal to Xit, independently and identically distributed across individuals (i) and over time (t), and is normalized to satisfy these constraints. Eq. (1) implies the following:

$$Q_Y\left(\tau|X_{it}\right) = \left(\alpha_{i+}\,\delta_i q\left(\tau\right)\right) + X_{it}^{'}\beta + Z_{it}^{'}\gamma\, q\left(\tau\right) \tag{8.3}$$

The independent variables in the current study include the natural logarithm of economic growth (LEG), the natural logarithm of information and communication technology (LICT), the natural logarithm of human capital (LHC), the natural logarithm of carbon dioxide emissions (LCO_2), and the natural logarithm of governance (LGVR). The scalar coefficient Xit—αi(τ) ≡ αi + δiq(τ) represents the quantile-specific fixed effect for individual i, and QY(τ|Xit) denotes the conditional quantile distribution of the dependent variable Yit (natural logarithm of fisheries production), which is contingent upon the location of the independent variable Xit. Unlike conventional least-squares fixed effects, the individual-level effects do not represent a simple intercept shift. Rather, these time-invariant parameters exhibit heterogeneous effects that can vary across the quantiles of the conditional distribution of the dependent variable Y. The τ-th sample quantile, denoted by q(τ), is obtained by solving the following optimization problem:

$$\min_q \sum_i \sum_t P_\tau \left(R_{it} - \left(\delta_i + Z_{it}^{'}\gamma \right) q \right) \tag{8.4}$$

Where the check function P_τ (A) = (τ—*1*) AI {A ≤ 0} + TAI {A > 0} is shown.

8.4 ICT Application in Fisheries Industry

Prior to estimating the unknown parameters, several common preliminary tests are utilized to determine the time-series characteristics of the variables. To begin, we test for the presence of cross-sectional dependence (CD) within the panel. Cross-sectional dependence can lead to biased estimates of true parameters, as this form of dependence, which occurs because of common, unobserved variables, can erode large portions of the efficiency gains realized when utilizing panel data [68]. Thus, it is crucial to address this problem in order to generate accurate coefficient estimates. We use the [69] CD test to test for the presence of CD within our panel. With the exception of fishery production, our other quantitative variables in Table 8.3 reveal strong CD between countries. To avoid potential size distortions as much as

Table 8.3 Cross-sectional dependence results

Variables	FP	GVR	ICT	CO_2	HC	EG
Pesaran (2004)	(18.82)***	(31.30)***	(43.28)***	(19.96)***	(99.27)***	(83.55)***

Remark: *** refer importance at the 1%, scale
Source: [61]

possible, our unit root tests, cointegration tests and panel estimation procedures will need to account for cross-sectional dependence.

We apply the [70] (LLC) and [71] (IPS) unit root tests to test the integrating properties of the variables concerned objectively. Panel unit root tests merge the information contained in the cross-section and time dimension, with the purpose of achieving better test efficiency. In order to compare the null hypothesis of a unit root against alternative hypotheses, several panel unit root tests have been proposed. Unlike the null of a unit root, [70] tests the null against the homogeneous alternative where all variables in the panel are stationary at the same rate of reversion. The alternative of at least one stationary variable in the panel is tested against the null of a unit root by [71]. In this case, the new test proposed in this paper addresses panel stationarity, thus leading us to focus on the LLC architecture. The panel unit root testing procedure (IPS) does relax the stronger assumption imposed by [70], allowing each individual to have a unit root as a specific AR process. It has been found that the LLC panel unit root testing strategy provided better power efficiency compared with other equivalent unit root tests for a comparatively small panel dataset size as used in this paper. These two tests are used to measure the effects of cross-sectional dependence on the panel unit root tests. As illustrated in Table 8.4, all of the unit root test specification variables have become stationary at the first difference and at the levels where they are non-stationary. That implies that all variables used in the estimation are of order I (1).

Whether there actually exists a long-run cointegrated relation is assessed using [62], the panel cointegration test, and [72], the bootstrapped panel cointegration test in Table 8.5 [62]. gives a comprehensive method for the panel cointegration test procedure built on Engle and Granger's two-step method. Pedroni's methodology removes person-specific deterministic patterns and short-run effects in the initial data processing to tackle the heterogeneity. From residual estimation results, Pedroni forms seven different statistical tests. These statistics are of either single-process "pooled" or "within-dimension" type or different process "grouped" or "between-dimension" types in Table 8.5. Under the null of no cointegration, four tests using the [72] methodology are carried out. The common factor restrictions imposed on residual-dynamics-based cointegration tests are abandoned, as evidence suggests that dynamics has a structural shape rather than being residual, and thus, the performances of some tests are weakened by the lack of common factor restrictions, as common factor restrictions are in reality based on the behavior of the residuals [73].

Since this restriction can be lifted, the long-run and short-run adjustment procedures do not have to be set to the same value [72] to eliminate this concern; the

Table 8.4 Panel unit root test results

	Difference		First difference	
Variable	LLC	IPS	LLC	IPS
FP	3.067	−3.700	−24.13***	−22.737***
GVR	−1.090	−0.254	−23.47***	−23.47***
ICT	−1.209	−3.942	−7.818***	−11.92***
CO_2	−5.000	−10.609	−29.637***	−37.388***
HC	−2.189	−5.018	−11.91***	−14.91***
EG	−1.571	−0.210	−18.74***	−68.36***

Remark: *** refer importance at the 1%, scale
Levin, Lin & Chu test (LLC), and Im, Pesaran, and Shin W-stat test (IPS)
Source: [61]

Table 8.5 Panel cointegration test for EU countries

Test		Without trend	With trend	
A) Pedroni Residual Cointegration Test:				
Alternative hypothesis: common AR coefficients (within dimension):				
Panel v-Statistic		−2.375 (0.991)	−1.823 (0.965)	
Panel rho-Statistic		4.241 (0.987)	1.697 (0.955)	
Panel PP-Statistic		−1.073* (0.085)	−6.305*** (0.000)	
Panel ADF-Statistic		−2.077** (0.018)	−7.329*** (0.000)	
Alternative hypothesis: common AR coefficients (between dimensions)				
Group rho-Statistic		−3.150	(0.966)	
Group PP-Statistic		−6.682***	(0.000)	
Group ADF-Statistic		−6.723***	(0.002)	
B) Westerlund (2007) Bootstrap Panel Cointegration:				
Statistics	Value	Z-Value	p-value	Robust p-value
Gt	−2.409	−1.095	0.095	0.081*
Ga	−7.225	3.100	0.999	0.985
Pt	−11.677	−1.408	0.080	0.074*
Pa	−5.673	1.569	0.942	0.999

Remark: ***, **, and * refer importance at the 1%, 5%, and 10% scales respectively
Values in parentheses are *p*-values
Source: [61]

bootstrap methodology we have developed allows us to control for the contaminating influence of cross-section dependence, allowing the generation of critical values that are robust to this problem. The empirical evidence shown in Table 8.5 reveals that we obtain cointegration at a high significance level from both the [62] test and the [72] bootstrapped cointegration test.

Table 8.6 reports the estimation results from the FMOLS, DOLS, and FEOLS estimation techniques. Although these estimation methods result in different significance levels for most of the coefficients, the estimated coefficient values for model 1 in Table 8.6 do not vary dramatically on average. Fisheries production in the EU region does appear to be robust with respect to all three estimators with respect to estimation coefficients and significance levels.

Information and communication technology in the EU region promotes the enhancement of sustainability by 0.82%, 0.05%, and 0.06% for every 1% change in ICT for FEOLS, FMOLS, and DOLS estimation techniques, respectively. As expected, this is positive and statistically significant for all estimations, and is in agreement with the works of [1–3]. A strong foundation of sustainable growth in marine information and communication technologies, alongside a marine environment, and the ability to provide for the development of the EU's blue economy are essential elements in achieving the EU's goal of a climate-neutral continent by 2050.

Furthermore, human capital in the EU region is high and has a significant and positive impact for all estimations. Every 1% increase in human capital leads to an increase in sustainability by 6.251%, 3.57%, and 2.86% for FEOLS, FMOLS, and DOLS estimation techniques, respectively. Our finding contradicts the evidence from [38, 39], but it is consistent with the modernization hypothesis. Strong and positive statistical evidence for the growth hypothesis supports the result from all the FMOLS, FEOLS, and DOLS specifications. The term human capital indicates the skill level and quality of staff operating the vessel as part of the fishing fleet (and can be viewed as a stock), while non-physical capital describes capital invested in non-tangible aspects such as ownership of a fishery right to operate in an individual quota fishery.

Economic growth contributes positively to the fisheries sector sustainability for the EU region. More specifically, an increase in economic growth leads to a sustainability enhancement of 0.74%, 0.26%, and 0.15% for FEOLS, FMOLS, and DOLS estimation techniques, respectively. Hence, we observe support for the growth hypothesis for the panel of fisheries-producing countries. Our finding is in agreement with the results from [44–46], which show that economic growth has a positive and inelastic impact on the fisheries sector development. A competitive and profitable market must be secured to maintain the competitiveness and long-term sustainability of the EU fisheries sector.

Table 8.6 Model 1. Panel estimation for the EU-27 region from 1990 to 2022

Long-run coefficient	DOLS		FMOLS		FEOLS	
	Coefficient	Std. error	Coefficient	Std. error	Coefficient	Std. error
ICT	0.064	(0.031)**	0.056	(0.019)***	0.825	(0.091)***
HP	2.860	(0.655)***	3.579	(0.858)***	6.251	(2.019)***
EG	0.153	(0.068)**	0.262	(0.101)**	0.740	(0.093)***
GVR	0.376	(0.217)*	0.467	(0.267)*	1.703	(0.391)*
CO_2	0.784	(0.150)***	0.520	(0.135)***	0.025	(0.121)

Note: ***, **, and * indicate significance at the 1%, 5%, and 10% levels respectively
Values in parentheses are P-values
Source: [61]

In more detail, the impact of governance on enhancing the fisheries sector sustainability has been calculated at 1.70%, 0.46%, and 0.37% for FEOLS, FMOLS, and DOLS estimation techniques, respectively. The findings of this estimation support the principal conclusion from [24–26]. The EU has become one of the central players of fisheries governance in the international scene, due to its position as the largest consumer of fish, its ability to import half of the total volume, and its role as a major fisheries player.

Carbon dioxide emissions positively impact the fisheries sector for the EU region. More specifically, an increase in carbon dioxide emissions leads to an enhancement in the sustainability of 0.52% and 0.78% for FMOLS and DOLS estimations, respectively, although the DOLS estimation results provide minimal evidence of its significance. Support for the growth hypothesis can be established for the panel of fisheries-producing countries. This result is in line with the results of [50–52] that reveal a positive and inelastic response of fisheries production increase to the carbon emissions increase. There is less emission to the atmosphere when there are more big fish like tuna, sharks, mackerel, and swordfish in the ocean, where dead fish in the ocean release the whole carbon content to the ocean floor.

According to the estimation results from panel quantile regression technique provided in Table 8.7, the impacts of information and communication technology outcomes on fisheries production have been significant and positive across all quantiles, and the magnitude of ICT's impacts tends to increase with higher quantiles, indicating that ICT contributes to achieving sustainability in the fisheries industry according to [1–3]. We also find substantial and positive effects of human capital outcomes on fisheries production for all quantiles (low to high), where magnitude impacts increase with lower quantiles, a finding consistent with [37–39], who argued that human capital enhances sustainability in the fisheries industry.

Also, the impact of economic growth, estimated using GDP, on fisheries production has a significant and positive impact for all quantiles, which implies that fisheries policies implemented by the EU have been effective in increasing the economic performance of high-production countries, where the magnitude of economic growth impacts tends to rise with higher quantiles according to [44–46]. It can be inferred that economic growth has a significant role in achieving sustainability in the fisheries sector in higher-performing countries.

In addition, the panel quantile estimation results illustrate that the impact of governance on fisheries sector sustainability has been statistically significant in all quantiles, apart from the lower quantile where the effect is relatively small. This means that EU marine fisheries governance is efficient in higher marine production countries, with the magnitude of governance's impact rising as we increase higher quantiles. Fisheries policies were found to be prioritized over growth in all countries, with support of the growth hypothesis only for the lower, median, and upper quantiles. Also, countries in the lower quantiles, due to the high cost of investment, could struggle to shift from a nonrenewable to a renewable resource in the fisheries industry, especially at the developing stage. In countries with the lower and highest levels of production, the impacts of governance on the fisheries industry are positive, which may result from the different techniques for extraction of marine

Table 8.7 Panel quantile estimation results for EU27 countries

Variables	Method of moments quantile regression with fixed effects										
	Location	Scale	Quantiles								
			0.1	0.2	0.3	0.4	0.5	0.6	0.7	0.8	0.9
ICT	(0.091)***	(0.069)***	(0.043)***	(0.713)***	(0.387)*	(0.319)***	(0.308)***	(0.327)***	(0.203)***	(0.233)***	(0.274)***
HP	(2.019)***	(0.485)	(9.017)***	(8.625)***	(4.480)*	(2.007)	(1.827)	(2.971)**	(2.742)**	(2.940)**	(7.054)
EG	(0.093)***	(0.668)***	(0.692)***	(0.745)***	(0.843)***	(0.714)***	(0.566)***	(0.427)***	(0.285)***	(0.231)***	(0.462)*
GVR	(0.391)*	(0.310)***	(0.858)	(1.277)***	(0.877)**	(1.296)***	(1.554)***	(1.994)***	(2.287)***	(2.163)***	(1.801)***
CO_2	(0.121)	(0.484)***	(1.293)***	(1.104)***	(1.031)***	(0.738)***	(0.677)***	(0.632)***	(0.564)***	(0.483)***	(0.566)***

Notes: *, **, and *** indicate significance at 10%, 5%, and 1% levels, respectively
Source: [61]

resources used in EU14 and emerging EU13. Our results support the growth hypothesis of [24–26] in many countries with strong evidence of the large contribution to overall production, which has significant governance implications. We also agree with [74], which found supporting evidence for the growth hypothesis, one of many popular theories regarding the exploitation of fisheries resources, and with [75], which provides empirical evidence that the growth hypothesis has been empirically validated for top-producing fisheries resource countries.

In all quantiles, the impacts of carbon dioxide emissions on fisheries production have been significant and positive, but the magnitude tends to decline with higher quantiles, which implies that the EU is embracing a strategy toward environmentally sustainable fisheries through the promotion of green markets [50–52].

The outcomes of FMOLS, DOLS, and FEOLS estimation approaches are described in Table 8.8. Although there exist certain differences between the significance levels of estimation approaches, on average the outcomes of estimates of coefficients between these 3 specifications are quite similar in Model 2 (Table 8.8). The information and communication technology among the EU14 developed countries exerts a statistically significant effect on fisheries industry sustainability, which increases the fisheries industry sustainability by 0.29%, 0.05%, and 0.12% as one percentage point increase in ICT among the FEOLS, FMOLS, and DOLS, respectively.

This result supports the findings [4–6]. The development of a sustainable blue economy for the future and marine technological transfer and innovation are at the heart of this strategy and are indispensable to meeting Europe's goal of being the first climate-neutral continent in 2050.

In the developed countries of the EU14, human capital leads to a positive impact across most of the estimations, where an increase in fisheries production for a percentage rise in human capital varies between 4.13% and 3.30% between the FMOLS and DOLS estimators, respectively. Nevertheless, the FEOLS coefficient seems inconsistent across specifications. This supports modernization theories ([40, 41]) and goes contrary to the study [48]. The growth hypothesis is supported by the estimates from most specifications and is robust across these specifications. Fish population size is determined by the ways human fishermen behave, and it is also determined over generations [40].

For example, a fish population that has been heavily hunted, particularly large fish, exerts evolutionary pressure for small-sized fish to prevail [40].

Table 8.8 Model 2. panel estimation for the EU-14 region from 1990 to 2022

Long-run coefficient	DOLS		FMOLS		FEOLS	
	Coefficient	Std. error	Coefficient	Std. error	Coefficient	Std. error
ICT	0.124	(0.051)**	0.055	(0.029)*	0.299	(0.152)*
HP	3.301	(1.508)**	4.137	(1.173)***	2.684	(8.391)
EG	0. 181	(0.104)*	0.194	(0.086)**	0.865	(0.341)**
GVR	0.320	(0.528)	0.531	(0.260)**	1.796	(1.226)
CO_2	0.379	(0.293)	0.465	(0.125)***	8.689	(0.088)***

Heavy pressure on fisheries resources could, in fact, put the question of the size of the fish population and the supply of resources at risk. The economic growth among the developed countries of the EU14 has a considerable influence on the fisheries sector, where one percentage of economic growth gives an impact of 0.86%, 0.19%, and 0.18% increase on fisheries production, based on the FEOLS, FMOLS, and DOLS estimations, respectively. This supports the growth hypothesis of the developed EU14 countries.

However, this result is opposite to [48], who shows an adverse and elastic impact on fisheries security from economic development and supports the main hypothesis of [47, 49]. Fisheries production contributes to other activities like other types of industries, such as shipping, fisheries, recreation, and aquaculture. For example, the cost of maintaining ships and equipment damaged by used fisheries gear represents an annual loss for EU countries.

Unfortunately, the coefficient regarding governance impact on the EU14 developed countries in the fisheries sector, in terms of magnitudes and statistical significance, gives an expected positive relation across only one specification, which increases by 0.53% for a percentage rise in governance in the fisheries industry, as seen among FMOLS, where governance shows inconsistent outcomes across FEOLS and DOLS specifications.

This confirms results derived from [28, 29, 36] and as for the EU communication it plays a crucial role in the EU's delivery of Sustainable Development Goal 14 on life below water and the 2030 Agenda for Sustainable Development of the UN, as the identifications of accessible readings for earlier commitments illustrates EU's commitment to ocean conservation, whereas the current agenda contributes to the blue component of the European Green deal. Based on the EU14 developed countries, information and communication technology has a positive impact of 8.68% on fisheries production in accordance with the FEOLS, whereas 0.46% based on FMOLS for a one percent rise in information and communication technology. Even though the results seem consistent with those in [53, 54], the outcome is still inconsistent regarding DOLS estimation, where it shows a positive relationship. Burning that fuel results in 1.8mtons of carbon emissions per year, or the amount generated by the whole of Malta.

EU countries' fishing fleets enjoy billions of euros in annual tax subsidies.

The panel quantile regression shows in Table 8.9 that information and communication technology has a statistically significant effect on fisheries sustainability in lower-to-above median quantiles. This confirms the use of technology transfer programs for the development of fisheries among developed countries EU14. The research confirms the growth hypothesis on the majority of quantiles on EU14 developed countries' fisheries, excluding two of the quantiles (0.2, 0.9), which implies that at high growth rates, EU14 developed countries give preference to production over technology development and transfer.

Nevertheless, the investment cost for developing alternative or renewable energy production technologies might lead to more challenges for EU14 developed countries at the lowest quantiles. Information and communication technology has more impact on higher-developing fisheries-producing countries of EU14 countries.

Table 8.9 Panel quantile estimation results for EU14 countries

Variables	Method of moments quantile regression with fixed effects										
	Location	Scale	Quantiles								
			0.1	0.2	0.3	0.4	0.5	0.6	0.7	0.8	0.9
ICT	(0.152)*	(0.079)***	(0.038)**	(0.164)	(0.241)**	(0.296)***	(0.323)***	(0.493)***	(0.336)***	(0.223)***	(0.013)
HP	(8.391)	(0.422)***	(0.175)	(5.360)	(8.666)*	(9.020)*	(12.77)***	(11.23)***	(14.18)***	(13.71)***	(9.380)***
EG	(0.341)**	(0.256)***	(0.167)*	(0.502)	(0.830)***	(1.036)***	(1.199)***	(1.204)***	(0.438)	(0.192)	(0.024)
GVR	(1.226)	(0.702)***	(1.740)	(1.098)	(1.934)**	(2.816)***	(2.767)***	(1.741)**	(0.217)	(0.508)	(0.694)
CO_2	(0.088)***	(0.249)***	(0.475)**	(0.572)**	(0.770)***	(0.974)***	(1.222)***	(0.962)***	(1.076)***	(1.113)***	(1.024)***

Notes: *, **, and *** indicate significance at 10%, 5%, and 1% levels, respectively
Source: [61]

Perhaps due to the heterogeneous structure of the technologies in the developed EU14 countries' fisheries sector [4], which tests for the growth hypothesis in the context of countries with an extremely high level of production (Fisheries Sector), leads us to predict that policy will therefore significantly focus on such a strategy.

In our case, the previous study supports the growth hypothesis, as [5] also confirmed the growth hypothesis as an appropriate frame for fisheries resource issues, and the results match the conclusions [6], which provided evidence for the growth hypothesis at the top-producing countries of the Fisheries Sector.

Furthermore, the development of human capital has a significantly positive correlation with median and higher quantiles and not so considerable an impact at the lowest quantiles. Human capital decreases its magnitude as the quantile increases. Economic growth (as per GDP) has a statistically significant effect on the fisheries sector at the lower and median quantiles, whereas on the higher quantiles the effect of economic expansion does not significantly affect fisheries production. These results indicate the fisheries strategic policy in generating economic progress in the highest producing countries of EU14.

This result supports the previous results indicating that at higher quantiles EU14 developed countries are prioritizing economic progress at the expense of marine environment conservation.

Moreover, governance has a strong positive impact on the median quantiles and is not so relevant for the higher and lower quantiles, whereas governance impacts more the higher quantiles as they increase. CO_2 emission, a component in our model, has a strong and positive correlation for the entire range of quantiles, and the same impact is more significant for the higher ones.

Table 8.10 in Model 3 exhibits the estimated coefficients from each of the several panel estimations, namely FMOLS, DOLS, and FEOLS. Each technique presents similar coefficient value estimates. The estimates demonstrate a clearly defined relationship between information and communication technology production and fisheries production through FEOLS, FMOLS, and DOLS techniques.

ICT is positively and significantly related to fisheries production through FEOLS, FMOLS, and DOLS estimates, increasing fisheries production by 0.24%, 0.13%, and 0.34% when ICT increases by 1%, respectively.

This result is similar to findings by [20, 21, 23, 54] showing that information and communication technology boosts fisheries production in developing countries. The empirical evidence shows a clear relationship between human capital and fisheries production in the EU13 developing countries, increasing fisheries production by 2.74%, 2.81%, and 5.57%, respectively, based on FEOLS, FMOLS, and DOLS estimation. This supports [42, 43, 54], suggesting human capital supports fisheries production in developing countries. Most importantly, economic growth has positive and profound impacts on fisheries sustainability in the EU13 developing countries.

For every 1% increase in GDP, fisheries production increases by 0.21%, 0.04%, and 0.06% under the FEOLS, FMOLS, and DOLS estimators, respectively.

The growth hypothesis in the EU13 developing countries has been established since the GDP coefficients were positive and statistically significant in all three techniques used, in line with previous literature on developing countries by [44–46].

Table 8.10 Model 3 panel estimation for the EU-13 region from 1990 to 2022

Long-run coefficient	DOLS		FMOLS		FEOLS	
	Coefficient	Std. error	Coefficient	Std. error	Coefficient	Std. error
ICT	0.345	(0.109)***	0.132	(0.025)***	0.240	(0.142)*
HP	5.574	(1.553)***	2.816	(0.010)***	2.745	(2.687)***
EG	0.069	(0.081)	0.048	(0.016)***	0.217	(0.088)**
GVR	1.604	(0.572)***	0.284	(0.023)***	0.316	(0.375)
CO_2	0.466	(0.368)	0.518	(0.021)***	0.373	(0.107)***

Long-run coefficient	DOLS		FMOLS		FEOLS	
	Coefficient	Std. error	Coefficient	Std. error	Coefficient	Std. error
ICT	0.345	(0.109)***	0.132	(0.025)***	0.240	(0.142)*
HP	5.574	(1.553)***	2.816	(0.010)***	2.745	(2.687)***
EG	0.069	(0.081)	0.048	(0.016)***	0.217	(0.088)**
GVR	1.604	(0.572)***	0.284	(0.023)***	0.316	(0.375)
CO_2	0.466	(0.368)	0.518	(0.021)***	0.373	(0.107)***

Note: ***, **, and * indicate significance at the 1%, 5%, and 10% levels respectively
Values in parentheses are P-values
Source: [61]

Governance had positive and strong impacts on fisheries production in the EU13 developing countries by increasing fisheries production by 0.28% and 1.60% for FMOLS and DOLS estimators, respectively. The governance coefficients being positive and significant confirm the growth hypothesis in EU13 countries under the FMOLS and DOLS methodologies but seem not to be consistent under the FEOLS method, as found in [31–33], stating that at the beginning of development, fisheries production is expected to increase as governance and ocean public policies develop.

The estimates confirm a positive relationship between CO_2 emissions and fisheries production in EU13 developing countries, where each unit increase of CO_2 emissions has a positive effect on the increase in fisheries production, by 0.37%, 0.51%, and 0.46%, by using FEOLS, FMOLS, and DOLS methodologies, respectively, as stated by [55, 57] for European countries and by [56] for developing EU countries where green fisheries market would promote fisheries production. Table 8.11 summarizes the method of moments quantile regression results. First, ICT showed a positive and strong impact on the sustainability of the fisheries at low and high quantiles, with the exceptions of the below-median and above-median quantiles (4th, 6th and 7th), where the effects of information and communication technologies are larger at higher quantiles.

Likewise, human capital input exhibits a positive and strongly significant effect on fisheries sustainability at all the quantiles (from the first to the ninth); however, its impact seems to get weaker with higher quantiles.

The table shows clearly that Fisheries sustainability is increasing with respect to Economic Growth and is decreasing as the quantile goes higher from 0.606 to 0.065 (from 1st to 9th). In other words, the positive effects of economic growth on fisheries sustainability have been evidenced through all the quantiles but the ninth one. A positive impact of governance can be demonstrated on the sustainability of the fisheries at lower and higher quantiles; however, not at below-median and above-median quantiles. According to the table, fisheries production varies significantly with respect to governance, and decreases as quantiles go higher from 2.033 to 0.144 (from the first to the ninth quantile). In other words, governance impacts fisheries sustainability; however, this effect is negligible and statistically not significant at the 4th, 6th, and 7th quantiles.

Lastly, CO_2 emissions' results were statistically negligible at low and above-median quantiles and were strongly and positively correlated to fisheries sustainability at the median and high quantiles; however, the impact gets lower with higher quantiles in emergent EU13 developing countries.

Panel FMOLS and DOLS estimators confirm the FEOLS panel estimations. As can be seen, panel FEOLS coefficients share the same sign and similar levels of significance as the panel FMOLS and DOLS coefficients. Thus, the results derived from panel FMOLS and DOLS are credible and allow the drawing of conclusions.

The panel FEOLS coefficient sign is the same as the one derived from panel FMOLS and DOLS estimates; the levels of significance are similar though slightly different.

Nevertheless, panel DOLS estimates are reliable and robust in the face of endogeneity and serial correlation issues. The EU27 countries were divided into two subgroups, the 14 developed countries and the 13 developing countries, in order to shed light on the impact of Information and communication technologies on Fisheries sustainability for developed and developing countries, respectively. Tables 8.8 and 8.9 show the potential impact of Information and communication technologies on Fisheries sustainability in the EU14 Developed Countries. Tables 8.10 and 8.11 show the estimated effect of Information and communication technologies on Fisheries sustainability in the EU13 developing countries. Results from Tables 8.8 and 8.10 suggest that Information and communication technologies are positively correlated with fisheries sustainability in both country subgroups.

Moreover, developed EU14 countries possess better access to ICT than developing EU13 countries and enhance their fisheries sustainability substantially. The exact value of ICT coefficients for EU14 and EU13 countries is 0.29 and 0.24, respectively. Thus, combined with green innovation technologies and dissemination mechanisms, developed EU14 countries have the potential to increase their Fisheries sustainability more substantially than the developing EU13 countries [76–81].

Tables 8.9 and 8.11 indicate that the EU13 developing countries, compared to the EU14 developing countries, had a distinct significant positive association with human capital input and Fishery sustainability, with a coefficient value of 2.745. Thus, with specialized institutions and training experience, developing EU13 countries can achieve a better increase in their Fishery sustainability than developed EU14 countries. Tables 8.8 and 8.9 show the impact of Fishery infrastructure on fishery sustainability in EU14 and EU13 developed countries, respectively. Results suggest that developed EU14 countries have a more significant impact on fisheries sustainability than developing EU13 Countries. The value of the fisheries sustainability coefficient is 0.865 for EU14 countries and 0.217 for EU13 countries.

Table 8.11 Panel quantile estimation results for EU13 countries

Variables	Method of moments quantile regression with fixed effects										
	Location	Scale	Quantiles								
			0.1	0.2	0.3	0.4	0.5	0.6	0.7	0.8	0.9
ICT	(0.142)*	(0.134)*	(0.054)***	(0.055)***	(0.054)***	(0.014)	(0.228)*	(0.163)	(0.160)	(0.726)***	(0.478)**
HP	(2.687)***	(0.355)***	(18.76)***	(26.80)***	(26.37)***	(20.06)***	(9.177)***	(10.27)***	(14.27)***	(7.776)**	(2.895)
EG	(0.088)**	(0.778)***	(0.606)***	(0.474)***	(0.463)***	(0.513)***	(0.599)***	(0.602)***	(0.482)***	(0.279)***	(0.065)
GVR	(0.375)	(0.893)*	(2.033)***	(1.300)**	(1.261)*	(1.153)	(1.598)*	(1.098)	(1.045)	(0.329)***	(0.144)***
CO2	(0.107)***	(0.084)**	(0.332)	(0.119)	(0.217)	(0.178)	(0.210)**	(0.261)***	(0.111)	(0.025)	(0.140)**

Notes: *, **, and *** indicate significance at 10%, 5%, and 1% levels, respectively
Source: [61]

It means that the wealthier EU14 countries achieve greater fisheries sustainability through infrastructure development, compared with developing EU13 countries. Results from Tables 8.10 and 8.11 indicate a positive association between carbon dioxide emissions and fisheries production for both country subgroups. More specifically, this positive correlation has been more remarkable in EU14 developed countries compared to the EU13 developing countries. For EU14 and EU13 countries, this coefficient is equal to 8.689 and 0.373, respectively.

This indicates that EU14 developed countries achieved an increase in fisheries sustainability through utilizing green gear technologies more readily than EU13 developed countries. This may be due to high Fishing production levels without adequate environmental regulation, which lead to high CO_2 emissions in EU14 developed countries.

8.5 Conclusion and ICT Solution in Fisheries Industry

We investigate the empirical relations among fishery production, information and communication technologies, human capital, economic growth, governance, and carbon dioxide emissions with the growth hypothesis as reference. Our contribution to the literature about blue growth concerning fisheries and sustainable development lies in this regard. We applied panel unit root, panel cointegration, and panel estimations to a dataset covering a 31-year span across 27 EU member countries. The quantitative variable governance (% confidence in the governance) was used to account for fishery governance and to measure how sustainable development is defined.

Panel unit root and panel cointegration techniques provided evidence that all variables were I(1) and shared a non-spurious long-run relationship. Though the magnitude differences across specifications were minimal, we found that standard estimation techniques using panel cointegration yielded different significant coefficients. For a detailed look into the relationships among the empirical relations, we utilized the MMQR technique to analyze the heterogeneous impact of explanatory variables across multiple quantiles of the conditional distribution of fishery production. The estimated coefficients for FMOLS, DOLS, and FEOLS supported the growth hypothesis in EU27-based fishery-producing economies.

The MMQR results, as summarized in Table 8.7, indicate that information and communication technology is significantly and positively related to fishery sustainability at quantiles 1 to 9, suggesting a consistent beneficial relationship between information and communication technology and fisheries sustainability at all quantiles.

In other words, ICT enhances both the early and advanced transitional growth of the EU27's fisheries sectors. To be specific, the human capital effect on fishery production decreases in importance as sustainability becomes more pronounced in higher quantiles. Hence, economies with significant dependence on traditional fisheries production methods and thus reliant on imports will need greater human capital investment to advance the fishing industry.

This implies dependence, but not necessarily abundance. Further supporting the resource curse from the deindustrialization process is the observation here that developed countries with better fisheries conditions are those dependent on imported fish products. The visual examination of Table 8.7 shows that MMQR estimates of economic growth were significantly and positively related from the lowest to the highest quantiles, peaking at the first quantile and gradually decreasing to the highest one. The results also indicated that quantiles where the influence of economic expansion on fisheries production was less prominent are the ones with the highest fisheries sustainability.

More concretely, governance was significantly and positively related from the 2nd to the 9th quantiles of the conditional distribution of fishery sustainability, and the coefficient peaked in the highest quantile, growing from the lowest to the highest quantile, meaning that the quantile where governance has the most significant influence on fishery production is where fishery sustainability reached maximum levels. It should be noted that the quantiles of fishery sustainability from 1 to 9 were statistically higher in the presence of carbon dioxide emissions. Fishery production economies having high CO_2 emissions have higher diversifications in economic sectors than emerging economies, where fish imports represent an important portion of the country's consumption. Therefore, we conclude that fishery products constitute an essential element of the marine natural resource base of developed economies and contribute less to the natural resources in developing economies.

Table 8.9 shows that information and communication technology were significantly and positively related at lower and higher quantiles of fisheries production for the developed economies of EU14 countries and developing economies of EU13, where environmental concerns regarding the reduction of ICT usage can lead to more efficient economies and significant contribution in development, and a variety of development opportunities for large-scale production in EU14 countries, while environmental regulations for those countries at less than or greater than the median are an issue at below or upper percentiles of the EU13 countries where the effects are negligent and a threat to the existing environment because effects are visible and could negatively affect them directly.

However, human capital was positive and significantly correlated to both lower and higher quantiles in the developed countries of EU14 and developing countries of EU13, respectively. Human capital's positively significant impact across all quantiles, as shown in Tables 8.9 and 8.11, supported the growth hypothesis. The results suggest potential reasons for these varying results among the different quantiles in countries where fisheries' production is successful.

One interpretation for lower quantiles in EU14 countries and higher quantiles in EU13 countries is that the use of human capital in fishery production is inefficient and hence other industries are substituted by an inappropriate human capital input, leading to greater dependence on imports and hence to a higher portion of economic activities consisting of the exchange of fish products. Similarly, due to a possibly high return on capital from the export of goods, high quantiles in EU13 countries and EU14 countries would experience higher capital flow from developing EU13 countries to developed EU14 countries, causing the fisheries production to collapse in EU13 economies.

Strictness of fisheries policies to industries dependent on marine resource inputs for their exports could mean the growth hypothesis directly applies to developed EU14 countries. Such an instance may force the fisheries sector to scale back its production and marine waste outputs. Numerous firms may choose to shift their activities to other, less developed countries with less rigorous fiscal policies and environmental regulation, and a greater abundance of marine resources, in order to profit more from such activity given the cost of hiring experienced specialists and introducing new technology diffusion in developing EU13 countries. Voters and policymakers may focus more on economic growth in Tables 8.9 and 8.11, which might have their roots in fisheries sector activities, based on sustainable development.

Given the prominence of other competitiveness drivers and a greater prevalence of the resource curse, the effect of economic growth is evident at some higher quantiles of EU13 developed and EU14 countries. The prominent and robust impact of economic growth at median quantiles suggests an improvement in sustainable fisheries development in EU14 and EU13 countries. Continued efforts by the government to avoid further development of overfishing and improper practices in the fisheries sector might be the cause for the insignificance of the positive effect of governance at lower quantiles in the EU14 developed countries. Given lesser production volumes in developing EU13 countries, regulation of the fisheries sector is of more concern for residents of EU14 developed countries (Tables 8.9 and 8.11).

Thus, better-performing governance, responding to voters demands, will tend to create tougher environmental policy toward overfishing and inappropriate fishing practices, and that explains the result in EU14 countries.

Based on these factors, it is important to enhance economic performance for countries that rely on the production from the fisheries sector to generate revenues, particularly from developing countries in this group, where a good portion of their economy might depend on it. All quantiles of fisheries production are substantially influenced positively by CO_2 emissions across Table 8.9; however, not in the upper quantiles (5th, 6th, and 9th quantiles) of Table 8.11. Although the lowest quantile (first in Table 8.9 and first in Table 8.11) is the most significant one in the impact of CO_2 on production, signifying the development of Fisheries tools and variables with increasing values from lower to higher quantiles of production, as can also be visible in marine habitat, marine environment, and marine ecology that results in increasing carbon sinks from lowest to higher quantiles.

Fisheries can continue to benefit from an increasing impact of information and communication technology (ICT), which suggests reliance on conventional methods in production activities in both EU14 and EU13 developing countries. However, more efforts need to be made in the fisheries sector to promote its move from traditional to greener and sustainable technology for improved levels of sustainable development in the fisheries sector in both the EU14 and EU13 countries. In simpler words, green technology should be utilized for promoting fisheries output in a sustainable and renewable way. Perhaps more pertinently, developing EU13 countries, unable to develop new green technology, may have to consider establishing environmental levies on their fisheries sector.

This implies moderate taxes on the fisheries sector, relying heavily on its marine resource inputs. These would contribute funds that may then be utilized for promoting the uptake of long-term environmentally friendly technologies to preserve this critical resource. Finally, this analysis could be supplemented by including other variables, such as those concerning tourism, globalization, and financial development, which are also known to affect the sustainability of the EU region's fisheries sector and its environment. In addition, by adopting new advanced econometric approaches, this study may be extended at the country level, for instance, to China and other country groups like ASEAN, BRICS, Group of 7, Next 11 countries, etc.

The application of models naturally comes with limitations, and this work is no exception.

Given the limitations in terms of available data, this work could not extend to address technology transfer benefits of fisheries. However, more research could be provided concerning these variables and using the present dataset, adding new variables like public-private investments in fisheries and in ocean energy. Furthermore, in this context, it would be of value to explore additional countries with significant pollution levels instead of focusing only on EU countries.

It is recommended that research also investigate the economics of fisheries technology transfer in a wider framework, exploring its changes across various time periods. The current research had its limitations, in spite of employing sophisticated and modern econometric techniques to confirm the proposed links. Firstly, the results found in this research were exclusively analyzed with econometric methods using a restricted list of variables over the period 1990–2022.

New and additional econometric techniques with the use of current variables and extra parameter statistics could be employed to investigate the impact of public–private investments in fisheries and in ocean energy upon production from the fisheries sector.

Second, given that the model used is only for EU countries, extra countries with significant emissions may be considered in future work. Last, the economic complexity of fisheries technology transfer must be assessed across different time frames in light of possible breaks in data.

References

1. Schwing FB (2023) Modern technologies and integrated observing systems are 'instrumental' to fisheries oceanography: a brief history of ocean data collection. Fish Oceanogr 32(1):28–69. https://doi.org/10.1111/fog.12619
2. Scarnecchia DL, Schooley JD (2022) Trophies, technology, and timescape in fisheries management, as exemplified through Oklahoma's world record paddlefish *Polyodon spathula*. Fisheries 47(9):381–394. https://doi.org/10.1002/fsh.10764
3. Ziegler R et al (2022) Social innovation for biodiversity: a literature review and research challenges. Ecol Econ 193:107336. https://doi.org/10.1016/j.ecolecon.2021.107336

4. Torres Cañete F, Oyanedel R, Gelcich S (2022) Adoption and impacts of fishing gear innovations: insights from a small-scale fishery in Chile. Fish Res 248:106200. https://doi.org/10.1016/j.fishres.2021.106200
5. Pathak S, Lavudya N, Kumar RM, Dash RK (2022) Use of innovative technologies for fisheries resource management. Just Agriculture 2(5):1–8
6. Kemp PS et al (2023) Future advances in UK marine fisheries policy: integrated nexus management, technological advance, and shifting public opinion. Mar Policy 147:105335. https://doi.org/10.1016/j.marpol.2022.105335
7. Kelkar N, Arthur RI (2022) A review of governance and tenure in inland capture fisheries and aquaculture systems of India. Online. https://books.google.com/books?hl=en&lr=&id=wL1qEAAAQBAJ&oi=fnd&pg=PP1&dq=Kelkar,+N.,+%26+Arthur,+R.+I.+(2022).+A+review+of+governance+and+tenure+in+inland+capture+fisheries+and+aquaculture+systems+of+India.+https://doi.org/10.4060/cb8615en&ots=43o6zSOa5g&sig=lUJVzKdVu4wfxr1-PtME1opbuaU. Accessed: May 20, 2026
8. Chudik A, Pesaran H (2015) Large panel data models with cross-sectional dependence: a survey. Online. https://academic.oup.com/edited-volume/41981/chapter/355335206?searchresult=1&utm_source=TrendMD&utm_medium=cpc&utm_campaign=Oxford_Academic_Books_TrendMD_0. Accessed 20 May 2026
9. Flaten O, Lien G, Koesling M, Løes A-K (2010) Norwegian farmers ceasing certified organic production: characteristics and reasons. J Environ Manag 91(12):2717–2726. https://doi.org/10.1016/j.jenvman.2010.07.026
10. Breitung J, Pesaran MH (2008) Unit roots and cointegration in panels. In: Mátyás L, Sevestre P (eds) The econometrics of panel data, Advanced Studies in Theoretical and Applied Econometrics, vol 46. Springer Berlin Heidelberg, Berlin, Heidelberg, pp 279–322. https://doi.org/10.1007/978-3-540-75892-1_9
11. Andrews DWK (2005) Cross-Section Regression with Common Shocks. Econometrica 73(5):1551–1585. https://doi.org/10.1111/j.1468-0262.2005.00629.x
12. Cerrato M, Sarantis N (2002) The cross sectional dependence puzzle. London Guildhall University. Online. http://fmwww.bc.edu/RePEc/res2002/Cerrato.pdf. Accessed 20 May 2026
13. Balsalobre-Lorente D, Shahbaz M, Roubaud D, Farhani S (2018) How economic growth, renewable electricity and natural resources contribute to CO2 emissions? Energy Policy 113:356–367
14. Bekun FV, Alola AA, Sarkodie SA (2019) Toward a sustainable environment: nexus between CO2 emissions, resource rent, renewable and nonrenewable energy in 16-EU countries. Sci Total Environ 657:1023–1029. https://doi.org/10.1016/j.scitotenv.2018.12.104
15. Isham J (2005) The varieties of resource experience: natural resource export structures and the political economy of economic growth. World Bank Econ Rev 19(2):141–174. https://doi.org/10.1093/wber/lhi010
16. Mehlum H, Moene K, Torvik R (2006) Institutions and the resource curse. Econ J 116(508):1–20. https://doi.org/10.1111/j.1468-0297.2006.01045.x
17. Leite MC, Weidmann J (1999) Does mother nature corrupt? Natural resources, corruption, and economic growth. International Monetary Fund. Online. https://books.google.com/books?hl=en&lr=&id=UFYYEAAAQBAJ&oi=fnd&pg=PA3&dq=Leite,C.,+Weidmann,+J.+(1999).+Does+Mother+Nature+Corrupt:+Natural+Resources,+Corruption,+and+Economic+Growth+IMF+Working+Papers.+International+Monetary+Fund.&ots=o29DeYoDjX&sig=TVLC-xIqLLL9VCD6nJ30s1TCTho. Accessed 20 May 2026
18. Machado JAF, Santos Silva JMC (2019) Quantiles via moments. J Econ 213(1):145–173. https://doi.org/10.1016/j.jeconom.2019.04.009
19. Koenker R (2004) Quantile regression for longitudinal data. J Multivar Anal 91(1):74–89. https://doi.org/10.1016/j.jmva.2004.05.006
20. Karagiannis D, Metaxas T (2023) 'Fishing' for winemaking perfection in cloudy waters…underwater wineries: innovation genius or another marketing trick? implications on wine & food tourism. In: Sigala M, Haller C (eds) Technology advances and innova-

tion in wine tourism. Springer Nature Singapore, Singapore, pp 231–250. https://doi.org/10.1007/978-981-19-8277-4_14
21. Nogueira LA, Lindeløv B, Olsen J (2023) From waste to market: exploring markets, institutions, and innovation ecosystems for waste valorization. Bus Strat Env 32(4):2261–2274. https://doi.org/10.1002/bse.3247
22. Li C-J, Kim H-H, Yang J-W (2021) A Comparative study of fishery industry competitiveness in China's coastal provinces. International journal of Advanced Smart Convergence 10(2):158–167. https://doi.org/10.7236/IJASC.2021.10.2.158
23. Fu X-M et al (2022) Evaluation of the innovation ability of China's marine fisheries from the perspective of static and dynamic. Mar Policy 139:105032. https://doi.org/10.1016/j.marpol.2022.105032
24. Chen X, Xu Q, Li L (2023) Illegal, unreported, and unregulated fishing governance in disputed maritime areas: reflections on the international legal obligations of states. Fishes 8(1):36. https://doi.org/10.3390/fishes8010036
25. Spijkers J, Mackay M, Turner J, McNeill A, Travaille K, Wilcox C (2023) Diversity of global fisheries governance: types and contexts. Fish Fish 24(1):111–125. https://doi.org/10.1111/faf.12713
26. Jarvis RM, Young T (2023) Pressing questions for science, policy, and governance in the high seas. Environ Sci Pol 139:177–184. https://doi.org/10.1016/j.envsci.2022.11.001
27. Guggisberg S, Jaeckel A, Stephens T (2022) Transparency in fisheries governance: achievements to date and challenges ahead. Mar Policy 136:104639. https://doi.org/10.1016/j.marpol.2021.104639
28. Okafor-Yarwood I, Kadagi NI, Belhabib D, Allison EH (2022) Survival of the richest, not the fittest: how attempts to improve governance impact african small-scale marine fisheries. Mar Policy 135:104847. https://doi.org/10.1016/j.marpol.2021.104847
29. Jentoft S, Chuenpagdee R (2022) Interactive learning and governance transformation for securing blue justice for small-scale fisheries. Administration & Society 54(7):1255–1282. https://doi.org/10.1177/00953997211073947
30. Dudayev R, Hakim LL, Rufiati I (2023) Participatory fisheries governance in Indonesia: are octopus fisheries leading the way? Mar Policy 147:105338. https://doi.org/10.1016/j.marpol.2022.105338
31. Acuña-Alonso C, Varandas S, Álvarez X, Martinho A (2023) Analysis of the evolution of a fisheries management plan based on environmental governance: living laboratory in the Olo River, Portugal. Fish Res 260:106595. https://doi.org/10.1016/j.fishres.2022.106595
32. Aguión A et al (2022) Establishing a governance threshold in small-scale fisheries to achieve sustainability. Ambio 51(3):652–665. https://doi.org/10.1007/s13280-021-01606-x
33. Linke S, Gillette MB, Jentoft S (2022) Institutionalizing injustice? aligning governance orders in swedish small-scale fisheries. In: Jentoft S, Chuenpagdee R, Bugeja Said A, Isaacs M (eds) Blue justice, MARE Publication Series, vol 26. Springer International Publishing, Cham, pp 529–547. https://doi.org/10.1007/978-3-030-89624-9_28
34. Ynacay-Nye A, Hisano S, Suryawan AS (2023) Alternatives to sustainable seafood certifications: transitions of small-scale fisheries governance in northeastern Japan. J Rural Stud 97:269–280. https://doi.org/10.1016/j.jrurstud.2022.12.002
35. Day S (2026) Pacific Salmon fisheries management: an (unusual) example of polycentric governance involving indigenous participation at multiple scales. The Center for Growth and Opportunity. Online. https://www.thecgo.org/books/the-environmental-optimism-of-elinor-ostrom/chapter-3-chapter-3-pacific-salmon-fisheries-management-an-unusual-example-of-polycentric-governance-involving-indigenous-participation-at-multiple-scales/. Accessed 20 May 2026
36. Sorensen JA, Kincl L, Weil R, Dzugan J, Christel D (2022) Fisheries governance and associated health implications: current perspectives from US commercial fishermen. Mar Policy 141:105119

37. Wang M, Janssen ABG, Bazin J, Strokal M, Ma L, Kroeze C (2022) Accounting for interactions between sustainable development goals is essential for water pollution control in China. Nat Commun 13(1):730. https://doi.org/10.1038/s41467-022-28351-3
38. Albasri H, Sammut J (2022) A comparative study of sustainability profiles between small-scale mariculture, capture fisheries and tourism communities within the Anambas Archipelago Small Island MPA, Indonesia. Aquaculture 551:737906. https://doi.org/10.1016/j.aquaculture.2022.737906
39. Ramos J, Drakeford B (2021) Blue economy and tourism: is there a potential to create sustainable jobs? In: Handbook of research on human capital and people Management in the Tourism Industry. IGI Global Scientific Publishing, pp 243–264. Online. https://www.igi-global.com/chapter/blue-economy-and-tourism/263358. Accessed 20 May 2026
40. Yıldırım DÇ, Yıldırım S, Bostancı SH, Turan T (2022) The nexus between human development and fishing footprint among mediterranean countries. Mar Pollut Bull 176:113426. https://doi.org/10.1016/j.marpolbul.2022.113426
41. Seisay M (2022) Human capital investment and financing for Africa's blue economy: conference on unlocking the full potential of the fisheries sector within the framework of a sustainable blue economy. Online. http://repository.au-ibar.org/handle/123456789/1175. Accessed 20 May 2026
42. Najib H, Nawangsari LC (2021) Effect of intellectual capital on organizational sustainability with employee innovative behavior as intervening variables in Pt. Jaya maritime services. EJBMR 6(1):158–163. https://doi.org/10.24018/ejbmr.2021.6.1.714
43. Nurlita L, Yanuar T, Syah R, Indradewa R, Hamdi E (2021) Human capital implementation strategy in the Marina company. American International Journal of Business Management (AIJBM) 4(8):152–156
44. Rahmizal M (2022) Analysis of Indonesia marine fisheries with economic growth, population and effort effectiveness. European Journal of Formal Sciences and Engineering 5(1):53–61. https://doi.org/10.26417/ejef.v1i1.p17-22
45. Setiyowati H, Nugroho M, Halik A (2022) Developing a blue economy in depok West Java, Indonesia: opportunities and challenges of Neon Tetra Fish cultivation. Sustainability 14(20):13028. https://doi.org/10.3390/su142013028
46. Coppola D, Lauritano C, Palma Esposito F, Riccio G, Rizzo C, De Pascale D (2021) Fish waste: from problem to valuable resource. Mar Drugs 19(2):116. https://doi.org/10.3390/md19020116
47. Abo-Taleb HA et al (2021) Effect of a new feed Daphnia magna (Straus, 1820), as a fish meal substitute on growth, feed utilization, histological status, and economic revenue of grey mullet, mugil cephalus (Linnaeus 1758). Sustainability 13(13):7093. https://doi.org/10.3390/su13137093
48. Shamsuddin M et al (2022) Effects of stocking larger-sized fish on water quality, growth performance, and the economic yield of Nile Tilapia (Oreochromis niloticus L.) in floating cages. Agriculture 12(7):942. https://doi.org/10.3390/agriculture12070942
49. Palomino E, Pardue J (2021) Alutiiq fish skin traditions: connecting communities in the COVID-19 era. Heritage 4(4):4249–4263. https://doi.org/10.3390/heritage4040234
50. Reza MS, Azad AK, Abu Bakar MS, Karim MR, Sharifpur M, Taweekun J (2022) Evaluation of thermochemical characteristics and pyrolysis of fish processing waste for renewable Energy feedstock. Sustainability 14(3):1203. https://doi.org/10.3390/su14031203
51. Saba GK et al (2021) Toward a better understanding of fish-based contribution to ocean carbon flux. Limnology & Oceanography 66(5):1639–1664. https://doi.org/10.1002/lno.11709
52. Colina M et al (2021) Trophic and non-trophic effects of fish and macroinvertebrates on carbon emissions. Freshw Biol 66(9):1831–1845. https://doi.org/10.1111/fwb.13795
53. McArley TJ, Sandblom E, Herbert NA (2021) Fish and hyperoxia—from cardiorespiratory and biochemical adjustments to aquaculture and ecophysiology implications. Fish Fish 22(2):324–355. https://doi.org/10.1111/faf.12522

54. Li Y, Kong J, Ji J (2022) Environmental regulation, technological innovation and development of marine fisheries—evidence from ten coastal regions in China. Fishes 7(1):20. https://doi.org/10.3390/fishes7010020
55. Chen C-Z, Li P, Liu L, Li Z-H (2022) Exploring the interactions between the gut microbiome and the shifting surrounding aquatic environment in fisheries and aquaculture: a review. Environ Res 214:114202. https://doi.org/10.1016/j.envres.2022.114202
56. Kamau JN et al (2021) Managing emerging fisheries of the North Kenya Banks in the context of environmental change. Ocean & Coastal Management 209:105671. https://doi.org/10.1016/j.ocecoaman.2021.105671
57. Chevallier A et al (2021) Diversity of small-scale fisheries in Chile: environmental patterns and biogeography can inform fisheries management. Environ Sci Pol 124:33–44. https://doi.org/10.1016/j.envsci.2021.06.002
58. World Bank (n.d.) Fish to 2030: prospects for fisheries and aquaculture
59. European Commission (2022) Joint research centre. *Clean Energy Technology Observatory, Ocean energy in the European Union: status report on technology development, trends, value chains and markets: 2022.* Publications Office, LU. Online. https://data.europa.eu/doi/10.2760/162254. Accessed 30 Dec 2025
60. European Commission (n.d.) The European Green Deal—European Commission. Online. https://commission.europa.eu/strategy-and-policy/priorities-2019-2024/european-green-deal_en. Accessed 08 June 2025
61. Alsaleh M, Yang Z (2023) The evolution of information and communications technology in the fishery industry: the pathway for marine sustainability. Mar Pollut Bull 193:115231. https://doi.org/10.1016/j.marpolbul.2023.115231
62. Pedroni P (2004) Panel cointegration: asymptotic and finite sample properties of pooled time series tests with an application to the PPP hypothesis. Economic Theory 20(03). https://doi.org/10.1017/S0266466604203073
63. Kao C, Chiang M-H (2000) On the estimation and inference of a cointegrated regression in panel data. In: Advances in Econometrics, vol 15. Emerald (MCB UP), Bingley, pp 179–222. https://doi.org/10.1016/S0731-9053(00)15007-8
64. Sarkodie SA, Strezov V (2019) Effect of foreign direct investments, economic development and energy consumption on greenhouse gas emissions in developing countries. Sci Total Environ 646:862–871. https://doi.org/10.1016/j.scitotenv.2018.07.365
65. Koenker R, Bassett G (1978) Regression quantiles. Econometrica 46(1):33. https://doi.org/10.2307/1913643
66. Binder M, Coad A (2011) From Average Joe's happiness to Miserable Jane and cheerful John: using quantile regressions to analyze the full subjective well-being distribution. J Econ Behav Organ 79(3):275–290. https://doi.org/10.1016/j.jebo.2011.02.005
67. Canay IA (2011) A simple approach to quantile regression for panel data. Econ J 14(3):368–386
68. Phillips PCB, Sul D (2003) Dynamic panel estimation and homogeneity testing under cross section dependence. Econ J 6(1):217–259. https://doi.org/10.1111/1368-423X.00108
69. Pesaran MH (2021) General diagnostic tests for cross-sectional dependence in panels. Empir Econ 60(1):13–50. https://doi.org/10.1007/s00181-020-01875-7
70. Levin A, Lin C-F, James Chu C-S (2002) Unit root tests in panel data: asymptotic and finite-sample properties. J Econ 108(1):1–24. https://doi.org/10.1016/S0304-4076(01)00098-7
71. Im KS, Pesaran MH, Shin Y (2003) Testing for unit roots in heterogeneous panels. J Econ 115(1):53–74. https://doi.org/10.1016/S0304-4076(03)00092-7
72. Westerlund J (2007) Testing for error correction in panel data*. Oxf Bull Econ Stat 69(6):709–748. https://doi.org/10.1111/j.1468-0084.2007.00477.x
73. Kremers JJM, Ericsson NR, Dolado JJ (1992) The power of cointegration tests. Oxf Bull Econ Stat 54(3):325–348. https://doi.org/10.1111/j.1468-0084.1992.tb00005.x
74. Yusuf M (2022) The competitiveness of Indonesian shrimp export in Malaysia and Singapore market. Online. http://repository.unimus.ac.id/5474/1/6.%20THE%20

COMPETITIVENESS%20OF%20INDONESIAN%20SHRIMP%20EXPORT%20IN%20MALAYSIA%20AND%20SINGAPORE%20MARKETS.pdf. Accessed 20 May 2026
75. Handaka AA, Lestari RP, Rizal A (2022) Analysis of competitiveness fisheries processing industry in West Java, Indonesia. AJFAR:8–19. https://doi.org/10.9734/ajfar/2022/v16i330372
76. Crouch GI, Ritchie JB (1999) Tourism, competitiveness, and societal prosperity. J Bus Res 44(3):137–152
77. Dwyer L, Kim C (2003) Destination competitiveness: determinants and indicators. Curr Issue Tour 6(5):369–414. https://doi.org/10.1080/13683500308667962
78. Anwar A, Siddique M, Dogan E, Sharif A (2021) The moderating role of renewable and non-renewable energy in environment-income nexus for ASEAN countries: evidence from method of moments quantile regression. Renew Energy 164:956–967
79. Ehigiamusoe KU, Lean HH, Smyth R (2020) The moderating role of energy consumption in the carbon emissions-income nexus in middle-income countries. Appl Energy 261:114215. https://doi.org/10.1016/j.apenergy.2019.114215
80. Dogan E, Seker F (2016) Determinants of CO2 emissions in the European Union: the role of renewable and non-renewable energy. Renew Energy 94:429–439. https://doi.org/10.1016/j.renene.2016.03.078
81. Apergis N, Payne JE (2010) The emissions, energy consumption, and growth nexus: evidence from the commonwealth of independent states. Energy Policy 38(1):650–655

Chapter 9
European Fisheries Intellectual Property Rights (IPR) in the EU

Acronyms

EU	European Union
NEF	New Economics Foundation
TAC	Total Allowable Catch
CO_2	Carbon Dioxide Emissions
MMQR	Method of Moments Quantile Regression
EUR	Euro
CD	Cross-sectional Dependency
FP	Fisheries Production
WDI	World Bank Development Indicators
GDP	Gross domestic product
PD	Population density
DOLS	Dynamic ordinary least squares
FMOLS	Fully modified ordinary least squares
FEOLS	Fixed-effects ordinary least squares
Europa	European data portal
Eurostat	European Commission's database
LLC	Lin & Chu Test
IPPS	Im, Pesaran, and Shin Test
EU27	27 European Union Member States
EU13	13 European Union Member States
BRICS	Brazil, Russia, India, China, South Africa
ASEAN	Association of Southeast Asian Nations
EU14	14 European Union Member States
OLS	Ordinary least squares
IPR	Intellectual property rights
PPP	Purchasing power parity
ICR	Intellectual capital rights
MRK	Market size
CFP	Common Fisheries Policy
MAGPs	Multi-Annual Guidance Programmes
TACs	Total Allowable Catches

M. Alsaleh, *Fisheries Sustainability in the EU*, Urban Sustainability,
https://doi.org/10.1007/978-981-92-4496-6_9

Preventing unauthorized exploitation of marine resources through the allocation of property rights can align economic incentives with environmental conservation objectives. Nevertheless, increasing exploitation of marine resources and the associated environmental pollution have accelerated changes in the living conditions of aquatic species within European waters. By incorporating fisheries productivity, intellectual property rights, intellectual capital rights, market size, governance, and economic growth over the period 1990–2022, this study investigates the dynamic influence of property-rights-related factors on the sustainability of the fisheries sector across 27 European countries. Employing the Method of Moments Quantile Regression (MMQR) with fixed effects, the findings reveal a significantly positive relationship between governance and fisheries sustainability, particularly at the higher quantiles. Furthermore, the effect of intellectual property rights in EU27 countries was found to be positive and statistically significant from the first through the ninth quantiles. The results further indicate that governance quality and intellectual capital rights are stronger in the EU14 developed economies than in the EU13 developing nations, thereby contributing more substantially to fisheries sustainability. Similarly, intellectual property rights, market size, and economic growth all demonstrate positive effects on fisheries sustainability across all quantiles, thereby supporting the growth hypothesis for fisheries-producing economies. More specifically, the positive influence of intellectual property rights, market size, and economic development on fisheries sustainability is greater in the EU13 developing countries than in the EU14 developed nations. These findings offer important policy implications for promoting property-rights-related mechanisms in both EU14 and EU13 countries through the adoption of effective green technologies within the fisheries sector to achieve sustainable development goals.

9.1 Fisheries and Intellectual Property Rights (IPR)

The application of property rights has been widely supported as a mechanism for fostering sustainable fisheries and strengthening marine environmental conservation initiatives [1]. The establishment of property rights is often regarded as a classical remedy for fisheries management, enthusiastically promoted by economists and broadly implemented by fisheries authorities [1]. Economists argue that clearly defined property rights enhance productivity, profitability, and the sustainable utilization of marine resources. Conversely, other scholars advocate for a broader fisheries policy framework that incorporates perspectives from ecology, sociology, biology, anthropology, and geography in order to address the diverse social and environmental challenges associated with fisheries governance [1]. Countries such as Ireland, Spain, the Netherlands, Portugal, and Germany ranked highest within the "Overfishing League," reflecting earlier findings published by the New Economics Foundation (NEF), which demonstrated that these countries exceeded scientifically recommended quotas by 24%, 35%, 23%, 23%, and 22%, respectively. Likewise, the United Kingdom, Denmark, and Spain recorded excess tonnage levels of 1.78 million t, 1.48 million t, and 1.04 million t, respectively.

The New Economics Foundation report also documented the Total Allowable Catch (TAC) for commercial fish species within EU waters, revealing that between 2001 and 2020, approximately six out of every ten TACs were established above scientific recommendations [2]. Although the proportion of TACs exceeding scientific advice declined from eight out of ten to five out of ten during this period, the decline remained relatively limited compared to the reduction in the extent to which TACs exceeded recommended levels, which fell from 39% to 10% across EU waters [2]. For example, France established quotas that exceeded scientific recommendations by 20% (771,000 t) between 2001 and 2020, ranking sixth in terms of excess tonnage [3]. Since 2018, Germany has led fisheries activity in the Baltic Sea, largely due to continued exploitation of cod and herring stocks, which has adversely affected both ecosystems and fishermen. Germany exceeded scientific recommendations by 22% (592,000 t) from 2001 to 2020, ranking ninth in excess tonnage [3]. Similarly, the Netherlands exceeded limits by 847,000 t during the same period, placing fourth in the Overfishing League Table [4]. Ireland surpassed scientific advice by an average of 24% (765,000 t), ranking second behind Spain [4]. Portugal ranked third, with quotas averaging 23% (212,000 t) above scientific recommendations [2], while Spain occupied the first position due to quotas averaging 35% above scientific advice [2] Sweden exceeded its recommended limits by 16% (552,000 t) between 2001 and 2020, ranking seventh in the Overfishing League Table [2].

The implementation of the EU Common Fisheries Policy has introduced numerous restrictions within the fisheries sector [5]. The economic operating space of EU fisheries enterprises is regulated through mechanisms such as Total Allowable Catches (TACs) and Multi-Annual Guidance Programmes (MAGPs), which are intended to constrain fisheries efforts [5]. These regulatory transformations demonstrate how restrictions have gradually evolved into rights allocated either to individual fishers or fishing communities. Consequently, restrictions and rights can be viewed as two dimensions of the same framework, as scarcity generated through limitations provides property rights with economic value. This study therefore seeks to examine whether lessons derived from intellectual property rights can contribute to fisheries sustainability [5]. Although naturally occurring living organisms are generally not patentable, products derived through human creativity and innovation that are "new" and "non-obvious" may, in principle, qualify for patent protection [6], thereby creating opportunities for patent applications within the fisheries sector. Depending on the technological innovation involved, breeder's rights, patents, trademarks, geographical indications, copyrights, industrial designs, or combinations of these intellectual property mechanisms may be used to protect innovations in livestock and fisheries industries [6]. Given the vast and largely unexplored potential of marine biological resources, together with the broader acceptance of biotechnology patents in industrialized nations, the patenting of aquatic living resources, including genetic material, is likely to expand significantly in the future. Such innovations possess substantial commercial potential in sectors including pharmaceuticals, diagnostic and analytical reagents, cosmetics, and food production [7].

This study addresses three principal research questions: (1) How do property-rights-related variables affect fisheries productivity, sustainability and the associated marine environment in European waters between 1990 and 2022? (2) Are EU member states more vulnerable to fisheries sustainability challenges resulting from property-rights-related issues? (3) Can the fisheries sector become more environmentally sustainable to achieve EU goals relating to marine and ocean sustainability by 2030? Accordingly, the study aims: (1) to examine fisheries sustainability in relation to property rights within EU seas between 1990 and 2022; (2) to comprehensively analyze the influence of property rights on fisheries sustainability and marine environmental quality between the EU13 developing and EU14 developed economies during 1990–2022; and (3) to evaluate the extent to which property-rights-related factors influence fisheries sustainability in both developed and developing countries over the same period.

Property rights are frequently regarded as a critical determinant of sustainable development potential and dynamics. Important policy reforms, including the EU Common Fisheries Policy (CFP), revisions to common agricultural and fisheries policies, and successive EU enlargements, have substantially reduced or eliminated trade barriers [8]. Consequently, the micro-marketing environment has undergone substantial transformation, intensifying competition among exporting nations. This evolving competitive landscape creates new risks and challenges for the fisheries sector. Simultaneously, concerns regarding food security, marine environmental sustainability, and ocean resource sustainability have become globally significant [8]. Because the fisheries sector must remain competitive while also responding to domestic and international concerns regarding marine ecosystems and sustainable blue food systems, property-rights-related governance has become increasingly complex [8]. The EU fisheries sector is no exception, as it operates within a highly competitive global market. Fisheries products account for a considerable share of the EU food market, while imports have expanded substantially over the past decade [8]. Due to the limited availability of relevant literature capable of adequately explaining the changing dynamics of property rights within the fisheries industry, investigating fisheries sustainability and the determinants of property rights may provide valuable insights for future policy formulation and marketing strategies.

A major weakness in previous studies concerning EU member states is the assumption that disturbance factors are independent and that no cross-sectional dependency (CD) exists among variables across countries. According to [9–13] these disturbances may arise from common shocks, unobserved spillovers, or combinations of both. Incorporating CD into empirical analysis is essential because EU economies are highly interconnected through economic integration. EU member states are therefore vulnerable to coordinated shocks affecting economic growth, property rights, intellectual property rights, and broader economic performance [12]., argued that failing to address CD may produce biased and inconsistent findings regarding the relationship between economic growth and intellectual property rights. The present study seeks to determine whether the growth-mitigating influence of natural resources similarly affects the property-rights profile of fisheries-producing countries. Earlier studies [14, 15] employed more aggregate exogenous resource variables without distinguishing between point-source and dispersed-source resources.

In contrast, the present research focuses specifically on fisheries sustainability as a specialized natural resource variable. Fisheries-producing EU economies were selected because point-source resources often generate stronger negative externalities compared with dispersed resources [16, 17]. Since effective governance has been shown to mitigate or even reverse the adverse growth effects associated with resource dependence [18], the sample includes both developed and developing economies alongside regulatory variables such as governance. Fisheries-producing countries within the EU27 include developed economies such as Germany, Italy, and Sweden, as well as emerging economies such as Estonia, Bulgaria, and Malta. The study covers the period from 1990 to 2022. Given the differing stages of economic development among these countries, this study adopts the advanced Method of Moments Quantile Regression (MMQR) developed by [19] to evaluate the validity of the economic-growth-induced resource sustainability nexus in the presence of fisheries production and other control variables across different quantiles of fisheries sustainability. By incorporating fixed effects, the MMQR approach captures the distributional heterogeneity of these relationships. Moreover, the method permits heterogeneous relationships between economic growth and fisheries productivity across varying conditional quantiles of blue resource sustainability, thereby identifying patterns that conventional mean regression models may fail to capture. Consequently, the growth hypothesis is comprehensively examined among fisheries-producing nations. To the best of our knowledge, this study is the first to analyze the influence of economic growth on resource sustainability in EU27 countries by incorporating fisheries productivity, intellectual property rights, intellectual capital rights, governance, market size, and economic growth. Evaluating the growth hypothesis across multiple conditional quantiles of fisheries sustainability is particularly important because conditional quantile estimators are more robust to outliers originating from the dependent variable than conditional mean estimators [20]. Additionally, conditional mean estimations fail to capture the complete distributional impact of property rights on fisheries sustainability. By dividing the distributional effects of explanatory variables into several quantile ranges, quantile regression provides more intuitive and detailed insights, particularly in panel data contexts. Consequently, conditional quantile estimates offer information unavailable through conditional mean estimation. The remainder of this study is organized as follows: Sect. 9.2 reviews the relevant literature concerning the growth hypothesis; Sect. 9.3 describes the data and methodology; Sect. 9.4 presents and discusses the empirical findings; and Sect. 9.5 concludes with policy recommendations derived from the results.

9.2 Historical Development of Fisheries IPR

The concept of fisheries industry sustainability refers to the capacity to generate and preserve inclusive wealth without undermining the sector's ability to maintain or expand existing wealth levels. For example, [2–4, 20] examined the sustainability of Indonesia's fisheries processing industry and found that Depok City ranked first in competitiveness, performing strongly in infrastructure, productivity, and the application of science and technology [2]. In Mexico [5], the influence of market competition on the fisheries industry was examined, and concluded that competitive markets substantially stimulate fisheries-related economic growth. Similarly, [7] reported that Malaysia experiences a considerable fish trade imbalance, thereby weakening the global competitiveness of its fisheries industry. [21, 22] further demonstrated that Shandong Province possesses the highest competitiveness among China's coastal fisheries regions and exhibits considerable potential for sustainable fisheries development. Their studies applied the Analytic Hierarchy Process (AHP) and Delphi methods to construct evaluation indices and assess the fisheries competitiveness of 11 coastal Chinese provinces.

A growing body of literature, including studies by [23–26]. It has emphasized that equitable and resilient policy innovations must coexist alongside intellectual property rights in order to facilitate socially, economically, and environmentally sustainable transformations within aquatic food systems. Likewise, [27–29] investigated future developments in European marine fisheries by integrating intellectual property rights, technological innovation, and public policy. These studies adopted the Fisheries-Energy-Environment Nexus resource management framework, highlighting the importance of systems thinking in balancing trade-offs and synergies among competing sectors to achieve both fisheries sustainability and environmental conservation [30, 31]. analyzed technological innovation and intellectual property rights development within marine fisheries across ten coastal Chinese regions and concluded that property rights are fundamental to supporting such progress. From another perspective, [32–34] confirmed that enhanced investment in intellectual property rights may partially offset the negative effects of fisheries decommissioning. In the Mediterranean region, [35] explored the relationship between intellectual capital development and fisheries footprints, concluding that low levels of intellectual capital rights negatively affect fisheries sustainability, whereas stronger intellectual capital rights produce positive outcomes. Similarly, [36, 37] investigated blue economy sustainability in Indonesia and proposed integrated, coordinated, and harmonized policy frameworks. According to [38, 39] sustainable fisheries involve the marketing and distribution of aquaculture and wild-catch products in ways that preserve species sustainability while safeguarding marine ecosystems over the long term. The role of market development in expanding the Asian fisheries sector was examined by [39–41], who concluded that effective fisheries management systems are essential for maintaining sustainable fisheries markets. Similarly, [42, 43] studied seafood market development in Europe and demonstrated that integrating domestic and international seafood markets can enhance both fisheries sustainability and market expansion.

Several studies, including [44–49], have investigated how weak governance and ineffective enforcement contribute to overfishing, emphasizing the importance of strong policy frameworks for sustainable fisheries management. Likewise, [8, 50], and related scholars examined the legal and policy structures governing fisheries in Indonesia and India, concluding that community partnerships and co-management approaches represent the most effective governance arrangements for supporting sustainable fisheries industries. Evaluations of fisheries management plans conducted by [51, 52] demonstrated significant improvements in fish abundance in Europe, increasing from 1229 in 2008 to 1781 in 2018, despite a decline observed in 2013 due to nearby wildfires [53]. Explored alternatives to sustainable seafood certification systems in Japan, noting that although major political and economic stakeholders advocate seafood certification as a strategy for addressing external shocks, local knowledge systems are often overlooked. In the United States, [54, 55] investigated fisheries management regulations aimed at conserving marine resources and concluded that enabling environments significantly influence fishers' welfare, particularly when combined with unpredictable weather conditions, unreliable labor availability, and fluctuations in market prices. Research conducted by [56–60] estimated that unsustainable fisheries practices lead to losses exceeding 20% of the global fisheries industry's value. Establishing a sustainable fisheries sector therefore offers substantial long-term economic, social, and environmental development potential [61, 62]. investigated optimal fish productivity growth levels in China and demonstrated that sustainable fisheries development can simultaneously improve economic performance and economic expansion. In East Asia, [62–64] examined the economic implications of fisheries productivity and trade, concluding that developing countries can significantly strengthen their economies through increased productivity and high-quality fisheries exports. Whereas earlier studies primarily concentrated on fisheries production itself, the present study investigates the broader role of property-rights-related factors in promoting fisheries sustainability and blue economic growth. Additionally, the study examines how property-rights variables influence fisheries productivity within EU seas to support sustainable development. While most previous research relied on time-series approaches, this study adopts both static and panel regression methods to analyze the effects of property rights and economic growth on fisheries sustainability. Several key contributions distinguish this research. First, it investigates the relationship among fisheries productivity, intellectual property rights, market size, governance, and economic growth in EU seas over the period 1990–2022. Second, it evaluates the varying effects of property-rights-related factors and economic development on the fisheries sector in EU27 waters in pursuit of sustainable blue growth. Third, the study categorizes the EU27 into two subgroups: the EU14 developed countries and the EU13 developing nations. Because the dataset spans 1990–2022 and exhibits random drift behavior, unit root and cointegration analyses are applied to assess stationarity. Furthermore, by utilizing the Method of Moments Quantile Regression (MMQR), the study explores the interrelationships among market size, intellectual capital rights, governance, intellectual property rights, and economic growth. Static and panel regression techniques are also employed to examine the determinants of economic development and property rights. Finally, the study forecasts how various property-rights-related variables may influence marine resource sustainability and fisheries development in the future.

9.3 Porter Theory and Fisheries IPR Application

Two key considerations are essential when developing fisheries sectors in developing economies: (1) the contribution of multinational economic activities to cluster and property-rights advantages, and (2) the role of interconnected networks among cluster participants. Although numerous studies on cluster development, property rights, and Porter's theory emphasize the global competitiveness of EU member states and regions, a new framework referred to as the "property rights of fisheries development" is proposed in response to the importance of these issues for sustainable fisheries development. The proposed framework builds upon the models of [65, 66], which themselves are derived from Porter's Diamond Model. However, three additional components are incorporated into the revised framework. First, the model highlights the significance of patents, trademarks, copyrights, and trade secrets as valuable strategic assets for all industries. Second, governance is identified as a major determinant of fisheries property rights within the EU13 developing countries, thereby making it a critical factor in achieving sustainable fisheries development. Third, because all forms of property rights—including intellectual property rights and intellectual capital rights—are interconnected within industrial clusters, it becomes essential to identify and demonstrate the nature and intensity of the relationships among them. The following model was partially developed based on the studies of [67–70].

$$FP_{it} = f\left(IPR_{it}, ICR_{it}, MRK_{it}, GVR_{it}, GDP_{it}\right)$$

In this framework, fisheries productivity (FP), measured as total fisheries output in metric tonnes, represents the dependent variable. Data for these variables were obtained from the World Bank's World Development Indicators database. The independent variables include governance (GVR), representing government effectiveness, and sourced from the official European data portal [71]. The present analysis utilizes annual data for EU economies spanning the period from 1990 to 2022, with all variables measured on a per-capita basis. Intellectual property rights (IPR) serve as a proxy for innovation ecosystem activities and are derived from the European Commission database [71]. Market size (MRK) is represented by the market-sector volume scaled using purchasing power parity (PPP) in current international U.S. dollars. Intellectual capital rights (ICR) refer to the average duration of education and learning within society. Gross domestic product (GDP), expressed in constant U.S. dollars and sourced from [71], measures economic development. To achieve the objectives of this study, annual data for all selected countries were collected from the World Bank's World Development Indicators (WDI), the European Commission database (Eurostat), and the official European data portal (Europa) for the period 1990–2022. Table 9.1 summarizes the datasets, measurement units, transformations, and data sources used in the analysis.

We shall review the results in this section. The statistical properties of each selected element, such as its maximum, minimum, mean, and standard deviation, are shown in Table 9.2. The following variables have mean values of 4.931, 1.867, 3.646, 4.408, 1.888, and 4.374, respectively. Each of the study's components has a standard deviation of 0.827, 0.069, 0.331, 0.229, 0.019, and 0.398. The descriptive qualities of the factors allow us to approach the unit root test.

For comparison purposes, this study employs fixed-effects ordinary least squares (FEOLS), dynamic ordinary least squares (DOLS), and fully modified ordinary least squares (FMOLS). The FEOLS model is complemented with Driscoll and Kraay standard errors, which are robust against common forms of cross-sectional dependence and autocorrelation up to a specified lag. According to [73], the principal challenges in estimating dynamic cointegrated panels arise from heterogeneity issues associated with variations in cross-sectional averages and cross-sectional adjustments toward the cointegrating equilibrium. To overcome these concerns,

Table 9.1 Data, units, transformation, and source

Data	Measurement unit	Transformation	Source
Fisheries production	metric tons	Natural logarithm transformation	WDI
Governance	% of confidence in governance	Natural logarithm transformation	WDI
Intellectual property rights	Trademark applications, direct nonresident	Natural logarithm transformation	Eurostat
Market size	GDP per PPP (current international $)	Natural logarithm transformation	WDI
Intellectual capital rights	The average number of years of education	Natural logarithm transformation	Eurostat
Economic growth	Constant US dollars	Natural logarithm transformation	Eurostat

Source: [72]

Table 9.2 Results of descriptive statistics

Variable	Observations	Mean	Std. dev.	Min	Max
FP	864	4.931	0.827	3.211	6.310
GVR	864	1.867	0.069	1.482	1.979
IPR	864	3.646	0.331	2.223	4.354
MRK	864	4.408	0.229	3.744	5.044
ICP	864	1.888	0.019	1.837	1.921
EG	864	4.371	0.398	3.130	5.248

Source: [72]

Pedroni's FMOLS framework incorporates individual-specific intercepts and accommodates different serial correlation properties of the error terms across panel units. Consistent with Monte Carlo simulation results, the DOLS estimator was found to be unbiased relative to both OLS and FMOLS estimators in finite samples. Consequently, [74] extended the DOLS estimator to panel data contexts. Furthermore, the DOLS estimator mitigates endogenous feedback effects through the inclusion of lead and lag differences, thereby correcting for endogeneity. Owing to the limitations of earlier estimation methods, a panel quantile regression approach was adopted to evaluate heterogeneous and distributional effects across quantiles [75].

[76] In their pioneering work, they introduced the panel quantile regression methodology. Unlike conventional least-squares regressions, which estimate the conditional mean of the dependent variable given specific values of the explanatory variables, quantile regressions estimate the conditional median or other conditional quantiles of the response variable. Quantile-based regressions are more robust in the presence of outliers. Moreover, they are particularly suitable in situations where there exists little or no relationship between the conditional means of variables [77]. Nevertheless, the present study applies the Method of Moments Quantile Regression (MMQR) with fixed effects proposed by [19]. Although traditional quantile regression is robust to outliers, it fails to account for unobserved heterogeneity across individuals within a panel. The MMQR framework allows individual-level effects to influence the entire distribution rather than merely shifting the mean, as highlighted in the works of [20, 78], and related studies. This feature enables the identification of conditional heterogeneous covariance effects among the determinants of fisheries sustainability. The MMQR estimation technique is especially relevant in panel data settings that incorporate individual effects alongside endogenous explanatory variables. In addition, the MMQR method is relatively easy to interpret because it generates non-crossing estimates of regression quantiles. For a location-scale specification, the conditional quantile function $QY(\tau|X)$ has the following form:

$$\mathrm{Y}_{\mathrm{it}} = \alpha_{\mathrm{i}+} \mathrm{X}_{\mathrm{it}}{}' \beta' + \left(\delta_{\mathrm{i}} + \mathrm{Z}_{\mathrm{it}} \gamma' \right) \mathrm{U}_{\mathrm{it}} \tag{9.1}$$

Where $P(\delta_i + Z_{it} \gamma' > 0) = 1.(\alpha, \beta', \delta, \gamma')$ are parameters that need to be evaluated, and the probability, Z is a k-vector of identifiable components of X that are differentiable transformations with element i provided by: $(\alpha_i, \delta_i), i = 1, \ldots, n$ represents the individual i fixed effects.

$$Z_l = Z_l\left(X\right), l = 1, \ldots, k \tag{9.2}$$

For each given iii, XitX_{it}Xit is independently and identically distributed and remains independent over time ttt. According to [19], UitU_{it}Uit is orthogonal to XitX_{it}Xit, independently and identically distributed across both individuals (i) (i)(i) and time (t)(t)(t), and normalized to satisfy the moment conditions, thereby eliminating the need for strict homogeneity assumptions. Equation (1) therefore implies the following conditional quantile representation:

$$Q_Y(\tau|X_{it}) = (\alpha_{i+}\ \delta_i q(\tau)) + X_{it}'\beta + Z_{it}'\gamma q(\tau) \tag{9.3}$$

In the present analysis, the independent variables include the natural logarithm of economic growth (LGDP), intellectual property rights (LIPR), intellectual capital rights (LICR), market size (LMRK), and governance (LGVR). The scalar coefficient Xit − αi(τ) = αi + δiq(τ)X_{it}-\alpha_i(\tau) = \alpha_i + \delta_i q(\tau) Xit − αi(τ) = αi + δiq(τ) represents the quantile-fixed effect for individual iii, while QY(τ|Xit)Q_Y(\tau|X_{it})QY(τ|Xit) denotes the quantile distribution of the dependent variable YitY_{it}Yit, namely the natural logarithm of fisheries productivity (FP), conditional upon the position of the explanatory variable XitX_{it}Xit. Unlike conventional fixed-effects least-squares models, the individual effect here does not merely reflect an intercept shift. Rather, the heterogeneous effects may vary across the quantiles of the conditional distribution of the endogenous variable YYY. These effects are treated as time-invariant parameters. The τ\tauτ-th sample quantile, denoted as q(τ)q(\tau) q(τ), is obtained by solving the following optimization problem:

$$\min_q \sum_i \sum_t P_\tau \left(R_{it} - \left(\delta_i + Z_{it}'\gamma \right) q \right) \tag{9.4}$$

Where $P_\tau(A) = (\tau - 1)AI\{A \leq 0\} + TAI\{A > 0\}$ denotes the check function.

9.4 The Outcomes of Fisheries IPR Application

Prior to estimating the unknown parameters, several preliminary tests were conducted to identify the time-series properties of the variables. First, the study examined whether cross-sectional dependence (CD) existed within the panel structure. Cross-sectional dependence may distort the true parameter values of coefficient estimates. If ignored, dependence arising from common unobserved factors can substantially diminish the efficiency gains associated with panel data estimation (Phillips & Sul, 2003). Therefore, accounting for this issue is essential to obtain reliable coefficient estimates. The [79] Pesaran's (2004) CD test was employed to assess cross-sectional dependence within the panel framework. The results presented in Table 9.3 indicate the presence of multiple forms of cross-sectional dependence across countries for all quantitative variables, with the exception of fisheries productivity. Consequently, to minimize potential size distortions, the unit root tests, cointegration tests, and panel estimation procedures applied in this

Table 9.3 Cross-sectional dependence results

Variables	FP	GVR	IPR	MRK	ICR	GDP
Pesaran (2004)	(18.82)***	(31.30)***	(43.28)***	(98.67)***	(99.27)***	(83.55)***

- Remark: *** refer importance at the 1%, scale
Source: [72]

study must adopt methodologies that are robust to cross-sectional dependence effects

To evaluate the integration properties of the variables under investigation in an unbiased manner, this study employs the [80] (LLC) and [79] (IPS) panel unit root tests. The primary objective of panel unit root testing is to enhance test efficiency through the combined use of cross-sectional and time-series observations. A variety of panel unit root procedures have been developed to test the null hypothesis of a unit root against alternative hypotheses. Levin, Lin, and Chu assess the null hypothesis of a unit root against a homogeneous alternative hypothesis in which every series within the panel is stationary and shares a common rate of mean reversion. In contrast, Im, Pesaran, and Shin test the null hypothesis of a unit root against the alternative that at least one series in the panel is stationary.

The testing framework adopted in this study focuses on the stationarity properties of the panel, thereby drawing attention to the LLC structure. However, the IPS (2003) framework relaxes this assumption by allowing each cross-sectional unit to possess its own autoregressive parameter. For relatively small panel datasets, such as the dataset utilized in the present study, the (LLC) panel unit root test has been shown to perform better than comparable unit root testing methods in terms of statistical power. Accordingly, both tests are employed to examine the influence of cross-sectional dependence on panel unit root testing outcomes.

The results reported in Table 9.4 indicate that all variables included in the unit root testing procedure are non-stationary at their level forms. However, after taking first differences, all variables become stationary. These findings confirm that all variables incorporated in the estimation process are integrated of order one, denoted as I(1).

To examine the existence of a long-run equilibrium relationship among the variables, this study applies the [73] panel cointegration test and the [81] bootstrapped panel cointegration test, as reported in Table 9.5. Following the framework of the Engle and Granger two-step procedure, [73] Developed a comprehensive methodology for panel cointegration analysis. Pedroni's approach accounts for heterogeneity by eliminating short-run disturbances and individual-specific deterministic trends during the first stage of estimation. Based on the estimated residuals, Pedroni constructs seven alternative test statistics. These statistics are categorized either as tests that assume a common process across the panel, generally termed "pooled" or "within-dimension" tests, or as tests that allow for individual processes, commonly referred to as "grouped" or "between-dimension" tests, as presented in Table 9.5.

In addition, the [81] approach introduces four new cointegration tests under the null hypothesis of no cointegration. Unlike residual-based methods, the Westerlund

Table 9.4 Panel unit root test results

	Difference		First difference	
Variable	LLC	IPS	LLC	IPS
FP	3.067	−3.700	−24.13***	−22.737***
GVR	−1.090	−0.254	−23.47***	−23.47***
IPR	−1.209	−3.942	−7.818***	−11.92***
MRK	−3.775	−3.465	−6.894***	−10.42***
ICR	−2.189	−5.018	−11.91***	−14.91***
GDP	−1.571	−0.210	−18.74***	−68.36***

– Remark: *** refer importance at the 1%, scale
– Levin, Lin & Chu test (LLC), and Im, Pesaran, and Shin W-stat test (IPS)
Source: [72]

Table 9.5 Panel cointegration test for EU countries

Test	Without trend	With trend		
A) Pedroni residual cointegration test:				
Alternative hypothesis: common AR coefficients (within dimension):				
Panel v-Statistic	−1.711(0.956)	−2.356(0.990)		
Panel rho-Statistic	2.600(0.995)	0.565(0.714)		
Panel PP-Statistic	−1.621*(0.052)	−7.871***(0.000)		
Panel ADF-Statistic	−2.011**(0.022)	−8.361***(0.000)		
Alternative hypothesis: common AR coefficients (between dimensions)				
Group rho-Statistic	−3.161	(0.999)		
Group PP-Statistic	−7.829***	(0.000)		
Group ADF-Statistic	−8.284***	(0.002)		
B) Westerlund (2007) bootstrap panel cointegration:				
Statistics	Value	Z-Value	p-value	Robust p-value
Gt	−2.409	−1.095	0.095	0.081*
Ga	−7.225	3.100	0.999	0.985
Pt	−11.677	−1.408	0.080	0.074*
Pa	−5.673	1.569	0.942	0.999

– Remark: ***, **, and * refer importance at the 1%, 5%, and 10% scales, respectively
– Source: [72].

procedure is founded on structural dynamics rather than residual dynamics, thereby avoiding the restrictive common factor assumptions imposed by residual-based cointegration tests. According to [82] violations of common factor restrictions may substantially reduce the effectiveness of residual-based cointegration analyses, which highlights the importance of adopting a structural dynamics framework.

With the relaxation of this restriction, the long-run and short-run adjustment mechanisms are no longer required to be identical. The bootstrap approach proposed by [81] Enables the reduction of distortionary effects arising from cross-sectional dependence while also providing robust critical values. The results presented in Table 9.5 indicate that both the [73] panel cointegration test and the

[81] bootstrapped cointegration test strongly support the existence of cointegration among the variables.

Table 9.6 reports the estimation results obtained from the FMOLS, DOLS, and FEOLS procedures. Although differences exist in the levels of statistical significance, Model 1 in Table 9.6 demonstrates that the coefficient estimates generated from the three estimation approaches are, on average, relatively comparable. In terms of both statistical significance and coefficient magnitude, fisheries productivity within the EU region appears to be the most stable variable across all three estimation specifications.

An increase of 1% in governance contributes to an increase in fisheries sector sustainability by approximately 0.19% under the FMOLS estimator and 0.88% under the DOLS estimator, while the FEOLS specification shows only marginal statistical significance. This finding is consistent with the results reported by [45, 46, 83]. Owing to its position as the world's largest seafood market, its dependence on imports for more than half of its seafood demand, and its role as a major fisheries power, the European Union has become a key actor in the governance of global fisheries.

Furthermore, intellectual property rights within the EU region exhibit a positive and statistically significant relationship with fisheries industry sustainability, as theoretically expected. Nevertheless, the coefficient estimates differ somewhat across the DOLS specifications. Specifically, a 1% increase in intellectual property rights enhances fisheries industry sustainability by approximately 0.03% and 0.04% under the FEOLS and FMOLS estimators, respectively. These findings are aligned with earlier studies conducted by [23–25]. Fisheries and marine-related intellectual property rights are fundamental for understanding marine ecosystems, advancing the future blue economy, and supporting Europe's ambition of becoming the first climate-neutral continent by 2050. Market size also exerts a substantial influence on fisheries sustainability within the EU region. Under the FEOLS, FMOLS, and DOLS estimators, a 1% increase in market size leads to increases of approximately 0.47%, 0.08%, and 1.41%, respectively, in fisheries sustainability. These results support the growth hypothesis for fisheries-producing nations. The findings are consistent with the conclusions of [38, 84], who reported that market size has a positive and relatively inelastic effect on fisheries industry development. The growth of the

Table 9.6 Model 1. Panel estimation for the EU-27 region from 1990 to 2022

Long-run coefficient	DOLS		FMOLS		FEOLS	
	Coefficient	Std. error	Coefficient	Std. error	Coefficient	Std. error
GVR	0.887	(0.498)*	0.192	(0.016)***	0.032	(0.184)
IPR	0.099	(0.084)	0.042	(0.011)***	0.039	(0.013)***
MRK	1.417	(0.377)***	0.088	(0.057)***	0.470	(0.121)***
ICR	9.816	(4.630)***	1.792	(0.089)***	7.318	(1.55)***
GDP	0.037	(0.219)	0.065	(0.016)***	0.234	(0.053)***

– Note: ***, **, and * indicate significance at the 1%, 5%, and 10% levels, respectively
– Values in parentheses are P-values. Source: Alsaleh et al. (2024)

European fisheries industry can largely be attributed to rising consumer expenditure on fish and shellfish products despite their relatively higher prices.

Similarly, intellectual capital rights are shown to have a positive and statistically significant effect across all estimation specifications. According to the FEOLS, FMOLS, and DOLS estimators, a 1% increase in intellectual capital rights development contributes approximately 7.31%, 1.79%, and 9.8%, respectively, to fisheries sustainability. Although this outcome contrasts with the modernization hypothesis advanced by [33, 34] it remains theoretically consistent with the broader growth hypothesis. The FEOLS, FMOLS, and DOLS models all provide strong statistical evidence in support of this hypothesis. Intellectual property rights reflect the efficiency and quality of labor contributions required in fisheries harvesting activities, whereas intangible capital refers to investments in non-physical assets, including fisheries ownership rights necessary for participation in rights-regulated fisheries systems.

Economic growth also exerts a significantly positive effect on fisheries sustainability within the EU region. More precisely, a 1% increase in economic growth improves fisheries sustainability by approximately 0.23% and 0.06% under the FEOLS and FMOLS estimators, respectively, although the coefficient appears less stable across DOLS specifications. These findings further support the growth hypothesis for fisheries-producing economies. The results are consistent with the major findings of [56–58], who documented a positive and relatively inelastic relationship between economic expansion and fisheries sector development. The fisheries industry in the EU region may therefore achieve sustainable long-term growth by establishing favorable conditions for a competitive and economically viable market environment.

The panel quantile regression estimates show that, except for the medium quantiles, government spending has a statistically significant impact on the fisheries industry (Table 9.7). At the medium quantiles (fourth and fifth quantiles), however, the influence of fisheries output is minimal. This result validates the effectiveness of marine laws pertinent to the fishing industry in higher marine-producing countries. The results show that the growth theory only applies to fisheries output from the earliest and most recent periods in the EU27. This implies that in countries with median incomes, development comes before maritime policy. Additionally, countries in the median quantiles may require assistance in switching their fishing industries from non-recycled to recycled supplies due to the related investment expenditures and reduced output levels. In countries with the highest and lowest production levels, governance favors the fishing industry. This could be brought on by the various extraction methods used in the developed EU14 nations and the expanding EU13 economy. These results indicate important implications for governance, according to [45, 46, 83], who evaluated the growth hypothesis for several countries that generate extraordinarily high levels of fisheries output. Furthermore, our findings are consistent with [1], recent empirical research, which backed the growth theory—another important source of fishing resources. The results also support the growth theory for the primary fisheries resource producer, as shown by [2]. Additionally, the effects of intellectual property rights on fisheries productivity are

Table 9.7 Panel quantile estimation results for EU27 countries

Variables	Method of moments quantile regression with fixed effects										
	Location	Scale	Quantiles								
			0.1	0.2	0.3	0.4	0.5	0.6	0.7	0.8	0.9
GVR	(0.184)	(0.520)	(1.734)***	(1.083)***	(0.546)	(0.060)	(0.498)*	(0.937)***	(1.206)***	(1.521)***	(1.721)***
IPR	(0.013)***	(0.549)***	(0.310)**	(1.006)***	(1.160)***	(1.128)***	(1.061)***	(0.952)***	(0.753)***	(0.563)***	(0.313)**
MRK	(0.121)***	(0.455)***	(0.081)	(0.041)	(0.166)***	(0.185)**	(0.354)***	(0.287)***	(0.219)***	(0.177)***	(0.487)
ICR	(1.55)***	(6.266)***	(0.075)	(0.060)	(0.035)***	(0.028)***	(0.038)***	(0.025)***	(0.030)***	(0.030)***	(0.280)*
GDP	(0.053)***	(0.198)	(0.862)***	(0.827)***	(0.886)***	(0.733)***	(0.611)***	(0.477)***	(0.271)***	(0.063)	(0.048)

Notes: *,**, and *** indicate significance at 10%, 5%, and 1% levels, respectively
Source: [72]

significantly and favorably correlated across all quantiles. The impact of intellectual property rights increases as quantiles rise. It is therefore in line with the conclusions of [23–25], who found that the fishing industry is encouraged to pursue more sustainable growth when intellectual property rights are protected. Furthermore, fisheries production in the medium and above-medium quantiles is strongly and favorably correlated with market-size outcomes. However, the impact of market size is increasing as quantiles go smaller. This corroborates the findings of [38, 84, 85], which demonstrated that market efficiency propels the fishing industry to achieve sustained competitiveness. Furthermore, the results of intellectual property rights and fisheries productivity have a substantial and positive link in the middle and most recent quantiles. In a similar vein, the influence of intellectual property rights is increasing at lower quantiles. This corroborates the conclusions drawn by [33, 34, 61] that the creation of intellectual property rights motivates the fishing industry to achieve sustainable objectives. Furthermore, the effect of GDP growth on fisheries output is significant in the middle and most recent quantiles but insignificant in the earlier quantiles. The effectiveness of fisheries policy in fostering economic growth in higher-producing countries is supported by this conclusion. According to [56–58], the fishing industry is motivated to pursue sustainable expansion by the effectiveness of economic performance.

The results of the FMOLS, DOLS, and FEOLS estimation procedures are shown in Table 9.8. Despite the variations in the degree of statistical significance between the coefficient estimates from the three specifications, Table 9.8 from Model 2 demonstrates that they are generally comparable. The EU14 developed countries have strong governance in terms of both statistical significance and coefficient size. For both the FMOLS and FEOLS estimators, an increase in governance has a positive percentage impact on fisheries output of 0.58% and 0.38%, respectively. The coefficients may be more uniform across DOLS specifications, though. This finding is consistent with the primary conclusions of [2–4]. The EU's implementation of Sustainable Development Goal 14 on life below water and the UN's 2030 Agenda for Sustainable Development depend on EU communication. The EU's strong commitment to the ocean is demonstrated by the availability of accessible translations of the preceding land. The European Green Deal's blue component is implemented in large part thanks to the updated agenda. All criteria indicate that there is a

Table 9.8 Model 2. Panel estimation for the EU-14 region from 1990 to 2022

Long-run coefficient	DOLS		FMOLS		FEOLS	
	Coefficient	Std. error	Coefficient	Std. error	Coefficient	Std. error
GVR	0.504	(0.365)	0.598	(0.236)**	0.388	(0.156)**
IPR	0.056	(0.021)	0.026	(0.013)	0.016	(0.018)
MRK	0. 607	(0.227)***	0.431	(0.137)***	0.391	(0.088)***
ICR	11.616	(3.238)***	10.301	(1.858)***	9.348	(1.251)***
GDP	1.039	(0.322)***	0.136	(0.080)*	0.091	(0.053)*

– Note: ***, **, and * indicate significance at the 1%, 5%, and 10% levels, respectively
– Values in parentheses are P-values. Source: Alsaleh et al., (2024)

statistically significant positive correlation between market size and the fishing sector in the developed EU14 countries. Fisheries production increases by 0.39%, 0.43%, and 0.60% for every percentage increase in market size, according to the FEOLS, FMOLS, and DOLS estimators. This study confirms previous research by [39–41]. Direct monitoring of the quantities, values, and prices of fishery goods from the point of first sale to the point of retail, including imports and exports, may lead to increased market intelligence, transparency, and efficiency of the sustainability of the fishing sector. With increases in fisheries productivity of 9.34%, 10.30%, and 11.61% for each percentage increase in intellectual capital rights for the FEOLS, FMOLS, and DOLS estimators, respectively, it is clear that intellectual capital rights are important and have a positive effect in the developed EU14 countries. This finding contradicts Seisay (2022), yet it is consistent with modernization hypotheses proposed by [86]. For every parameter, the growth hypothesis is very statistically significant. Fish population numbers and generation-to-generation evolution are influenced by human fisheries practices. Fisheries that primarily target large fish put lesser species under evolutionary pressure to survive, according to studies. Overfishing can eventually lead to issues with the food supply. Furthermore, the level of development of the EU14's fishing industry is positively correlated with economic growth. A percentage increase in economic growth increases by 0.09%, 0.13%, and 1.03% for the FEOLS, FMOLS, and DOLS estimators, respectively. The growth theory is endorsed by the panel of developed EU14 countries. In contrast, [35] found that the security of fisheries was negatively and inelastically impacted by economic expansion. Additionally, it is consistent with the main findings of [59–61]. Fisheries production affects other economic activities, such as shipping, fishing, aquaculture, and recreation. For example, replacing boats and other equipment damaged by abandoned fishing gear is a significant yearly expense for the fishing industry.

The panel quantile regression estimates in Table 9.9 show that the effects of governance on fisheries sustainability are statistically significant from the aforementioned medium and latest quantiles. This result validates the effectiveness of maritime policy related to fisheries production in the developed EU14 countries. The results show that the growth theory is supported by the above-medium and most current quantile fishing industries in the developed EU14 countries. This implies that among the 14 developed countries of the EU, maritime policy comes before fisheries output. Due to their level of fisheries output, EU14 developed countries at earlier quantiles may find it more challenging to transition from conventional to recycled sources due to the associated investment requirements. In the 14 EU countries with the highest level of fisheries production, the government has a less positive effect on the fishing industry. The different extraction methods utilized in the fishing sectors of the EU14 developed countries could be the reason for this. These results have important policy implications and are consistent with [3] findings, which supported the growth hypothesis for several industrialized countries and produced extremely high fisheries productivity. Our findings also corroborate a recent empirical study by [4], which provided evidence in favor of the growth hypothesis—another important source of fishery resources. The results are consistent with

Table 9.9 Panel quantile estimation results for EU14 countries

Variables	Method of moments quantile regression with fixed effects										
	Location	Scale	Quantiles								
			0.1	0.2	0.3	0.4	0.5	0.6	0.7	0.8	0.9
GVR	(0.156)**	(0.143)*	(0.774)	(0.434)	(0.320)	(0.319)	(0.404)	(1.135)***	(1.627)***	(1.861)***	(2.131)***
IPR	(0.018)	(0.301)*	(0.011)**	(0.017)	(0.014)**	(0.015)**	(0.016)***	(0.015)***	(0.020)**	(0.025)	(0.035)
MRK	(0.088)***	(0.199)***	(0.653)*	(0.021)	(0.112)**	(0.094)**	(0.229)***	(0.263)***	(0.195)***	(0.183)***	(0.213)***
ICR	(1.251)***	(0.303)	(3.170)	(0.703)	(1.371)*	(0.125)	(1.054)***	(1.386)***	(1.438)***	(1.322)***	(1.573)***
GDP	(0.053)*	(0.189)***	(2.599)*	(0.213)	(0.406)*	(0.525)	(0.424)***	(0.523)***	(0.845)**	(0.711)**	(0.777)***

Notes: *, **, and *** indicate significance at 10%, 5%, and 1% levels, respectively
Source: [72]

those of [2], who demonstrated that the growth hypothesis for the primary fisheries resource producer was valid. Additionally, in the medium and above-medium quantiles, intellectual property rights have a strong and positive correlation with the fisheries industry's outcomes, although their impact is minimal at earlier and later quantiles. In a similar vein, the impact of intellectual property rights grows with greater quantiles. While the impact of market size is minimal at the second quantile, the outcomes of market size are likewise strongly and favorably connected with the fisheries industry throughout all quantiles. However, the market size effect is getting smaller at larger quantiles. Furthermore, in the medium and most recent quantiles, there is a strong and positive correlation between the results of intellectual property rights and the fishing sector. On the other hand, in earlier quantiles, the effect of human capital development is minimal. Except for the second quantile, there was a substantial and positive correlation between market size and the fishing industry across all quantiles; however, the influence decreased with increasing quantiles. The majority of quantiles indicate that GDP-based fisheries productivity is significantly impacted by economic growth. The consequences of economic expansion are negligible in the second and fourth quantiles. The effectiveness of fisheries policy in fostering economic growth in higher-producing countries is supported by this conclusion. This suggests that in the EU14 developed nations, economic growth takes precedence over the sustainability of marine resources.

Table 9.10 lists the results of the different panel estimate methods utilized in Model 3, including FMOLS, DOLS, and FEOLS. These methods yield nearly identical coefficient values. The findings show that the productivity of fisheries in the EU13 developing countries is positively and significantly impacted by governance. In the case of DOLS, fisheries productivity rises by 0.85% for every 1% increase in the enabling environment. The growth hypothesis in the EU13 developing countries is confirmed by the positive and significant sign of the coefficients of governance in the DOLS approach. The coefficient appears to be contradictory across the FMOLS and FEOLS specifications, though. This outcome is in line with the findings of [51, 52, 87], who assert that fisheries productivity increases as governance and ocean public policy increase from the start of development. The production of intellectual property rights is intimately linked to fisheries productivity in the FEOLS, FMOLS, and DOLS techniques. For every 1% increase in IP rights, fisheries output rises by

Table 9.10 Model 3. Panel estimation for the EU-13 region from 1990 to 2022

Long-run	DOLS		FMOLS		FEOLS	
coefficient	Coefficient	Std. error	Coefficient	Std. error	Coefficient	Std. error
GVR	0.855	(0.766)**	0.330	(0.472)	0.097	(0.327)
IPR	0.376	(0.166)**	0.083	(0.050)*	0.078	(0.026)***
MRK	0.692	(0.436)***	0.527	(0.304)*	0.429	(0.195)**
ICR	0.772	(0.958)*	6.776	(4.036)*	6.209	(2.708)**
GDP	0.550	(0.701)**	0.300	(0.139)**	0.272	(0.081)***

– Note: ***, **, and * indicate significance at the 1%, 5%, and 10% levels, respectively
– Values in parentheses are P-values. Source: Alsaleh et al., (2024)

0.07% for FEOLS, 0.08% for FMOLS, and 0.37% for DOLS. This outcome supports claims made by [28–31] that the productivity of fisheries in developing countries rises when intellectual property rights are protected. A negative relationship between market size and fisheries output in EU13 developing countries is explained by the empirical data. For FEOLS, FMOLS, and DOLS, a 1% increase in market size results in increases in fisheries output of 0.42%, 0.52%, and 0.69%, respectively [42, 43, 88]. for EU developing countries confirmed this finding, which implies that market efficiency increases fisheries productivity. Similarly, the empirical findings clarify a negative relationship between fisheries productivity in EU13 developing countries and intellectual capital rights. If intellectual capital rights increase by 1%, fisheries output increases by 6.20%, 6.77%, and 0.77% for FEOLS, FMOLS, and DOLS, respectively. According to [36, 37] for developing countries, and Nielsen et al. (2022) for EU developing countries, intellectual capital rights increase fisheries productivity. The conclusion shows that the sustainability of fisheries in EU13 developing countries is favorably and significantly impacted by economic expansion. Fisheries output increases by 0.27%, 0.30%, and 0.55% for FEOLS, FMOLS, and DOLS, respectively, for every 1% increase in GDP. The existence of the growth hypothesis in EU13 rising nations is demonstrated by the positive and significant sign of the GDP coefficients in all three techniques. For developing countries, this finding is consistent with [63, 64, 89, 90]. Table 9.11 displays the results of the Movement Quantile regression method. First, it is possible to show that governance has a positive effect on the sustainability of fisheries for medium and later quantiles, but not for below-medium quantiles. The table makes it abundantly evident that there is significant volatility in fisheries output as a result of governance, rising from 0.806 to 0.380 as the quantile rises (at first and from fifth to third). However, the effect of governance is statistically insignificant in quantiles from second to fourth. Furthermore, there is a significant and positive association between the outcomes of intellectual property rights and the sustainability of fisheries across all quantiles, with the exception of the medium quantiles (the fifth, sixth, and seventh quantiles). In a similar vein, the influence of intellectual property rights increases with larger quantiles. Furthermore, there is a robust and positive association between market-size outcomes and fisheries sustainability across all quantiles, with the exception of the most recent (eighth and ninth). On the other hand, for the EU13 emerging countries, the effect of market size on higher quantiles is decreasing. In a similar vein, intellectual property rights contribute positively to fisheries sustainability across all quantiles; this relationship is statistically significant between quantiles one and nine. As in previous cases, higher quantiles result in a reduction in the impact of inputs related to intellectual property rights. Stated differently, the sustainability of fisheries is positively impacted by economic growth. The table demonstrates how economic development has significantly improved fisheries sustainability, rising from 0.483 to 0.513 as the quantile rises (from fifth to ninth). Both the newest and medium quantiles exhibit this.

The estimates from the panel FEOLS have been confirmed by Panel FMOLS and DOLS. It is evident that the panel FMOLS and DOLS coefficients share the same sign and level of significance as the panel FEOLS coefficients. This proves

Table 9.11 Panel quantile estimation results for EU13 countries

Variables	Method of moments quantile regression with fixed effects										
	Location	Scale	Quantiles								
			0.1	0.2	0.3	0.4	0.5	0.6	0.7	0.8	0.9
GVR	(0.327)	(0.252)	(0.806)**	(0.221)	(0.156)	(0.166)	(0.904)***	(0.973)***	(1.204)***	(1.501)***	(1.380)***
IPR	(0.026)***	(0.435)**	(0.675)***	(0.614)***	(0.583)***	(0.525)***	(0.210)	(0.129)	(0.027)	(0.566)***	(0.885)***
MRK	(0.195)**	(0.903)**	(2.060)***	(2.440)***	(1.844)***	(1.792)***	(1.611)***	(1.488)***	(0.979)***	(0.237)	(0.477)
ICR	(2.708)**	(0.170)***	(0.426)***	(0.443)***	(0.399)***	(0.424)***	(0.400)***	(0.367)***	(0.287)***	(0.125)***	(0.104)***
GDP	(0.081)***	(0.120)	(0.249)	(0.251)	(0.275)	(0.638)	(0.483)**	(0.429)**	(0.398)**	(0.334)*	(0.513)*

Notes: *, **, and *** indicate significance at 10%, 5%, and 1% levels, respectively
Source: [72]

that panel FMOLS and DOLS data are trustworthy and useful for drawing conclusions. Although their significance levels differ slightly, the coefficients from FEOLS have the same sign as those from the panel FMOLS and DOLS. However, panel DOLS estimates may be considered dependable and free of endogeneity and serial correlation problems. In order to examine the impact of property rights on the sustainability of fisheries based on economic development levels, the EU 27 countries were divided into two groups: 14 developed countries and 13 developing countries. Tables 9.8 and 9.9 present the anticipated impact of property rights on the sustainability of fisheries in the 14 developed EU member states. Tables 9.10 and 9.11 illustrate the estimated impact of property rights on the sustainability of fisheries in the EU14 developing nations. The results of Tables 9.8 and 9.10 demonstrate that property rights significantly improve fisheries sustainability. The results also imply that governance significantly improves the sustainability of fisheries in developed EU14 countries but not in less developed EU13 countries. EU14 and EU13 countries had impact magnitudes of 0.097 and 0.388, respectively. Fisheries sustainability may be improved more in EU14 developed countries than in EU13 undeveloped countries if ocean governance and marine ecological taxation are implemented. However, the results also indicate that intellectual property rights significantly improve fisheries sustainability in EU13 developing countries but not in EU14 developed countries. The impact magnitudes are 0.078 and 0.016 for EU13 and EU14 countries, respectively. This suggests that EU13 poor countries may considerably increase fisheries sustainability more than EU14 developed countries by utilizing green technology innovation and scientific patent and trademark approaches on intellectual property rights in the fisheries sector. From an alternative perspective, the data also indicate that the EU13 developing countries benefit more from market size in terms of fisheries sustainability than do the EU14 developed countries. EU13 and EU14 countries had impact magnitudes of 0.429 and 0.391, respectively. This demonstrates that, in comparison to developed countries in the EU14, developing countries in the EU13 may achieve a significant boost in fisheries sustainability by using a market-size approach. The current changes to tax legislation in the wealthier EU14 countries may lead to tax exemptions for seafood consumption and demand. The results, however, also indicate that the EU14 rich countries benefit more from intellectual capital rights input on the sustainability of fisheries than do the EU13 developing countries. The EU14 and EU13 countries have impact magnitudes of 9.348 and 6.209, respectively. This demonstrates that by utilizing specialized institutions and educated talents for intellectual capital rights in the fisheries industry, wealthy EU14 countries may considerably boost fisheries sustainability more easily than undeveloped EU13 countries. The results, however, also demonstrate that the significant and positive influence of economic development on fisheries sustainability is greater in the developing countries of the EU13 than in the developed countries of the EU14. EU13 and EU14 countries had impact magnitudes of 0.272 and 0.091, respectively.

This demonstrates how emerging EU13 countries may significantly improve fisheries sustainability in comparison to developed EU14 countries by investing in fisheries infrastructure and fleets.

9.5 Conclusion and Solutions of IPR in Fisheries Sector

As a contribution to the literature on marine and ocean resources on blue growth in the context of fisheries and sustainable development, this study investigates the empirical relationships between fisheries productivity, intellectual property rights, intellectual capital rights, governance, market size, and economic growth while taking the growth hypothesis into consideration. The researchers used panel unit roots testing, panel cointegration, and panel estimation techniques using data collected over 31 years across 27 EU member states. The quantitative variable governance (% of confidence for governance) was used in the model to account for fisheries governance and assess the level of sustainable development. Using panel unit root testing and panel cointegration approaches, it was demonstrated that all the variables under consideration followed an I(1) process and had a non-spurious long-run link. Conventional panel cointegration estimation approaches revealed variances in coefficient significance, even if the sizes produced from the various requirements are not distinct. We employed the MMQR approach, which allows the assessment of the different effects of the exogenous determinants over many quantiles of the conditional distribution of marine rubbish, for a more comprehensive analysis of the empirical connection. The FMOLS, DOLS, and FEOLS coefficient estimates validated the growth assumptions for the economies of the EU27 member states that engage in fishing. Governance is positively significant throughout all quantiles of the conditional distribution of fisheries sustainability, from the first to the ninth quantiles, but it is most noticeable at the third and fourth medium quantiles, according to Table 9.7's MMQR results. All quantiles of the conditional distribution of fisheries output show a positive correlation between intellectual property rights and fisheries sustainability; however, this relationship is only statistically significant between the first and ninth quantiles. This may indicate that both the early and advanced phases of the transitional growth of the fisheries sector in the EU27 countries benefit from intellectual property rights. As market size increased, fisheries sustainability statistically increased from the third to the eighth quantiles. It is crucial to stress that industrialized countries that depend on fish imports are less economically diversified than emerging nations that produce more fish. This implies that a much larger proportion of the marine resources in rising nations may come from their fishing businesses. A visual analysis of Table 9.7 shows that the MMQR estimate of economic growth is significantly positive from the first to seventh quantiles. The economic growth estimate's coefficient was greatest at the lowest quantile. From the lowest to the highest quantile, the economic growth coefficient decreases. This implies that fisheries sustainability is highest in the quantile where economic expansion has the least impact on fisheries productivity. Similarly, in quantiles where fisheries sustainability was highest, the effects of intellectual property rights on fisheries productivity were most pronounced. This indicates that countries whose fishing industries rely less on traditional methods are creating greater intellectual property rights. Dependency in this case does not necessarily imply plenty. This offers additional evidence of the deindustrialization-induced resource curse. The insignificance of the positive effects of governance in comparison to the preceding quantiles in Tables 9.9 and 9.11 may be related to the present government fisheries initiatives designed to prevent the further rise of overfishing and develop fisheries practices. Residents of the EU14 developed countries prioritize fisheries policy and regulation, in contrast to the lower productivity of the EU13 underdeveloped countries. Therefore, an effective administration that conforms with popular desire would enact additional regulations against fisheries and improper actions to secure the support of the electorate. The importance of intellectual property rights at medium fisheries productivity quantiles (from fourth to sixth quantiles in Tables 9.9 and 9.11) stems from the potential that policies that restrict intellectual property rights could be advantageous to growth and that the large-scale fisheries productivity in the EU13 developing nations may also present numerous development opportunities. At the greatest quantiles in EU14 nations and middle quantiles in EU13 countries, where intellectual property rights have a negligible beneficial impact on fisheries sustainability, environmental restrictions may cause citizens of these EU countries excessive anxiety. This is due to the immediate impacts of environmental degradation.

The electorate and elected officials in Tables 9.9 and 9.11 may place greater importance on the activities of the fishing industry, which may rely on economic growth and sustainable development. The influence of economic growth may be observed at specific lower quantiles in EU13 developing and EU14 developed countries due to the predominance of other competitiveness drivers with a more severe resource curse presence. The better sustainable fisheries development of the EU14 and EU13 countries points to the notable and significant influence of economic expansion on higher quantiles. The experiences of Mexico, Ecuador [6], Malaysia [7], China [22, 91] demonstrate that growing economic growth will boost fisheries sustainability. As a result, we support improving economic performance in fish-producing countries, especially in the developed EU14 countries with substantial fisheries output. Tables 9.9 and 9.11 show a positive and significant relationship between intellectual capital rights and fisheries sustainability across all quantiles. However, in the industrialized EU14 countries, it is only significant at lower quantiles (from the first to the fourth quantile). The growth notion in the corpus of earlier research is supported by this finding. In the context of EU13 and EU14 fisheries-producing nations, the data may offer multiple explanations for different quantiles. The influence of weak intellectual property rights on fisheries productivity could be one reason for decreased quantiles in the EU14. This situation arises when all other fishing methods based on tradeable economic sectors are replaced by ineffective intellectual property rights. They thus become more dependent on imports, which means that fish imports and exports would account for a sizable amount of their economic growth. In an attempt to increase profit levels, the labor-intensive export-oriented fisheries sector may migrate from EU13 developing economies to EU14 developed economies. This could tip the dynamics of economic growth in favor of imports, which would consequently lower the spate of fisheries productivity in EU13 countries at higher quantiles in EU13 and EU14. The growth hypothesis in the developed EU14 countries may also be directly impacted by the stringent fisheries policy on businesses that depend on marine resources and are export oriented. This situation might compel the fishing industry to cut back on both marine waste and output. Due to the cost of hiring qualified professionals and innovative technology in the EU13 emerging countries, many corporations may move their operations to less developed countries with less burdensome taxes, environmental regulations, and marine resources to make more money. Market size has a positive and significant impact on all medium quantiles of fisheries output (from the third to the seventh quantiles) in Tables 9.9 and 9.11. However, it was shown that the impact of market size on fisheries sustainability is greatest at the lowest quantile. The domestic fisheries market, the international fisheries market, and the fisheries trade can all demonstrate an improvement in market tools and determinants from the lowest to the greatest quantile. It also implies the result. The fact that property rights and the fishing industry still have a very positive association may suggest that conventional practices are still employed in the fisheries of EU14 developed countries and EU13 developing countries. This finding implies that in order to increase the level of sustainable development in the fisheries sector in EU14 and EU13 countries, more substantial efforts need be undertaken to transition fishing operations from conventional techniques to sustainable friendly practices. To put it another way, it is crucial to increase fish yield from sustainable and regenerative sources. Even better, the EU13 underdeveloped countries might implement small environmental tax rules for the fishing industry if they lack the resources to embrace sustainable and renewable sources. The fishing industry, which depends on marine waste, is subject to light taxes under these laws. Long-term infrastructure development for sustainable fisheries should be funded with the proceeds from these levies. Lastly, additional elements that affect the sustainability of the EU region's fishing industry and marine environment, such as tourism, globalization, and financial expansion, might be added to this research to make it more thorough. Additionally, utilizing state-of-the-art econometric approaches, this research can be conducted at the level of a country, like the Republic of China, or in other geographic areas, like ASEAN, BRICS, the Group of Seven, or the Next 11 nations. There are definitely limitations to this type of modeling, and this study is no different. Further research is necessary because the modeling was unable to assess the potential benefits of fishery property rights due to data limitations. Furthermore, given the scope of the current study, the model might be investigated in the contexts of other countries to gain a comprehensive contextual understanding. Despite using contemporary econometric techniques to show these associations, the study had a number of limitations. For example, the current findings only encompassed a limited set of characteristics from 1990 to 2022 and were only assessed using econometrics. Future studies may look at how public–private investments in the fishing industry and ocean energy affect fisheries production using the current dataset and additional parameter variables. Second, the current framework was limited to EU countries; this might be resolved by incorporating a number of other countries with significant emissions. Thirdly, researchers might analyze the economic complexity of fishery property rights over a number of time periods to account for structural gaps in the data.

References

1. Yusuf M (2022) The competitiveness of Indonesian shrimp export in Malaysia and Singapore markets. Available: http://repository.unimus.ac.id/5474/1/6.%20THE%20COMPETITIVENESS%20OF%20INDONESIAN%20SHRIMP%20EXPORT%20IN%20MALAYSIA%20AND%20SINGAPORE%20MARKETS.pdf. Accessed May 20, 2026
2. Handaka AA, Lestari RP, Rizal A (2022) Analysis of competitiveness fisheries processing industry in West Java, Indonesia. AJFAR:8–19. https://doi.org/10.9734/ajfar/2022/v16i330372
3. Suryana AAH, Amalia DH (2021) Analysis of competitiveness fisheries processing industry in Indonesia. AJFAR:30–42. https://doi.org/10.9734/ajfar/2021/v15i530343
4. Suryana AAH, Perdiansyah (2021) Analysis of human resources competitiveness of capture fisheries in Indonesia. AJFAR:40–46. https://doi.org/10.9734/ajfar/2021/v15i230326
5. Peñaloza-Talavera MF, Martínez-Arroyo JA, Valenzo-Jiménez MA (2022) Formation of a fishing and aquaculture cluster as a tool for regional competitiveness. In: Terzioğlu MK (ed) Advances in econometrics, operational research, data science and actuarial studies. Springer International Publishing, Cham, pp 267–283. https://doi.org/10.1007/978-3-030-85254-2_16
6. Menéndez Delgado ER, Rodríguez Ríos EB, Bernal Casasola D, Cerbán Jiménez MDM (2021) Artisanal fishery in Ecuador. A case study of Manta city and its economic policies to improve competitiveness of the sector. Mar Policy 124:104313. https://doi.org/10.1016/j.marpol.2020.104313
7. Soh BH, Lim G-T, Chua SY (2021) Competitiveness of Malaysian fisheries exports: a constant market share analysis. MJES 58(2):175–198. https://doi.org/10.22452/MJES.vol58no2.1
8. Kelkar N, Arthur RI (2022) A review of governance and tenure in inland capture fisheries and aquaculture systems of India. [Online]. Available: https://books.google.com/books?hl=en&lr=&id=wL1qEAAAQBAJ&oi=fnd&pg=PP1&dq=Kelkar,+N.,+%26+Arthur,+R.+I.+(2022).+A+review+of+governance+and+tenure+in+inland+capture+fisheries+and+aquaculture+systems+of+India.+https://doi.org/10.4060/cb8615en&ots=43o6zSOa5g&sig=lUJVzKdVu4wfxr1-PtME1opbuaU. Accessed May 20, 2026
9. Chudik A, Pesaran H (2015) Large panel data models with cross-sectional dependence: a survey. Accessed: May 20, 2026. [Online]. Available: https://academic.oup.com/edited-volume/41981/chapter/355335206?searchresult=1&utm_source=TrendMD&utm_medium=cpc&utm_campaign=Oxford_Academic_Books_TrendMD_0
10. Flaten O, Lien G, Koesling M, Løes A-K (2010) Norwegian farmers ceasing certified organic production: characteristics and reasons. J Environ Manag 91(12):2717–2726. https://doi.org/10.1016/j.jenvman.2010.07.026
11. Breitung J, Pesaran MH (2008) Unit roots and cointegration in panels. In: Mátyás L, Sevestre P (eds) The econometrics of panel data, Advanced Studies in Theoretical and Applied Econometrics, vol 46. Springer Berlin Heidelberg, Berlin, Heidelberg, pp 279–322. https://doi.org/10.1007/978-3-540-75892-1_9
12. Andrews DWK (2005) Cross-section regression with common shocks. Econometrica 73(5):1551–1585. https://doi.org/10.1111/j.1468-0262.2005.00629.x
13. Cerrato M, Sarantis N, (2002) The cross sectional dependence puzzle. London Guildhall University. [Online]. Available: http://fmwww.bc.edu/RePEc/res2002/Cerrato.pdf. Accessed May 20, 2026

14. Balsalobre-Lorente D, Shahbaz M, Roubaud D, Farhani S (2018) How economic growth, renewable electricity and natural resources contribute to CO2 emissions? Energy Policy 113:356–367
15. Bekun FV, Alola AA, Sarkodie SA (2019) Toward a sustainable environment: nexus between CO2 emissions, resource rent, renewable and nonrenewable energy in 16-EU countries. Sci Total Environ 657:1023–1029. https://doi.org/10.1016/j.scitotenv.2018.12.104
16. Isham J (2005) The varieties of resource experience: natural resource export structures and the political economy of economic growth. World Bank Econ Rev 19(2):141–174. https://doi.org/10.1093/wber/lhi010
17. Mehlum H, Moene K, Torvik R (2006) Institutions and the resource curse. Econ J 116(508):1–20. https://doi.org/10.1111/j.1468-0297.2006.01045.x
18. Leite MC, Weidmann J (1999) Does mother nature corrupt? Natural resources, corruption, and economic growth. International Monetary Fund. [Online]. Available: https://books.google.com/books?hl=en&lr=&id=UFYYEAAAQBAJ&oi=fnd&pg=PA3&dq=Leite,C.,+Weidmann,+J.+(1999).+Does+Mother+Nature+Corrupt:+Natural+Resources,+Corruption,+and+Economic+Growth+IMF+Working+Papers.+International+Monetary+Fund.&ots=o29DeYoDjX&sig=TVLC-xIqLLL9VCD6nJ30s1TCTho. Accessed May 20, 2026
19. Machado JAF, Santos Silva JMC (2019) Quantiles via moments. J Econ 213(1):145–173. https://doi.org/10.1016/j.jeconom.2019.04.009
20. Koenker R (2004) Quantile regression for longitudinal data. J Multivar Anal 91(1):74–89. https://doi.org/10.1016/j.jmva.2004.05.006
21. Kangas HL, Ollikka K, Ahola J, Kim Y (2021) Digitalisation in wind and solar power technologies. Renew Sust Energ Rev 150:111356. https://doi.org/10.1016/j.rser.2021.111356
22. Li Y, Kong J, Ji J (2022) Environmental regulation, technological innovation and development of marine fisheries—evidence from ten coastal regions in China. Fishes 7(1):20. https://doi.org/10.3390/fishes7010020
23. Schwing FB (2023) Modern technologies and integrated observing systems are 'instrumental' to fisheries oceanography: a brief history of ocean data collection. Fish Oceanogr 32(1):28–69. https://doi.org/10.1111/fog.12619
24. Scarnecchia DL, Schooley JD (2022) Trophies, technology, and timescape in fisheries management, as exemplified through oklahoma's world record paddlefish *Polyodon spathula*. Fisheries 47(9):381–394. https://doi.org/10.1002/fsh.10764
25. Ziegler R et al (2022) Social innovation for biodiversity: a literature review and research challenges. Ecol Econ 193:107336. https://doi.org/10.1016/j.ecolecon.2021.107336
26. Pathak S, Lavudya N, Kumar RM, Dash RK (2022) Use of innovative technologies for fisheries resource management. Just Agri 2(5):1–8
27. Kemp PS et al (2023) Future advances in UK marine fisheries policy: integrated nexus management, technological advance, and shifting public opinion. Mar Policy 147:105335. https://doi.org/10.1016/j.marpol.2022.105335
28. Karagiannis D, Metaxas T (2023) 'Fishing' for winemaking perfection in cloudy waters…underwater wineries: innovation genius or another marketing trick? Implications on wine & food tourism. In: Sigala M, Haller C (eds) Technology advances and innovation in wine tourism. Springer Nature Singapore, Singapore, pp 231–250. https://doi.org/10.1007/978-981-19-8277-4_14
29. Nogueira LA, Lindeløv B, Olsen J (2023) From waste to market: exploring markets, institutions, and innovation ecosystems for waste valorization. Bus Strat Env 32(4):2261–2274. https://doi.org/10.1002/bse.3247
30. Li C-J, Kim H-H, Yang J-W (2021) A comparative study of fishery industry competitiveness in China's coastal provinces. Int J Adv Smart Conv 10(2):158–167. https://doi.org/10.7236/IJASC.2021.10.2.158
31. Fu X-M et al (2022) Evaluation of the innovation ability of China's marine fisheries from the perspective of static and dynamic. Mar Policy 139:105032. https://doi.org/10.1016/j.marpol.2022.105032

32. Wang M, Janssen ABG, Bazin J, Strokal M, Ma L, Kroeze C (2022) Accounting for interactions between sustainable development goals is essential for water pollution control in China. Nat Commun 13(1):730. https://doi.org/10.1038/s41467-022-28351-3
33. Albasri H, Sammut J (2022) A comparative study of sustainability profiles between small-scale mariculture, capture fisheries and tourism communities within the Anambas Archipelago Small Island MPA, Indonesia. Aquaculture 551:737906. https://doi.org/10.1016/j.aquaculture.2022.737906
34. Ramos J, Drakeford B (2021) Blue economy and tourism: is there a potential to create sustainable jobs? In: Handbook of research on human capital and people management in the tourism industry. IGI Global Scientific Publishing, pp 243–264. [Online]. Available: https://www.igi-global.com/chapter/blue-economy-and-tourism/263358. Accessed May 20, 2026
35. Seisay M (2022) "Human capital investment and financing for Africa's blue economy: conference on" unlocking the full potential of the fisheries sector within the framework of a sustainable blue economy. [Online]. Available: http://repository.au-ibar.org/handle/123456789/1175. Accessed May 20, 2026
36. Nurlita L, Yanuar T, Syah R, Indradewa R, Hamdi E (2021) Human capital implementation strategy in the Marina company. Am Int J Bus Manage (AIJBM) 4(8):152–156
37. Najib H, Nawangsari LC (2021) Effect of intellectual capital on organizational sustainability with employee innovative behavior as intervening variables in Pt. Jaya Maritime services. EJBMR 6(1):158–163. https://doi.org/10.24018/ejbmr.2021.6.1.714
38. Climent-Pérez P, Galán-Cuenca A, García-d'Urso NE, Saval-Calvo M, Azorin-Lopez J, Fuster-Guillo A (2023) Automatic fish size estimation from uncalibrated fish market images using computer vision and deep learning. In: García Bringas P, Pérez García H, Martinez-de-Pison FJ, Villar Flecha JR, Troncoso Lora A, De La Cal EA, Herrero Á, Martínez Álvarez F, Psaila G, Quintián H, Corchado Rodriguez ES (eds) 17th International Conference on Soft Computing Models in Industrial and Environmental Applications (SOCO 2022) (in Lecture Notes in Networks and Systems, vol. 531), vol 531. Springer Nature Switzerland, Cham, pp 319–329. https://doi.org/10.1007/978-3-031-18050-7_31
39. Malto MAD, Dumilag RV (2023) Market fish resources in magallanes, Sorsogon, Philippines. Philippine J Sci 152(1) [Online]. Available: https://www.researchgate.net/profile/Richard-Dumilag/publication/365890886_Market_Fish_Resources_in_Magallanes_Sorsogon_Philippines/links/63883b6abbdef30dc9887097/Market-Fish-Resources-in-Magallanes-Sorsogon-Philippines.pdf. Accessed May 20, 2026
40. Shimose T, Kanaiwa M (2022) Influence of the body color and size on the market value of wild captured coralgroupers (Serranidae, Plectropomus): implications for fisheries management. Fish Res 248:106223. https://doi.org/10.1016/j.fishres.2021.106223
41. Macusi ED, Morales IDG, Macusi ES, Pancho A, Digal LN (2022) Impact of closed fishing season on supply, catch, price and the fisheries market chain. Mar Policy 138:105008. https://doi.org/10.1016/j.marpol.2022.105008
42. Ewald C-O, Haugom E, Kanthan L, Lien G, Salehi P, Størdal S (2022) Salmon futures and the fish pool market in the context of the CAPM and a three-factor model. Aquac Econ Manag 26(2):171–191. https://doi.org/10.1080/13657305.2021.1958105
43. Nielsen M et al (2022) Can small-scale fisheries survive market-based management? Nordic evidence. Fish Fish 23(1):256–272. https://doi.org/10.1111/faf.12614
44. Chen J, Glibert PM, Cai W-J, Huang D (2023) Editorial: eutrophication, algal bloom, hypoxia and ocean acidification in large river estuaries, volume II. Front Mar Sci 10:1225903. https://doi.org/10.3389/fmars.2023.1225903
45. Spijkers J, Mackay M, Turner J, McNeill A, Travaille K, Wilcox C (2023) Diversity of global fisheries governance: types and contexts. Fish Fish 24(1):111–125. https://doi.org/10.1111/faf.12713
46. Jarvis RM, Young T (2023) Pressing questions for science, policy, and governance in the high seas. Environ Sci Pol 139:177–184. https://doi.org/10.1016/j.envsci.2022.11.001

47. Guggisberg S, Jaeckel A, Stephens T (2022) Transparency in fisheries governance: Achievements to date and challenges ahead. Mar Policy 136:104639. https://doi.org/10.1016/j.marpol.2021.104639
48. Okafor-Yarwood I, Kadagi NI, Belhabib D, Allison EH (2022) Survival of the Richest, not the Fittest: how attempts to improve governance impact African small-scale marine fisheries. Mar Policy 135:104847. https://doi.org/10.1016/j.marpol.2021.104847
49. Jentoft S, Chuenpagdee R (2022) Interactive learning and governance transformation for securing blue justice for small-scale fisheries. Admin Soc 54(7):1255–1282. https://doi.org/10.1177/00953997211073947
50. Dudayev R, Hakim LL, Rufiati I (2023) Participatory fisheries governance in Indonesia: Are octopus fisheries leading the way? Mar Policy 147:105338. https://doi.org/10.1016/j.marpol.2022.105338
51. Agui6n A et al (2022) Establishing a governance threshold in small-scale fisheries to achieve sustainability. Ambio 51(3):652–665. https://doi.org/10.1007/s13280-021-01606-x
52. Linke S, Gillette MB, Jentoft S (2022) Institutionalizing injustice? Aligning governance orders in swedish small-scale fisheries. In: Jentoft S, Chuenpagdee R, Bugeja Said A, Isaacs M (eds) Blue justice (in MARE Publication Series, vol. 26), vol 26. Springer International Publishing, Cham, pp 529–547. https://doi.org/10.1007/978-3-030-89624-9_28
53. Ynacay-Nye A, Hisano S, Suryawan AS (2023) Alternatives to sustainable seafood certifications: transitions of small-scale fisheries governance in northeastern Japan. J Rural Stud 97:269–280. https://doi.org/10.1016/j.jrurstud.2022.12.002
54. Day S (2026) Pacific Salmon fisheries management: an (unusual) example of polycentric governance involving indigenous participation at multiple scales. The Center for Growth and Opportunity. [Online]. Available: https://www.thecgo.org/books/the-environmental-optimism-of-elinor-ostrom/chapter-3-chapter-3-pacific-salmon-fisheries-management-an-unusual-example-of-polycentric-governance-involving-indigenous-participation-at-multiple-scales/. Accessed May 20, 2026
55. Sorensen J, Kincl L, Weil R, Dzugan J, Christel D (2022) Fisheries governance and associated health implications: current perspectives from US commercial fishermen. Mar Policy 141:105119. https://doi.org/10.1016/j.marpol.2022.105119
56. McKinley E (2022) A blue economy for whom?: linking marine social sciences with blue economy discourse. In: Blue economy. Routledge, pp 13–25. [Online]. Available: https://www.taylorfrancis.com/chapters/edit/10.4324/9781003280248-4/blue-economy-emma-mckinley. Accessed May 20, 2026
57. Nunes AJP, Masagounder K (2022) Optimal levels of fish meal and methionine in diets for juvenile litopenaeus vannamei to support maximum growth performance with economic efficiency. Animals 13(1):20. https://doi.org/10.3390/ani13010020
58. Starkweather KE, Keith MH, Zohora FT, Alam N (2023) Economic impacts and nutritional outcomes of the 2017 floods in B angladeshi S hodagor fishing families. Am J Hum Biol 35(1):e23826. https://doi.org/10.1002/ajhb.23826
59. Karimi MS, Khezri M, Khan YA, Razzaghi S (2022) Exploring the influence of economic freedom index on fishing grounds footprint in environmental Kuznets curve framework through spatial econometrics technique: evidence from Asia-Pacific countries. Environ Sci Pollut Res 29(4):6251–6266. https://doi.org/10.1007/s11356-021-16110-8
60. Alharthi M, Hanif I (2020) Impact of blue economy factors on economic growth in the SAARC countries. Maritime Bus Rev 5(3):253–269
61. Wang Q, Wang S (2022) Carbon emission and economic output of China's marine fishery—A decoupling efforts analysis. Mar Policy 135:104831. https://doi.org/10.1016/j.marpol.2021.104831
62. Peng D, Yang Q, Yang H-J, Liu H, Zhu Y, Mu Y (2020) Analysis on the relationship between fisheries economic growth and marine environmental pollution in China's coastal regions. Sci Total Environ 713:136641. https://doi.org/10.1016/j.scitotenv.2020.136641

63. Suh D, Pomeroy R (2020) Projected economic impact of climate change on marine capture fisheries in The Philippines. Front Mar Sci 7:232. https://doi.org/10.3389/fmars.2020.00232
64. Fisu AA, Didiharyono D, Bakhtiar (2020) Economic & financial feasibility analysis of tarakan fishery industrial estate masterplan. IOP Conf Ser: Earth Environ Sci 469(1):012002. https://doi.org/10.1088/1755-1315/469/1/012002
65. Crouch GI, Ritchie JB (1999) Tourism, competitiveness, and societal prosperity. J Bus Res 44(3):137–152
66. Dwyer L, Kim C (2003) Destination competitiveness: determinants and indicators. Curr Issue Tour 6(5):369–414. https://doi.org/10.1080/13683500308667962
67. Anwar A, Siddique M, Dogan E, Sharif A (2021) The moderating role of renewable and non-renewable energy in environment-income nexus for ASEAN countries: evidence from method of moments quantile regression. Renew Energy 164:956–967
68. Ehigiamusoe KU, Lean HH, Smyth R (2020) The moderating role of energy consumption in the carbon emissions-income nexus in middle-income countries. Appl Energy 261:114215. https://doi.org/10.1016/j.apenergy.2019.114215
69. Dogan E, Seker F (2016) Determinants of CO2 emissions in the European Union: the role of renewable and non-renewable energy. Renew Energy 94:429–439. https://doi.org/10.1016/j.renene.2016.03.078
70. Apergis N, Payne JE (2010) The emissions, energy consumption, and growth nexus: evidence from the commonwealth of independent states. Energy Policy 38(1):650–655
71. European Commission (2022) Joint research centre, clean energy technology observatory, ocean energy in the European Union: status report on technology development, trends, value chains and markets: 2022. LU: Publications Office. Accessed: Dec. 30, 2025. [Online]. Available: https://data.europa.eu/doi/10.2760/162254
72. Alsaleh M, Abdul-Rahim AS, Liu R, Sun Q (2024) RETRACTED: nature of property rights and motivation for blue growth: an empirical evidence from the fisheries industry. Nat Res Forum 48(1):184–210. https://doi.org/10.1111/1477-8947.12317
73. Pedroni P (2004) Panel cointegration: asymptotic and finite sample properties of pooled time series tests with an application to the ppp hypothesis. Economic Theory 20(03). https://doi.org/10.1017/S0266466604203073
74. Kao C, Chiang M-H (2000) On the estimation and inference of a cointegrated regression in panel data. In: Advances in econometrics, vol 15. Emerald (MCB UP), Bingley, pp 179–222. https://doi.org/10.1016/S0731-9053(00)15007-8
75. Sarkodie SA, Strezov V (2019) Effect of foreign direct investments, economic development and energy consumption on greenhouse gas emissions in developing countries. Sci Total Environ 646:862–871. https://doi.org/10.1016/j.scitotenv.2018.07.365
76. Koenker R, Bassett G (1978) Regression quantiles. Econometrica 46(1):33. https://doi.org/10.2307/1913643
77. Binder M, Coad A (2011) From Average Joe's happiness to Miserable Jane and Cheerful John: using quantile regressions to analyze the full subjective well-being distribution. J Econ Behav Organ 79(3):275–290. https://doi.org/10.1016/j.jebo.2011.02.005
78. Canay IA (2011) A simple approach to quantile regression for panel data. Econ J 14(3):368–386
79. Im KS, Pesaran MH, Shin Y (2003) Testing for unit roots in heterogeneous panels. J Econ 115(1):53–74
80. Levin A, Lin C-F, James Chu C-S (2002) Unit root tests in panel data: asymptotic and finite-sample properties. J Econ 108(1):1–24. https://doi.org/10.1016/S0304-4076(01)00098-7
81. Westerlund J (2007) Testing for error correction in panel data*. Oxf Bull Econ Stat 69(6):709–748. https://doi.org/10.1111/j.1468-0084.2007.00477.x
82. Kremers JJM, Ericsson NR, Dolado JJ (1992) The power of cointegration tests. Oxf Bull Econ Stat 54(3):325–348. https://doi.org/10.1111/j.1468-0084.1992.tb00005.x
83. Chen X, Xu Q, Li L (2023) Illegal, unreported, and unregulated fishing governance in disputed maritime areas: reflections on the international legal obligations of states. Fishes 8(1):36. https://doi.org/10.3390/fishes8010036

84. Schrobback P, Hoshino E, Pascoe S, Curtotti R (2022) Market integration of domestic and imported seafood: Insights from the Sydney fish market*. Aus J Agri Res Econ 66(1):216–236. https://doi.org/10.1111/1467-8489.12452
85. Pincinato RBM, Gasalla MA, Garlock T, Anderson JL (2022) Market incentives for shark fisheries. Mar Policy 139:105031. https://doi.org/10.1016/j.marpol.2022.105031
86. University of montenegro, faculty of maritime studies kotor, Šekularac - Ivošević S (2021) Fostering maritime innovations th rough human capital: exploring the status Quo of the Adriatic Universities. Naše more 68(3):189–198. https://doi.org/10.17818/NM/2021/3.6
87. Acuña-Alonso C, Varandas S, Álvarez X, Martinho A (2023) Analysis of the evolution of a fisheries management plan based on environmental governance: living laboratory in the Olo River, Portugal. Fish Res 260:106595. https://doi.org/10.1016/j.fishres.2022.106595
88. Łuczyńska J, Łuczyński MJ, Nowosad J, Kowalska-Góralska M, Senze M (2022) Total mercury and fatty acids in selected fish species on the polish market: a risk to human health. IJERPH 19(16):10092. https://doi.org/10.3390/ijerph191610092
89. Ayilu RK, Fabinyi M, Barclay K (2022) Small-scale fisheries in the blue economy: review of scholarly papers and multilateral documents. Ocean Coastal Manage 216:105982. https://doi.org/10.1016/j.ocecoaman.2021.105982
90. Nunan F (2020) The political economy of fisheries co-management: challenging the potential for success on Lake Victoria. Glob Environ Chang 63:102101
91. Kang B et al (2021) Climate change impacts on China's marine ecosystems. Rev Fish Biol Fish 31(3):599–629. https://doi.org/10.1007/s11160-021-09668-6

Chapter 10
Fisheries Governance and EU Blue Governance Management

Acronyms

EU	European Union
WFD	Water Framework Directive
OSPAR	Convention for the Protection of the Marine Environment of the North-East Atlantic
CO_2	Carbon Dioxide Emissions
MMQR	Method of Moments Quantile Regression
EUR	Euro
CD	Cross-sectional Dependency
FP	Fisheries Production
WDI	World Bank Development Indicators
GDP	Gross domestic product
DOLS	Dynamic ordinary least squares
FMOLS	Fully modified ordinary least squares
FEOLS	Fixed-effects ordinary least squares
Europa	European data portal
Eurostat	European Commission's database
LLC	Lin & Chu Test
IPPS	Im, Pesaran, and Shin Test
EU27	27 European Union Member States
EU13	13 European Union Member States
BRICS	Brazil, Russia, India, China, South Africa
ASEAN	Association of Southeast Asian Nations
EU14	14 European Union Member States
OLS	Ordinary least squares
VOA	Voices, accountability, and social development
PSAV	Political and economic stability
GOE	Public sector efficiency and quality
REQ	Regulatory quality and private sector quality
CFP	Common Fisheries Policy

M. Alsaleh, *Fisheries Sustainability in the EU*, Urban Sustainability,
https://doi.org/10.1007/978-981-92-4496-6_10

COC	Elite management, private interests, and corruption control
LGDP	Natural logarithm of economic growth
LVA	Natural logarithm of social development
LPS	Natural logarithm of economic stability
LRL	Natural logarithm of property rights
LCC	Natural logarithm of private interest

LCC: Natural-logarithm of private interest. The current EU leadership role has an opportunity to lead the international agenda on global governance into healthier oceans protection and sustainability, a policy goal as mentioned by the European Green Deal, a model of sustainability for Europe globally. This chapter tries to review the dynamic impact of global governance on the fisheries industry sustainability in EU27 countries and how fisheries production and global governance dimensions have influenced it during the period of 1996–2022. It provides an inquiry into issues about ocean protection and marine health for living creatures, as well as global ocean governance to mitigate the threats to ocean resources by assessing the impacts of key dimensions on fisheries sustainability in EU27 countries from 1996 to 2022, using the Method of Moments Quantile Regression (MMQR) together with the use of fixed effects for fisheries production and global governance features. For both EU13 and EU14 countries, public sector quality, economic growth, economic stability, and fisheries sustainability had a positive, substantial relation in all quantiles. For EU27 countries at upper quantiles, private sector quality and property rights variables had a positive, substantial relation with fisheries sustainability. At the same time, private interest variables have a negative, substantial effect on fisheries sustainability in the EU27 at upper quantiles, but have a positive, substantial impact at median quantiles in the EU27 on social development and fisheries sustainability. Notably, the strong positive effect of economic stability, economic growth, property rights, public sector quality, and private sector quality on fisheries sustainability is greater in EU13 developing countries than EU14 developed countries; this result favors the growth hypothesis for fisheries-producing countries. Nevertheless, developed EU14 countries have a greater positive impact on social development, and fisheries sustainability is greater than that of developing EU13 countries. Using appropriate technology and green environmental investment will make marine protection and governance effectively promoted by policymakers on fisheries protection, and also help achieve sustainable development goals in EU14 and EU13 countries.

10.1 Fisheries Governance in the EU Background

Sustainable development goal 14 (life below water). The SDG on the oceans, seas, and marine resources focuses on the need to protect and utilize the oceans, seas, and marine resources for sustainable development. The oceans and seas form the largest system in the world on which we depend for our survival and for life on Earth [1].

These bodies of water, covering 70% of the planet's surface, supply humanity with food and water, generate oxygen, and regulate the climate.

The ocean captures a quarter of our annual global emissions of carbon dioxide (CO_2) to delay the full effects of climate change, which highlights its importance for our planet. Protection and use of the oceans and seas in a sustainable way are fundamental to our world. But the oceans and seas-our most unique environment-are under pressure, and their ability to continue supporting hundreds of millions of lives is at risk [1]. As set out in the UN's report on the Sustainable Development Goals, two primary drivers threatening the sustainable use of oceans and marine ecosystems are: (1) increasing ocean acidification and warming oceans, which threaten marine life and ecosystem services [2].

By 2021, an amount of more than 17 million t of plastic, making up 85% of all marine litter, was entering the world's oceans [4] (3) Only 8% of the ocean and coastal waters of the world were covered by a protected marine area in 2021 2 More than one-third (35.4%) of the world's fish stocks were fished more heavily than can be caught on a sustainable basis, in comparison to 34.2% in 2017 and 10% in 1974.

By 2040, plastic waste entering our oceans each year could potentially triple. (5) Half of the world population relies partially on small-scale fishing as an alternative source for income, whereas the sector of small-scale fishing comprises 90% of the world's fishing labor [2]. However, the rate of increase in loss has stabilized in recent years.

Overfishing: This is the most serious threat to marine environments, and it impacts humans and marine ecosystems alike. Overfishing occurs when too many fish are removed from a specific fishing area at one time to allow the remaining fish to reproduce and rebuild their numbers [3]. Furthermore, overfishing is often a symptom of unsustainable fishing practices, which result in an excess of bycatch (catching many non-targeted animals), most of which are discarded dead or injured [3].

One third of the world's fish stocks, as many of these fisheries are not properly explored, are in danger due to years of heavy fishing [3].

Indeed, 40% of the North-East Atlantic fish stocks are still being over-fished (caught at a rate faster than they can multiply), and recent estimates concerning the Mediterranean stocks show that more than 90% of those caught by the EU alone are over-exploited and going beyond sustainable levels, namely Maximum Sustainable Yield (MSY) [3]. Marine Plastic Litter: The UN estimates there are currently between 75 and 199 million tonnes of plastic litter in the seas. According to a 'business as usual' scenario, the annual 9–14 million tonnes of plastic entering our aquatic environment would increase to between 23 and 37 million tonnes per year by 2040 [4]. Some 80% of the waste that ends up in marine waters is said to have its origin on land, but this varies depending on the regions of the world. For example, for the Northeast Atlantic, sea-based sources such as shipping and fishing are a key source of marine debris.

For marine waste sources, in addition to sea-based sources, which are receiving particular attention, such as lost and abandoned fishing gear that may damage marine life and impact fisheries and the economy, waste can impose serious negative impacts on local communities, tourism, shipping, and fishing.

Therefore, with the accumulation and distribution of marine waste, this is one of the fastest-growing marine problems [4]. Ocean Governance Strategy, the EU has laid down an international ocean governance agenda with the overall mission to foster healthy, productive, safe, secure, and resilient oceans based on an effective cross-sectoral and rules-based international approach [4]. Correspondingly, marine environmental protection legislation is being gradually developed and is in place over many policy areas in the EU.

For instance, the CFP manages fisheries, and WFD aims at limiting the inflow of nutrients and other chemicals into the waters [4]. Nevertheless, they have certain limitations as they only target specific sources of pressure, and the scope for conservation of the marine environment is relatively fragmented. Aquatic resources and fisheries are significantly pressured by human activities.

The marine environment and marine resources in the EU Region are faced with various pressure sources, such as fishing, illegal fishing, climate change, and marine pollution.

With the world population forecast to reach 9 to 10 billion by 2050, the pressure on global resources such as food, water, and raw materials will be higher [1]. In this context, by managing EU fisheries and marine resources properly, the EU will contribute to the global objectives for ocean governance and will meet its international obligations for sustainable development, climate change, and the protection of biological diversity. The EU intends to pursue these activities for the management of the oceans and seas to be safe, secure, clean, and sustainable. The present research questions are: (1) How does global governance influence the sustainability of fisheries production and associated marine ecology in European seas between 1996 and 2022?

(2) Are EU member states more susceptible to effects on fisheries sustainability caused by global governance issues?

(3) Can the fisheries sector achieve better governance to meet EU objectives for ocean and marine sustainable development by 2030? The main objective of the research will be to analyze the fisheries production sustainability regarding global governance in the European Union seas from 1996 to 2022 (2) To investigate global governance effects between EU13 developing and EU14 developed countries on fisheries production sustainability and marine ecology between 1996 and 2022 comprehensively (3) To explore the extent to which the impact of global governance factors on fisheries production sustainability, as it exists between developing and developed countries from 1996 to 2022.

Many argue that due to recent governance developments, they have now become a key determinant in the sustainable development potential and dynamics. Due to the major policy developments in the Common Fisheries Policy, the Common Agriculture and Fisheries Policy reforms, and the most recent EU expansion, significant improvements have been made in reducing and, in some cases, eliminating the effects of overfishing, depleted resources, and the causes of marine pollution [1]. In return, this implies that the macro marketing environment is changing significantly, intensifying rivalry for export countries to an even greater extent. The environmental circumstances in which the fisheries sector is forced to exist are more threatening than ever. What's more, there is the dilemma concerning sustainable food security, depleted resources, and the sustainability of both marine resources and the environment on a global level [1]. As a consequence of this, governance is becoming a complicated topic as the "blue" food has to remain competitive and meet the local and international organizations' expectations about the sustainable use of marine ecosystem resources and of ocean resources and the sustainability of the blue food system. In the case of the EU fishery industry, there are no different scenarios because they face intense competition in a globalized world. In EU markets, the sector accounts for quite a large amount, and the growth of imports has accelerated over the past decade [1].

Since recent developments in fisheries industry governance are inadequately addressed in the literature on sustainability in this industry and influencing global governance, studies focusing on fisheries industry sustainability and global governance can help to derive appropriate future marketing policies and strategies. There are various ways of analyzing the numerous institutions and the complexity of fisheries governance within modern Europe. To better understand why current conditions persist, one needs to examine the nature of the systems that must be managed and the preconditions for efficient governance. Other influences on a system's development include historical, biophysical, social, cultural, and political factors. Where the region had strong historical institutions with a history of robust cooperation, a stable democracy, and an economically prosperous condition, the institutions would have developed better. Because the Baltic and North Seas score higher on these factors, they have better-developed institutions than the Mediterranean and the Black Seas. Moreover, given that all Contracting Parties in both conventions (Helsinki and OSPAR) are also member states of the EU, it is natural to assume that the EU will be better equipped to impose more rigid regulations in the former areas compared to the Mediterranean and especially the Black Sea, where the influence of the EU has so far been limited.

Even though the differences are huge, the similarities were great enough to draw important lessons from the experiences in various EU countries. Above all, the nature of the environmental problems that face the EU member states in managing fisheries and ensuring the sustainability of their marine ecosystem seems largely to be the same, regardless of where in the EU they are located. Except for the Black Sea, chemical pollution is posing serious problems to both fish stocks and marine organisms. Alien species, shipping's impact, and eutrophication cause problems for all member states. Many of the EU's stocks are subject to overfishing and often, as mentioned, combine badly with eutrophication to expand areas of marine environments that lack dissolved oxygen.

The problem of non-independence of disturbance factors or, in other words, non-cross-sectional dependence (CD) between the above variables over countries has been a major problem in previous works on the EU countries. According to [5–9], these factors may include the common or idiosyncratic shock or spillovers, or both. Given the increasing economic integration, it becomes necessary to include CD in our framework because there will be some sort of mutual interdependence amongst the group of EU economies in our sample. Being an open economy, the EU will be prone to global shocks that affect global governance and economic conditions, such as political stability or institutions. As pointed out by [8], failure to account for CD will lead to biased and inconsistent estimates in measuring the impact of economic growth on institutional performance.

In this work, we are thus interested in identifying whether the growth-mitigating effect of natural resources would be the same for fisheries-producing countries' global governance profile. Unlike previous works that use broader exogenous variables that make no distinction between point-source and diffuse-source resources [10, 11], this study uses a narrower natural resource as a measure for sustainable utilization of natural resources, namely, sustainable production of fisheries resources. The decision to select the EU countries is because, in general, point-source resources tend to have greater detrimental effects than diffuse-source ones [12, 13]. Moreover, in this study we use samples that include both developing and developed economies, and we added institutional quality and regulation control variables which may alleviate or remove the negative impacts of the resource-led growth process on institutional quality [14].

Our sample includes EU27 economies that produce fish as part of the group of developed countries (e.g., Germany, Italy, and Sweden) and emerging countries (e.g., Estonia, Bulgaria, and Malta) and covers the time span from 1996 to 2022. Given that there is variation in economic development among our sample countries, we use a cutting-edge Method of Moments Quantile Regression of [15] to investigate the validity of the economic growth-driven natural resource sustainability nexus in the presence of fisheries production and other controls across different quantiles of the conditional distribution of fisheries sustainability. Using fixed-effects estimation, this approach offers a better understanding of how far a distributional heterogeneous relationship can be explained. It captures effects at different quantiles of the economic growth-fisheries production relationship in blue resource sustainability, which cannot be revealed through simple mean regressions. Hence, we thoroughly test the growth hypothesis in fisheries-producing countries.

To the best of our knowledge, this study is the first to utilize distributional heterogeneity to test the extent to which EU27 countries' natural resources' sustainability can support economic growth considering fisheries production, political stability, government effectiveness, regulatory quality, rule of law, corruption control, and economic growth. The evaluation of the growth hypothesis at different quantiles of the fisheries sustainability conditional distribution is justified by the presence of numerous identified reasons. One of the most appealing properties of conditional quantile estimation is that, unlike mean estimation, it is not sensitive to the effect of outliers in the dependent or independent variables [16]. The latter may bias or eliminate the influence of global governance on fisheries sustainability when estimated by simple means. Instead, with quantile estimation, it seems quite convenient, especially in the case of panel regressions, to decompose the effects of variables on the dependent variables according to different ranges of quantiles, facilitating their categorization by group [4]. Thus, conditional quantile estimates result in information that the mean estimation technique does not have [4].

10.2 Fisheries Governance Historical Development

Fisheries governance is the system of laws, regulations, social customs, economics, and politics through which fisheries are managed. Fisheries governance is an open system composed of laws and policies that create enforceable regulations along with social rules and customs and has dimensions at global, national, and local levels. For example of different levels of individuality [1, 2] explore the network resilience and EU and USA policies in engagement in the third countries and explore for instance the capacity of Brexit governance within the frame of fisheries policy of Scotland respectively where it found the sectors absorbable as being fisheries management, fisheries operationalisation, and Fisheries knowledge and where the require further capacity strengthening is for both communication, relational and politically at high level. [1, 4] investigate how women experience and are shaping and influential in small-scale fisheries governance with in different scales, such as interactions with the existing social norms, customary practices and relationships; in co-management type participatory and as part of various other social, political and institutional relationships; in relations and arrangements that govern resource access and management; interactions between traditional and formal law in a range of institutional arrangements and in relations and discourses with others involved in fisheries governance. Ocean governance brings diverse actors together to work on ecosystem protection and the sustainable management and use of marine and water resources for sustainable, just and inclusive communities.

In China, [17] investigates whether self-governance has addressed the water commons issue by offering causal evidence of water pollution treatment's impact on the health of aquaculture farmers; it found that to properly tackle the water commons issue, one needs self-governance, as Individual treatment cannot prevent this issue from occurring. [18] in India analyze the functioning of ocean governance in sustainable development of marine resources, stating that adequate legal and fisheries practice can achieve a better state of the ocean from an economic, social, and environmental view. Fisheries management policy is argued by [19] on natural resources in Bangladesh to achieve the sustainable development goals from an economic, social, and environmental viewpoint. [20] investigate effective ways to strengthen governance in Africa's small-scale marine fisheries, showing that poor fisheries management, a variety of governance shortcomings in fisheries, an ever-increasing number of overfished stocks, and increasing fish caught illegally result in rising water pollution. Others [21] in UN Ocean [8] in UN Oceans and [22] in China investigate ocean governance's impact on fisheries in general as well as on water pollution. The EU will support global efforts for the transition toward a sustainable agri-food system and will promote the sustainable management of fish and seafood products, strengthen ocean governance, marine cooperation and coastal management, which is why it has zero tolerance for illegal, unregulated and unreported (IUU) fishing and overfishing. [23, 24] explore the role of fisheries management systems in food security and development promotion, arguing that an enabling environment fostered by agricultural and fisheries policy is what could lead to a self-perpetuating cycle for food security and nutrition. [25, 26] demonstrate the roles that the fisheries industry and aquaculture industry play in ensuring food security and the achievement of sustainable development goals in Africa by making sure to establish regulatory policy providing secured property rights to their practitioners, which African countries did not perfectly do for the fishery industry.

There is an extensive body of research by [20, 27–31] that focuses on the negative impact of poor governance and enforcement practices on overfishing and the importance of good management policy, as well as different means to strengthen and maintain them. [18, 32], among others, have investigated the existing legal and political status of the fisheries in Indonesia and India, respectively, and stated that community partnership and co-management practices are the best way to ensure a suitable atmosphere for the fish sector in Indonesia and India.

Recent literature indicates that, as evidenced by evaluations of fisheries management plans, by [33–35], a fish abundance trend is visible that demonstrates that effective marine governance is associated with an improved ecosystem health in Europe, showing that fish abundance levels have improved from 1229 in 2008, 1142 in 2013, and 1781 in 2018 (where the slight decrease in 2013 may be due to a wildfire located nearby). [36], despite the fact that local knowledge is in principle accepted as part of a governance approach to sustainable seafood, they have analyzed how alternatives to sustainable seafood certifications for seafood in Japan are advocated by influential political and economic actors when coping with external shocks, in which external impacts are analyzed from a broader, more policy-, market-, and economically oriented perspective and not from local knowledge perspectives. In the US, the studies conducted by [37, 38] focus on the underlying structure of policies designed for fisheries management so as to ensure their sustainability and health and highlight the existence of enabling environment factors outside of direct control of the fishermen that severely affects their quality of life when together with variable weather conditions, ability to access reliable crew, and volatility of fish price changes in the market. On the other hand, an important topic discussed by academics and economists in the fisheries sector related to the environmental issue, specifically the marine water contamination issue, is fisheries-led development.

Since [39] first proposed a Fisheries-led development policy, significant literature has grown on the subject. Various economic theories on development have been presented by economists, although one needs to account for the fisheries sector in the issues relating to marine and water pollution. [40] studied the effect of antibiotics and the waste of fish feeds on the ecosystem of benthic marine environment in Mediterranean Sea, arguing that fish feed altered habitat characteristics and resulted in macro-invertebrates loss and significantly affecting the microbial community organization, while [41, 42] evaluated the impact of fisheries waste on the survival and health of the marine environment with a key influence of fishing gear that impacts the marine pollution and the survival rate. [43] states that using fishery waste (fisheries products) is an environmentally friendly way of achieving the marine conservation goal through its transformation into a product that provides health, social, economic, and marine benefits to human communities. Developing a fisheries industry contributes to enhancing a wide variety of opportunities for economic, social, and environmental development in the long term.

For example, the relationship between sustainable fisheries production and economic development has been addressed by [44–46], showing that fishery production plays a significant role in the development of several South Asian Countries. [47] also evaluated the impact of substitution of partial or whole fish production on growth performance, feed utilization, histological condition, and economic status.

In [48], the researcher tested the effects of the initial stocking sizes of Nile tilapia (Oreochromis niloticus) on the quality of water, growth performance, and economic income. The researchers concluded that medium stocking fish size had the highest growth performance and return. Likewise [49] evaluated how Arctic fish skin crafts contributes to the sustainable development in Southwest Alaska's coastal regions, suggesting how the utilization of fish skin from Arctic peoples might be an option to promote sustainable development including poverty reduction, gender equality, improve health and well-being, access to water and sanitation, decent work and economic growth, reduced inequality, tackle climate change, ocean conservation, and responsible consumption and production. Numerous studies such as [50–54] et al., have estimated total financial losses of unsustainable fisheries at more than 20% of total global fisheries sector value.

By adopting a sustainable Fisheries sector, it increases the opportunities for long-term economic, social, and environmental development. [55–57] examine the optimal level for the production of fish for China to support high performance and economic efficiency, which demonstrate that fishery industry development could increase economic growth.

In East Asia, the impacts of fish production and trade economics have been extensively investigated by scholars such as [58–60] to conclude that developing countries are able to significantly contribute to the economy by dedicating more emphasis to fish production and superior fisheries products exports. Compared with previous literature confined solely to the fisheries sector, this study looks into several global governance indicators' effects on the sustainable development of the fisheries industry to reach sustainable blue growth. Meanwhile, we analyze the impacts of global governance factors on fisheries production in EU seas toward sustainable development.

As most studies adopted time-series analysis for their research, we utilize both static and panel regression models for analyzing the influences of global governance indicators and economic development on the sustainability of fisheries production.

First of all, the most important focus of the paper is analyzing the correlations of fisheries production, social development, economic stability, public sector quality, private sector quality, property rights, private interest, and economic development in EU waters from 1996 to 2022. Furthermore, more significantly, we analyzed how global governance drivers and economic development affect the fisheries sector in EU27 seas for obtaining sustainable blue growth. The EU27 region is broken down into two contrasting groups: EU14 countries (developed) andEU13 countries (developing countries). We employ a unit root and cointegration analysis, as our dataset covers 1996–2022 and shows non-stationary behavior, to analyze the stationarity of the data.

In addition, through a new method based on the MMQR method, called the method of moments quantile regression, we studied interactions between social development, economic stability, public/private sector quality, property rights, and private interests.

Static and panel regressions on issues affecting economic development and competitiveness on the global scene were carried out as well. At the end, we have made predictions on how several aspects of global governance will contribute to sustainable oceanic resources and fisheries.

10.3 Fisheries Governance Theory Modeling

For the purpose of this chapter, we shall adopt the fisheries governance system designed by T.S. Grey, and this approach contains market, hierarchical, and participatory governance. The participatory governance is further divided into six types, such as: social development (VOA), economic stability (PSVA), public & private sector quality, rule of law (ROL) & private interest (COC) governance. We analyze three participatory governance types—co-management, community collaboration, environmental protection, fisheries sector & management guideline policy. We use the EU-regulation subtype, since it is implemented on a practical and legal level in the EU fisheries sector. We develop the following model based on recent studies by [61–64]:

$$FP_{it} = f\left(VOA_{it}, PSVA_{it}, GOE_{it}, REQ_{it}, ROL_{it}, COC_{it}, GDP_{it}\right)$$

Fisheries production (FP) is our dependent variable measured in metric tonnes (total catch), where total catch represents all species caught in a particular fishing zone within a specific time period. Data on all variables for selected EU Member States were obtained from the European Commission [65]. The dependent variable comprises: (VOA) Voice and accountability; and social development according to the World Bank [66]. We analyze the EU economy from 1996 to 2022. All variables are presented per capita. (PSAV) Political and economic stability, which is proxied by the World Bank Database [66]. (GOE) Public sector efficiency and quality. (REQ) Regulatory quality and private sector quality, which is found in the database of the World Bank. (ROL) Rule of law and property rights. (COC) Elite management, private interests, and corruption control. (GDP) Gross domestic product, which is measured in 2017 US$ and collected from the European Commission [67].

In order to fulfill the objective of this paper, we have gathered an annual data series of selected countries from the World Bank's World Development Indicators (WDI), from the European Commission's data portal (Eurostat), and from the official European Union portal (Europa) for the period 1996 to 2022. Table 10.1 presents variables, units and measurement, source, and transformations.

Table 10.1 Data, units, transformation, and source

Data	Measurement unit	Transformation	Source
Fisheries production (FP)	Metric tons	Natural logarithm transformation	Eurostat
Social development (VOA)	% of confidence in governance	Natural logarithm transformation	WDI
Economic stability (PSAV)	% of confidence in governance	Natural logarithm transformation	WDI
Public sector quality (GOE)	% of confidence in governance	Natural logarithm transformation	WDI
Private sector quality (REQ)	% of confidence in governance	Natural logarithm transformation	WDI
Property rights (ROL)	% of confidence in governance	Natural Logarithm transformation	WDI
Private interest (COC)	% of confidence in governance	Natural logarithm transformation	WDI
Economic growth (GDP)	constant US dollars	Natural logarithm transformation	Eurostat

Source: [68, 69]

Let us examine the results presented in this section. Each of the components has characteristics shown in Table 10.2 concerning the maximum, minimum, average, and standard deviation of these factors. These components show the values for average: 4.929, 1.183, 1.847, 1.857, 1.751, 1.868, and 1.862, respectively. Each of the components in this study presents outstanding values on standard deviation of 0.825, 0.737, 0.124, 0.107, 0.123, 0.842, 0.103, and 0.701, respectively. The descriptive attributes of the components enable us to approach the unit root test.

For the purpose of this study, to enable comparisons and robustness checks, we apply fixed-effects OLS (FEOLS), dynamic OLS (DOLS), and fully modified OLS (FMOLS). Besides the FEOLS procedure, Driscoll and Kraay (1998) standard errors, which prove effective robustness against many types of cross-sectional dependence and autocorrelation within certain lag orders, are also considered. The main difficulties when estimating dynamic cointegrated panels involve heterogeneity across countries related to differences in cross-sectional averages and to differential cross-sectional adjustments to the cointegrating equilibrium. Therefore, in the FMOLS estimation of Pedroni (2000), each panel member receives individual intercepts, and each individual has serial correlation property in error terms. Based on monte-carlo results indicating DOLS estimation being unbiased relative to both the FMOLS and OLS estimations in finite samples, [71] extend the DOLS estimator in panels. Moreover, DOLS estimators correct for potential endogenous feedbacks through inclusion of lead and lag values of differences of the regressors.

Given the shortcomings of the methods mentioned earlier to estimate such dynamic panels, we use a panel quantile regression technique to examine the distributional and heterogeneous effects across quantiles [72]. The panel quantile regression method was established in the foundational work of [73]. Quantile regression estimates are derived in contrast to ordinary least-squares regression, which

Table 10.2 Results of descriptive statistics

Variable	Observations	Mean	Std. Dev.	Min	Max
FP	864	4.929	0.825	3.211	6.310
GDP	864	1.183	0.737	2.596	5.530
COC	864	1.847	0.124	1.178	1.990
GOE	864	1.857	0.107	1.206	1.921
PSAV	864	1.751	0.123	1.268	1.975
REQ	864	1.868	0.842	1.388	1.965
ROL	864	1.862	0.103	0.281	1.979
VOA	864	1.868	0.701	1.682	1.911

Source: [68, 69]

estimates the conditional mean of the dependent variable conditional upon values of the independent variable, typically applied to estimate the median, or other quantile, of the conditional distribution of the dependent variable. Furthermore, quantile regressions are much less sensitive to outliers than OLS regression. Also, they are particularly suitable if no (or negligible) relation between conditional means of two variables exists [74].

However, the researchers apply the Method of Moments Quantile Regression (MMQR) as estimated by [15] in fixed-effects estimation in the current work. Despite the robust nature of quantile regressions to the influence of outliers, they don't cater for heterogeneity between individuals in panel data. MMQR estimation can capture individual-specific effects that strongly influence the distribution of Y(i)conditional on X (i), rather than shifting its mean, which has important implications for identifying conditional heterogeneous covariance effects of fisheries sustainability determinants, as indicated in [16, 75]. When individual fixed effects exist, and an equation contains endogeneity, MMQR estimation methodology may be suitable. A further benefit is that MMQR is simple to implement, yielding non-crossing estimates of the regression quantiles. The equation for estimating the conditional quantiles QY(|X)in the context of a location-scale model specification is:

$$Y_{it} = \alpha_{i+} X_{it}^{/'} \beta + \left(\delta_i + Z_{it}' \gamma' \right) U_{it} \tag{10.1}$$

Where parameters requiring estimation and the probability, P(δi + Zitγ >0) = 1. (α, β, transformations with element l given by: (αi, δi), i = 1,..., n, represents the individual i fixed effects).

$$Z_l = Z_l(X), l = 1, \ldots, k \tag{10.2}$$

For each given i, Xit is i.i.d across i and independent across t. By [15], Uit is orthogonal to Xit, i.i.d across i and across t, and normalization criteria are applied, one of which is the relaxation of strict exogeneity. Equation (1) results in the Following,

$$Q_Y(\tau|X_{it}) = (\alpha_{i+}\, \delta_i q(\tau)) + X_{it}^{'}\beta + Z_{it}^{'}\gamma q(\tau) \tag{10.3}$$

The independent variables in the current paper were the natural logarithm of economic growth (LGDP), natural logarithm of social development (LVA), natural logarithm of economic stability (LPS), natural logarithm of private sector quality (LRQ), natural logarithm of property rights (LRL), and natural logarithm of private interest (LCC).

The scalar term is the i.i.d, while the i.i.d across t is the quantile-specific fixed effect for the individual i, and QY(τ|Xit) is the conditional quantile distribution of the dependent variable Yit (natural logarithm of fisheries production), depending on the values of the independent variable Xit.

The individual-level effect does not work like the standard fixed effects in OLS models that represent a constant vertical shift for each i. Instead, these effects do not vary across time for a given individual i and vary across quantiles of the conditional distribution of Yit. The-th sample quantile q(τ) is calculated by the following optimization problem.

$$\min_q \sum_i \sum_t P_\tau \left(R_{it} - \left(\delta_i + Z_{it}^{'}\gamma \right) q \right) \tag{10.4}$$

Where the check function P_τ (A) = (τ—1) AI {A ≤ 0} + TAI {A > 0} is shown.

10.4 The Outcome of Fisheries Governance in The EU

The unknown parameters will be estimated after several typical pre-estimation tests for time-series properties of the variables. Firstly, we will check whether there exists cross-sectional dependence (CD) in the panel data. Cross-sectional dependence will give us biased parameter estimates. For instance, if cross-sectional dependence, as a result of some unobserved variables common to countries, is not controlled, some gains of panel data efficiency would be erased, even greatly [76]. Thus, it will be beneficial to take CD into consideration to achieve efficient parameter estimation. CD can be tested by the [77] CD test in the panel context. Other quantitative variables in Table 10.3 have cross-sectional dependence strongly across countries, except for fisheries production. Therefore, to mitigate possible size distortions, unit root and cointegration tests and panel estimation procedures should use tests that are robust to the presence of cross-sectional dependence.

In order to provide an objective perspective of the integrating property of variables in the investigation, we have employed the [78] (LLC) and [79] (IPS) unit root tests. In order to improve test efficiency by combining both cross-sectional and

Table 10.3 Cross-sectional dependence results

Variable	FP	GDP	COC	GOE	PSAV	REQ	ROL	VOA
Pesaran CD (2004)	(8.653) ***	(9.663) ***	(1.642) ***	(2.207) ***	(4.765) ***	(14.38) ***	(3.697) ***	(3.292) ***

– Remark: *** refer importance at the 1%, scale
Source: [68, 69]

time-series data, a number of panel unit root test statistics have been constructed to compare the null of a unit root with alternative hypotheses. [78] considers testing the alternative of all panels being stationary at the same speed of reversion against the unit root null. [79] tests the alternative that at least one panel is stationary against the unit root null hypothesis. A feature of the new test developed in this paper is that it tests the hypothesis of stationarity of the panel, which naturally turns our focus to the architecture of the LLC test. In the (IPS) test panel, this restriction is relaxed, so that individual panels can be allowed to have individual autoregressive parameters. The power of the (LLC) panel unit root test was demonstrated to be superior to others on a very small sample of panel data such as what this chapter is using. These two tests are chosen to test for CD effects on the panel unit root test. All the variables unit root test specifications are shown in Table 10.4 to be stationary in the first difference but non-stationary in level, which implies they are integrated of order one, or I (1).

To ascertain whether there is a non-spurious long-run relationship between the variables, we employ the [70] panel cointegration test and the [80] bootstrapped panel cointegration test in Table 10.5. [70] published a thorough methodology for panel cointegration testing based on the Engle and Granger two-step method. By eliminating short-term variables and person-specific deterministic patterns in the initial processing stage, Pedroni's approach accounts for heterogeneity. From estimated residuals, Pedroni produces seven alternative test statistics, some of which may be tests that typically presume a single process. In Table 10.5, the process is referred to as "pooled" or "within-dimension" tests, or tests that presume individual processes, also known as "grouped" or "between-dimension" tests.

The null hypothesis is that there is no cointegration, and four new tests are presented using the [80] technique. The test removes the common factor constraints that apply to tests based on residual dynamics since these dynamics are structural rather than residual. The effectiveness of residual-based cointegration investigations may be significantly reduced by the failure of common factor restrictions, requiring structural dynamics [81].

With this restriction removed, the long-term and short-term adjustment processes no longer need to provide the same view. Using [80], bootstrap methodology, we can obtain durable critical values and lessen the distortionary impacts of cross-sectional reliance. Both the bootstrapped cointegration testing method of [80] and the [70] testing analysis provide strong support for cointegration, as shown by Table 10.5. The results of the FMOLS, DOLS, and FEOLS estimation procedures are shown in Table 10.6. We can observe from Model 1 in Table 10.6 that the

Table 10.4 Panel unit root test results

Variable	Difference		First difference	
	LLC	IPS	LLC	IPS
FP	3.067	−3.700	−24.13***	−22.737***
GDP	−1.090	−0.254	−23.47***	−23.47***
COC	−1.209	−3.942	−7.818***	−11.92***
GOE	1.964	1.897	7.144***	11.235***
PSAV	4.055	3.788	12.122***	13.123***
REQ	3.124	3.963	7.452***	10.499***
ROL	2.987	3.801	11.203***	13.166***
VOA	2.245	3.877	10.823***	12.887***

– Remark: *** refer importance at the 1%, scale
– Levin, Lin & Chu test (LLC), and Im, Pesaran, and Shin W-stat test (IPS)
– Source: [68, 69]

Table 10.5 Panel cointegration test is EU countries

Test	Without trend	With trend		
A) Pedroni residual cointegration test:				
Alternative hypothesis: common AR coefficients (within dimension):				
Panel v-Statistic	−0.921(0.758)	−2.464(0.896)		
Panel rho-Statistic	0.728(0.995)	0.892(0.714)		
Panel PP-Statistic	2.326**(0.010)	−2.511***(0.006)		
Panel ADF-Statistic	−1.895**(0.022)	−2.893***(0.000)		
Alternative hypothesis: common AR coefficients (between dimensions)				
Group rho-Statistic	−1.971	(0.975)		
Group PP-Statistic	−4.088***	(0.000)		
Group ADF-Statistic	−3.892***	(0.002)		
B) Westerlund (2007) bootstrap panel cointegration:				
Statistics	Value	Z-Value	p-value	Robust p-value
Gt	−2.409	−1.095	0.095	0.081*
Ga	−7.225	3.100	0.999	0.985
Pt	−11.677	−1.408	0.080	0.074*
Pa	−5.673	1.569	0.942	0.999

Source: [68, 69]
– Remark: ***, **, and * refer importance at the 1%, 5%, and 10% scales respectively
– Values in parentheses are *p*-values

coefficient estimates generated from all three specifications are, on average, rather similar, despite differences when taking into account their statistical significance level. Fish production in the EU region is the most consistent across all three criteria in terms of statistical significance and coefficient size. The sustainability of the fishing industry is positively impacted by a 0.96% improvement in public sector quality, according to the FEOLS estimate. Nonetheless, there is now statistically strong support for FMOLS and DOLS estimates. The results of [1, 2] support this conclusion.

Table 10.6 Model 1. Panel estimation for the EU27 region from 1996 to 2022

Long-run coefficient	DOLS		FMOLS		FEOLS	
	Coefficient	Std. error	Coefficient	Std. error	Coefficient	Std. error
COC	−1.203	(0.600)**	−0.918	(0.350)***	−0.612	(0.764)
GOE	0.452	(0.453)	0.143	(0.329)	0.962	(0.926)*
PSAV	0.146	(0.191)	0.053	(0.138)	2.043	(0.279)***
REQ	0.812	(0.431)*	0.709	(0.334)**	3.081	(0.675)***
ROL	1.424	(0.566)**	1.528	(0.420)***	0.987	(0.827)
VOA	0.871	(0.755)	0.932	(0.419)**	2.525	(0.957)***
GDP	0.149	(0.710)**	0.223	(0.055)***	0.517	(1.013)***

– Note: ***, **, and * indicate significance at the 1%, 5%, and 10% levels, respectively
– Values in parentheses are P-values
Source: [68, 69]

The EU has become an important player in international fisheries regulation because of its size as the world's largest market for seafood, its capacity to import more than half of its needs, and its position as a major fisheries power. Additionally, the sustainability of the fishing industry is positively correlated with the level of private sector quality in the EU region, as expected; increases in the sustainability of the fishing industry correspond to increases in private sector quality in the ranges of 3.08%, 0.70%, and 0.81% for the FEOLS, FMOLS, and DOLS estimators, respectively. The results of [3, 17] are supported by this study. High-quality fisheries and the private marine sector are essential to protecting the marine environment, growing the blue economy in the future, and achieving Europe's objective of becoming the first climate-neutral continent by 2050. Economic stability has a significant impact on the sustainability of the fishing industry in the EU. The FMOLS and DOLS estimation specifications have little statistically meaningful support, much like a percentage increase in economic stability. This bolsters the growth argument for the panel of fishing-prod. This outcome supports the key findings of [1, 82], who found that market and economic stability had a positive and inelastic impact on the expansion of the fishing industry. The stability of the European fish industry is flourishing due to a number of factors, including the fact that Europeans are spending more on fish and shellfish despite their higher prices. According to the estimated specifications, social improvement in the EU region is regarded as significant and beneficial. A percentage increase in social development increases by 2.52% and 0.93%, respectively, for the FEOLS and FMOLS, while its coefficient appears to be inconsistent between DOLS specifications. This outcome is consistent with the findings of but contradicts the modernization theory proposed by [18, 19]. The growth hypothesis is strongly supported statistically by the FMOLS and FEOLS specifications. The sustainability of blue growth and the preservation of the maritime environment are human responsibilities, according to social development indicators. The goals of blue growth sustainability may be achieved by the freedom of expression, freedom of association, freedom of the press, and the right of individuals to select their government. The sustainability of the fishing industry in the EU region is positively impacted by economic growth. Additionally, economic growth has increased by 0.51%, 0.22%, and 0.14% points, respectively, according to the FEOLS, FMOLS, and DOLS estimators. This bolsters the growth argument for the panel of fishing-producing countries. This outcome supports the important findings of [44, 45, 50], who found that economic expansion had a positive and inelastic impact on the growth of the fishing industry. By establishing the conditions necessary for a competitive and economically sustainable market, the fishing industry in the EU region may experience long-term growth. Remarkably, the FEOLS estimation specification shows a modest, statistically insignificant correlation between property rights and private interests in the EU region and the fishing industry. Nonetheless, there is a strong statistically significant correlation between the FMOLS and DOLS estimation specifications. This outcome contradicts [83], but it confirms Charter's primary conclusion (2022).

The panel quantile regression estimates in Table 10.7 show that the effects of public sector quality on the fisheries industry are statistically significant at all quantiles, from the first to the ninth. For the most recent quantiles, the public sector quality impact is increasing at the same time. This result validates the effectiveness of marine laws pertinent to the fishing industry in higher marine-producing countries. The results show that the growth theory only applies to fisheries output from the earliest and most recent periods in the EU27. This implies that in countries with lower quantiles, development comes before fisheries policy. Additionally, countries in the lower quantiles may require assistance in converting their fishing sectors from non-recycled to recycled supplies due to the related investment costs and lower output levels. In countries with the lowest and most significant output levels, the quality of the public sector favors the fishing industry. This could be brought on by the various extraction methods used in the developed EU14 nations and the expanding EU13 economy. These results predict major consequences for the quality of the public sector, according to [50, 51], who confirmed the growth theory for several countries that generate extraordinarily high levels of fishery production. Our results are consistent with recent empirical research by [53], which verified the growth hypothesis as another important source of fishing resources. The results are consistent with those of [54], who showed that the growth hypothesis was valid for the top producer of fishery resources. Furthermore, the results of private sector quality and property rights and fisheries output have a strong and positive link in the most recent quantiles. However, personal interest is significantly and negatively correlated with fisheries productivity in the most recent quantiles. However, there is a strong and favorable association between fisheries productivity and social development in the middle quantiles. It concurs with [18] that societal development motivates fisheries to seek more sustainable growth. Additionally, there is a significant and positive association between fisheries output and the outcomes of economic development and stability across all quantiles. However, at earlier quantiles, the influence of stability and economic growth is growing. According to [44–46], the fishing industry is motivated to pursue sustainable expansion by the efficiency of economic performance.

The results of the FMOLS, DOLS, and FEOLS estimation procedures are shown in Table 10.8. Model 2's Table 10.8 demonstrates that while the coefficient

Table 10.7 Panel quantile estimation results for EU27 countries

Variables	Method of moments quantile regression with fixed effects										
	Location	Scale	Quantiles								
			0.1	0.2	0.3	0.4	0.5	0.6	0.7	0.8	0.9
COC	(0.764)	(2.378)**	(−0.510)	(−1.375)	(−0.063)	(1.601)	(1.310)	(1.273)	(0.744)	(0.088)	(0.793)*
GOE	(0.926)*	(0.276)***	(1.382)***	(1.979)***	(1.748)***	(1.528)***	(1.863)***	(1.788)***	(1.937)***	(1.641)**	(1.979)**
PSAV	(0.279)***	(0.068)	(3.594)***	(2.348)***	(1.884)***	(1.603)***	(1.455)***	(1.145)***	(1.040)***	(0.098)***	(1.028)***
REQ	(0.675)***	(0.704)**	(1.238)	(1.988)	(0.527)	(0.086)	(0.063)	(0.405)	(0.449)	(0.982)*	(0.946)**
ROL	(0.827)	(2.301)*	(0.698)	(0.297)	(1.007)	(1.738)**	(1.000)	(0.675)	(0.784)	(0.064)	(1.985)**
VOA	(0.957)***	(0.645)	(−1.095)	(0.061)	(2.005)*	(2.226)**	(2.078)***	(1.408)**	(1.031)	(0.950)	(0.445)
GDP	(1.013)***	(0.844)	(0.750)***	(0.710)***	(0.682)***	(0.579)***	(0.416)***	(0.362)***	(0.338)***	(0.292)***	(0.290)***

Notes: *,**, and *** indicate significance at 10%, 5%, and 1% levels, respectively
Source: [68, 69]

Table 10.8 Model 2. Panel estimation for the EU-14 region from 1996 to 2022

Long-run coefficient	DOLS		FMOLS		FEOLS	
	Coefficient	Std. error	Coefficient	Std. error	Coefficient	Std. error
COC	−3.265	(1.751)*	−0.285	(0.040)***	−0.235	(0.229)
GOE	1.286	(0.668)*	0.517	(0.037)***	0.564	(0.220)**
PSAV	0. 116	(0.161)	0.137	(0.040)	0.023	(0.023)
REQ	0.520	(0.522)	0.244	(0.042)***	0.243	(0.243)
ROL	1.807	(1.090)*	0.793	(0.039)***	0.681	(0.681)**
VOA	2.349	(0.559)***	1.376	(0.028)***	1.385	(1.385)***
GDP	0.186	(0.098)**	0.220	(0.064)***	0.095	(0.095)*

– Note: ***, **, and * indicate significance at the 1%, 5%, and 10% levels, respectively
– Values in parentheses are P-values
Source: [68, 69]

estimates from the three specifications vary in statistical significance, they are generally comparable. The public sector quality in the EU14 developed countries is consistent across two specifications in terms of statistical significance and coefficient size. A percentage increase in public sector quality has a positive impact on fisheries output by 0.56%, 0.51%, and 1.28%, respectively, according to the FEOLS, FMOLS, and DOLS estimators. This outcome is consistent with the key conclusions of [20, 21]. The effectiveness of the public sector is crucial to the EU's implementation of the 2030 Agenda for Sustainable Development, especially Sustainable Development Goal 14 on life below water. The EU's strong commitment to the ocean is demonstrated by looking for easily available translations of the prior land. Achieving the blue component of the European Green Deal depends on the EU countries' updated agenda. In the industrialized EU14 countries, the increase in fisheries production for a percentage increase in property rights for the FEOLS, FMOLS, and DOLS estimators, respectively, ranges from 0.68%, 0.79%, and 1.80%. Across all criteria, there is a positive and statistically significant correlation between property rights and the fishing sector. The results of this study are in line with those of [8, 22]. The sustainability of the technological sophistication, transparency, and efficiency of the fishing industry can be enhanced by closely monitoring fisheries technology patents, capital property rights, and intellectual property rights from the planning and implementation stages, including marketing and distribution. With the FEOLS, FMOLS, and DOLS estimators finding increases in fisheries production of 1.38%, 1.37%, and 2.34% for a percentage increase in social development, respectively, all of the estimation specifications demonstrate that social development has a positive impact and is significant in the EU14 developed countries. This finding contradicts [48], but it validates the modernization theories of [23, 84]. For every parameter, the growth hypothesis is very statistically significant. Fish populations, marine life, and the maritime environment have been affected by social development tactics for generations. Research indicates that when fisheries target marine species, other marine species are under evolutionary pressure to survive. An imbalance in marine life could eventually result from fisheries pressure. Furthermore, the level of development of the EU14's fishing industry is positively correlated with economic growth. A percentage increase in economic growth increases by 0.09%, 0.22%, and 0.18% for the FEOLS, FMOLS, and DOLS estimators, respectively. The growth theory is endorsed by the panel of developed EU14 countries. This finding runs counter to the findings of [48], who documented the detrimental and inelastic impact of economic expansion on fisheries security. It does, however, corroborate the main conclusions of [47, 49] and others. Fisheries output affects other economic activities such as shipping, fishing, aquaculture, and recreation. For example, replacing boats and other equipment damaged by abandoned fishing gear is a significant yearly expense for the fishing industry. Remarkably, the FEOLS estimation specification reveals a weak, statistically insignificant relationship between the fishing industry, economic stability, and the quality of the private sector. On the other hand, there is a high statistically significant correlation between the FMOLS and DOLS estimation specifications. Although this result contradicts [43], it confirms [40] primary finding. The impact of public sector quality on the sustainability of fisheries is statistically significant in the earlier and more recent quantiles, as shown by the panel quantile regression estimates in Table 10.9. This result demonstrates the effectiveness of a marine strategy that is pertinent to the sustainability of fisheries in the developed EU14 countries. The results show that the growth notion is supported by both the older and more recent quantile fishing businesses in the industrialized EU14 countries. This implies that among the 14 developed countries of the EU, maritime policy comes before fisheries output. Due to a high level of fisheries output, EU14 wealthy states in early quantiles may find it more challenging to transition from conventional to recycled sources due to the investment costs involved. The quality of the public sector has a more positive effect on the fishing industry in the 14 EU countries with the largest fisheries output. The different extraction methods utilized in the fishing sectors of the EU14 developed countries could be the reason for this. These results are consistent with those of [21], who confirmed the growth hypothesis for several developed countries with exceptionally high levels of fisheries output, and they have important policy implications. Our findings are consistent with a recent empirical study conducted by [22], which provided evidence in favor of the growth theory, another important source of fisheries resources. The findings are also consistent with those of [8], who showed that the growth hypothesis for the top producer of fishery resources was valid. Furthermore, the outcomes of economic stability and private sector quality are significantly and favorably correlated with the total quantiles of the fishing business. The amplitude of the effects of private sector quality and economic stability decreases with increasing quantiles. The fishing industry is significantly and negatively correlated with the outcomes of personal interests in the early and middle quantiles. Personal interests, however, have very little effect on the final quantile. Interestingly, the impact of personal interest rises with greater quantiles. Conversely, the outcomes of property rights and social development are positively and significantly correlated with the fishing industry. However, there is no relationship between property rights, social development, and the sustainability of fisheries in the first and eighth quantiles, respectively. In a similar vein, the impact of market size decreases as quantiles rise.

Table 10.9 Panel quantile estimation results for EU14 countries

Variables	Method of moments quantile regression with fixed effects										
	Location	Scale	Quantiles								
			0.1	0.2	0.3	0.4	0.5	0.6	0.7	0.8	0.9
COC	(0.764)	(1.969)***	(−7.901)***	(−2.064)	(−3.259)***	(−2.411)***	(−1.969)***	(−2.368)***	(−1.745)	(−0.224)	(−0.951)
GOE	(0.926)*	(1.277)	(6.430)*	(0.966)	(0.594)	(1.315)	(1.277)	(1.807)	(3.836)***	(4.538)***	(4.022)**
PSAV	(0.279)***	(1.571)***	(2.965)***	(1.692)***	(1.569)***	(1.596)***	(1.571)***	(1.420)***	(1.465)***	(1.376)***	(1.395)***
REQ	(0.675)***	(3.036)***	(11.35)**	(9.317)*	(4.361)***	(3.985)***	(3.036)***	(2.294)***	(1.812)***	(1.550)***	(2.522)***
ROL	(0.827)	(2.565)***	(5.510)***	(4.860)**	(3.975)***	(3.589)***	(2.565)***	(2.916)***	(3.782)***	(2.252)	(1.756)**
VOA	(0.957)***	(3.003)***	(6.543)	(4.064)***	(4.699)***	(3.755)***	(3.003)***	(2.753)***	(2.837)***	(2.678)***	(2.576)**
GDP	(1.013)***	(0.267)	(1.265)	(0.207)	(0.012)	(0.167)	(0.267)	(0.194)	(0.191)	(0.190)	(0.209)

Notes: *, **, and *** indicate significance at 10%, 5%, and 1% levels, respectively
Source: [68, 69]

The results obtained from the different panel estimation approaches in Model 3—namely FMOLS, DOLS, and FEOLS—are reported in Table 10.10. These estimation techniques generate coefficient values that are largely similar across specifications. The findings indicate that the quality of the public sector has a positive and statistically significant effect on fisheries production in EU13 developing countries. The consistently positive coefficients across FEOLS, FMOLS, and DOLS confirm the validity of the growth hypothesis for EU13 developing economies. Specifically, a 1% improvement in public sector quality leads to an increase in fisheries production of 2.40% under FEOLS, 0.24% under FMOLS, and 0.81% under DOLS. This outcome aligns with the findings of [38], which show that during early development stages, strong ocean governance and effective public sector quality significantly enhance fisheries production.

In all three estimation frameworks (FEOLS, FMOLS, and DOLS), private sector quality is also strongly associated with fisheries output. A 1% rise in private sector quality results in increases in fisheries production of 1.39% under FEOLS, 0.65% under FMOLS, and 2.26% under DOLS. These results are consistent with [19, 24], who argue that improvements in private sector efficiency enhance fisheries productivity in developing economies.

The empirical results for EU13 developing countries also reveal a negative association between economic stability and fisheries production, although the estimated effects differ across methods. A 1% increase in economic stability is associated with changes in fisheries output of 2.56%, 0.11%, and 0.51% under FEOLS, FMOLS, and DOLS, respectively. Overall, this indicates that stronger economic performance supports fisheries production, a conclusion supported by [58], Suh and [59, 60] for developing EU countries. Similarly, the findings indicate a negative relationship between property rights and fisheries production in EU13 developing countries, although FEOLS and FMOLS suggest that improvements in property rights increase fisheries output by 3.85% and 1.08%, respectively, while DOLS estimates remain inconsistent. This supports the argument that stronger intellectual property rights can enhance fisheries production, in line with [35, 36] for developing EU economies. Across FEOLS, FMOLS, and DOLS estimators, social development and

Table 10.10 Model 3. Panel estimation for the EU-13 region from 1996 to 2022

Long-run coefficient	DOLS		FMOLS		FEOLS	
	Coefficient	Std.error	Coefficient	Std. error	Coefficient	Std. error
COC	−0.530	(0.566)	−0.837	(0.045)***	−0.828	(0.765)
GOE	0.816	(0.628)	0.248	(0.040)***	2.401	(0.573)***
PSAV	0.516	(0.244)**	0.115	(0.044)***	2.567	(0.422)***
REQ	2.268	(0.869)***	0.651	(0.047)***	1.395	(0.754)*
ROL	0.550	(0.522)	1.087	(0.038)***	3.851	(0.990)***
VOA	2.722	(1.289)**	0.433	(0.033)***	4.570	(1.319)***
GDP	0.404	(0.116)***	0.124	(0.019)***	2.387	(0.258)***

– Note: ***, **, and * indicate significance at the 1%, 5%, and 10% levels, respectively
– Values in parentheses are P-values
Source: Alsaleh and Wang (2025) and Alsaleh et al. (2024)

fisheries production are also closely linked. A 1% increase in social development leads to rises in fisheries output of 4.5% under FEOLS, 0.43% under FMOLS, and 2.72% under DOLS. This finding aligns with [18, 32], who emphasize that social development enhances fisheries productivity in less developed contexts. More importantly, the results confirm that economic development has a strong and positive effect on fisheries viability in EU13 developing countries. A 1% increase in GDP raises fisheries output by 2.38%, 0.12%, and 0.40% under FEOLS, FMOLS, and DOLS, respectively. The consistently positive and significant GDP coefficients across all models validate the growth hypothesis in EU13 economies. This is consistent with [55–57].

Table 10.11 presents the findings from the Method of Moments Quantile Regression (MMQR). The results indicate that public sector quality positively influences fisheries sustainability, but this effect is primarily concentrated in the middle quantiles rather than at the lower and upper extremes. As the quantiles increase (from the first to the seventh), the contribution of economic stability to fisheries sustainability becomes increasingly significant, rising from 1.753 to 2.688. However, this effect becomes statistically insignificant at the eighth and ninth quantiles. Furthermore, all quantiles except the highest (eighth and ninth) display a significant and positive relationship between social development, private sector quality, and fisheries sustainability. In a similar pattern, the influence of both social development and private sector quality is stronger at lower quantiles. Overall, economic expansion is confirmed to have a positive effect on fisheries sustainability. The results also show that economic growth has a significant and increasing impact on fisheries sustainability at middle quantiles, with the effect rising from 0.632 to 0.686 as the distribution moves from the second to the eighth quantiles. Finally, property rights exhibit a positive and statistically significant effect on fisheries sustainability from the first to the fourth quantiles, with the magnitude of the effect increasing as quantiles rise, following a similar pattern observed in previous results.

The estimates from the FEOLS panel have been confirmed by FMOLS and DOLS. It is evident that the panel FMOLS and DOLS coefficients have the same sign and degree of significance as the panel FEOLS coefficients. This proves that panel FMOLS and DOLS data are trustworthy and useful for drawing conclusions. Although their significance levels are marginally different, the coefficients from FEOLS show the same sign as those from the panel FMOLS and DOLS. However, panel DOLS estimates may be considered dependable and free of endogeneity and serial correlation problems. To evaluate the impact of global governance elements on the sustainability of fisheries in each country depending on economic development levels, the EU27 countries were split into two groups: 14 developed countries and 13 developing countries. Tables 10.8 and 10.9 show the expected impact of global governance issues on the sustainability of fisheries in the developed EU14 countries. However, the results of the anticipated impact of international governance parameters on fisheries sustainability in the EU14 developing countries are shown in Tables 10.10 and 10.11. The results presented in Tables 10.8 and 10.10 show that aspects of global governance have a significant positive impact on the sustainability of fisheries. The results also show that public sector quality has a

Table 10.11 Panel quantile estimation results for EU13 countries

Variables	Method of moments quantile regression with fixed effects										
	Location	Scale	Quantiles								
			0.1	0.2	0.3	0.4	0.5	0.6	0.7	0.8	0.9
COC	(0.765)	(−1.407)	(−1.009)	(−0.345)	(−0.262)	(−1.371)	(−1.407)	(−0.519)	(−1.105)	(−0.060)	(−1.672)*
GOE	(0.573)***	(2.229)***	(0.277)	(0.471)	(1.045)	(1.961)***	(2.229)***	(2.309)***	(1.810)***	(0.863)	(0.903)
PSAV	(0.422)***	(2.941)***	(1.753)***	(2.467)***	(3.013)***	(3.268)***	(1.941)***	(2.733)***	(2.688)***	(1.014)	(0.340)
REQ	(0.754)*	(4.632)***	(2.387)***	(3.337)***	(4.325)***	(5.190)***	(4.632)***	(3.617)**	(2.202)*	(1.488)	(1.003)
ROL	(0.990)***	(2.626)	(2.446)***	(2.447)**	(2.883)**	(2.730)*	(2.626)	(2.356)	(2.714)	(1.743)	(1.368)
VOA	(1.319)***	(4.698)***	(4.414)***	(5.430)***	(6.455)***	(5.656)***	(4.698)***	(3.986)***	(2.744)*	(1.840)	(1.439)
GDP	(0.258)***	(0.847)***	(0.319)	(0.632)***	(0.892)***	(0.918)***	(0.847)***	(0.888)***	(0.922)***	(0.686)**	(0.527)

Notes: *,**, and *** indicate significance at 10%, 5%, and 1% levels, respectively
Source: [68, 69]

much greater positive impact on fisheries sustainability in EU13 poor countries than in EU14 developed countries. Specifically, the impact magnitudes for EU13 and EU14 countries are 0.573 and 0.564, respectively. This suggests that EU13 developing countries using ocean governance and marine governance strategies in the fisheries sector may achieve a significantly higher level of fisheries sustainability than EU14 developed countries. Similarly, the results also show that the major positive impact of economic stability on the sustainability of fisheries is greater in EU13 poor countries than in EU14 rich countries. The impact magnitudes are 0.422 and 0.023 for EU13 and EU14 countries, respectively. This indicates that developing countries in the EU13 may boost fisheries sustainability more significantly than wealthy countries in the EU14 by using market tools and a market taxation policy on local and international markets of the fisheries industry. From a different perspective, the results indicate that compared to the EU14 developed countries, the EU13 developing countries see a greater advantage of private sector quality on fisheries sustainability. EU13 and EU14 countries had impact magnitudes of 0.754 and 0.243, respectively. This indicates that EU13 poor countries may significantly improve fisheries sustainability more than EU14 developed countries when considering fisheries sustainability from a private sector development perspective. This could be due to low interest rates and tax advantages for private investments in loans for the fishing sector in the EU13 developing nations. The results also demonstrate that the strong positive influence of property rights on the sustainability of fisheries is greater in the developing countries of the EU13 than in the industrialized countries of the EU14. EU13 and EU14 countries had impact magnitudes of 0.990 and 0.681, respectively. This demonstrates that EU13 developing countries may considerably improve fisheries sustainability more than EU14 developed countries by utilizing intellectual property rights and capital property rights for technology and innovation in the fisheries industry. However, the results also show that social development has a greater positive impact on fisheries sustainability in EU14 developed countries than in EU13 developing countries. The impact magnitudes for EU13 and EU14 countries are 1.319 and 1.385, respectively. This indicates that by investing in education development and establishing a healthy work environment for coastal cities, EU14 rich states may considerably improve fisheries sustainability in comparison to EU13 developing countries. The results, however, also demonstrate that the significant and positive influence of economic development on fisheries sustainability is greater in the developing countries of the EU13 than in the developed countries of the EU14. EU13 and EU14 countries had impact magnitudes of 0.258 and 0.095, respectively. This demonstrates how emerging EU13 countries may significantly improve fisheries sustainability in comparison to developed EU14 countries by investing in fisheries infrastructure and fleets.

10.5 Conclusion and Solution of Blue Governance

This study investigates the empirical relationships among fisheries production, public sector quality, private sector quality, economic stability, social development, property rights, and private interests while incorporating the growth hypothesis framework. It contributes to the literature on blue growth in the context of fisheries production and sustainable development. Using data spanning more than 31 years across 27 EU nations, we employ panel unit root tests, panel cointegration techniques, and panel estimation approaches.

The model includes the enabling environment variable (% confidence in governance) to assess fisheries governance and evaluate the level of sustainable development. The results show that all variables are integrated of order one, I(1), and exhibit a valid long-run relationship based on both panel unit root and panel cointegration methods. Although coefficient magnitudes are broadly similar across specifications, traditional panel cointegration estimators reveal differences in statistical significance. To provide a more comprehensive empirical analysis, we apply the MMQR approach, which allows examination of heterogeneous effects of explanatory variables across different quantiles of the conditional distribution of marine sustainability.

The coefficient estimates from FMOLS, DOLS, and FEOLS support the growth hypothesis for EU27 fisheries-producing economies. The MMQR results presented in Table 10.7 show that public sector quality is positively significant across all quantiles from the first to the ninth quantiles, indicating a consistent effect across the full conditional distribution of fisheries sustainability.

Similarly, economic stability and economic growth are positively significant across all quantiles of the conditional distribution of fisheries production. This suggests that economic growth and stability support both early and mature stages of fisheries sector development in EU27 nations.

In contrast, private sector quality and property rights become statistically significant mainly at higher quantiles of fisheries sustainability. This suggests that emerging fisheries-producing economies tend to be more economically diversified compared to industrialized nations, which rely more heavily on fish imports and exhibit stronger private sector and property rights effects. This implies that the marine resource structure of emerging economies represents a relatively larger share of national marine resource dependence. On the other hand, private interest is negatively significant at the highest quantile, indicating that adverse private interest factors such as corruption affect fisheries sustainability at advanced stages of development in EU27 countries. The MMQR estimate of social development is positively significant from the third to sixth quantiles, as shown in Table 10.7. The lowest quantile presents the highest coefficient for social development, while the magnitude declines as quantiles increase. This indicates that the influence of social development is stronger at lower levels of fisheries sustainability and weaker at higher levels.

The limited significance of public sector quality in Tables 10.9 and 10.11 may be attributed to existing government fisheries policies aimed at preventing overfishing and unethical practices. In EU13 developing countries, where fisheries may be relatively more sustainably managed than in EU14 developed countries in some respects, greater emphasis is placed on fisheries policy and regulation. Therefore, effective public sector governance tends to implement stricter rules against overfishing and unethical practices in response to public demand.

The significance of economic stability and private sector quality at earlier and medium fisheries production quantiles (from first to seventh quantiles) in Table 10.11 arises from the fact that improvements in macroeconomic stability and private sector development can support growth in EU13 countries. At the same time, higher levels of fisheries production in EU13 economies create additional development opportunities.

In contrast, Table 10.9 shows that economic stability and private sector quality remain significant across all quantiles in EU14 countries, reflecting more stable structural conditions. At higher quantiles, environmental regulations may reduce the marginal positive impact of economic stability and private sector development on fisheries sustainability in EU countries, as environmental degradation becomes more immediately observable. As a result, fisheries operations are increasingly influenced by both economic growth and sustainability policies in Tables 10.9 and 10.11. Due to the dominance of other competitiveness determinants and a stronger presence of resource-curse dynamics, the effect of economic growth is observed across all quantiles in EU14 developed countries. In contrast, improved sustainable fisheries development in EU13 countries reflects the stronger positive effect of economic growth at medium quantiles (from the second to the eighth quantiles).

Evidence from China, as described by [55] East Asia, as described by [58], and Africa, as described by [48], supports the conclusion that higher economic development enhances fisheries sustainability. Therefore, improving economic performance is strongly recommended for fisheries-producing countries, particularly EU14 developed nations with high fisheries output. In Tables 10.9 and 10.11, social development shows a positive and statistically significant effect overall; however, Table 10.11 indicates that in EU13 developing countries, social development exhibits a positive but statistically insignificant relationship with fisheries sustainability at the highest quantiles (eighth and ninth quantiles). Similarly, Table 10.9 shows that in EU14 developed countries, social development at the lowest quantile (first quantile) also has a positive but insignificant association with fisheries sustainability. This outcome is consistent with the growth theory in existing literature. These findings may be explained by differences across quantiles within fisheries-producing EU13 and EU14 countries. One possible explanation for lower quantiles in EU14 economies is inefficient investment in human welfare and poor working conditions in the fisheries sector. Such conditions emerge when unsafe labor environments and weak human capital investment replace other productive fisheries activities linked to tradable economic sectors. Consequently, these economies become more dependent on imports, meaning that a considerable share of their economic activity is driven by fish imports and exports.

The relocation of labor-intensive, export-oriented fisheries industries from EU13 developing countries to EU14 developed countries in pursuit of higher profits may shift economic growth patterns toward import reliance and significantly reduce fisheries production levels in EU13 countries at higher quantiles across both EU13 and EU14 groups. Additionally, stringent fisheries regulations imposed on marine resource-dependent and export-oriented firms may also directly influence the validity of the growth hypothesis in EU14 developed economies. Such policies may lead to reduced fisheries production and lower marine environmental pressure. As a result, many firms may relocate operations to less developed countries with lower taxation, weaker environmental restrictions, and abundant marine resources in order to maximize profits, particularly due to the high costs associated with skilled labor, experts, and advanced technologies in EU13 developing countries.

However, property rights demonstrate a positive and statistically significant relationship with fisheries sustainability at early and mid-range quantiles (first and fourth quantiles) in EU13 developing countries (see Table 10.11). In contrast, property rights show a positive and significant relationship with fisheries production across all quantiles in EU14 countries (see Table 10.9). The effect of property rights on fisheries sustainability is strongest at lower quantiles and weakest at higher quantiles in EU14 developed economies (Table 10.9). Conversely, in EU13 developing countries, the influence of property rights on fisheries sustainability appears strongest at higher quantiles and weakest at lower quantiles. This pattern suggests an improvement in capital asset formation and structural determinants across quantiles, which may be reflected in patents related to fisheries technology and ecosystem innovation. It also implies a shift in the effectiveness of intellectual property rights from higher to lower quantiles. Continued strong positive effects of global governance on fisheries may indicate that traditional fisheries practices remain dominant in both EU14 developed and EU13 developing countries. This implies that greater efforts are required to transition fisheries operations from conventional methods to more sustainable and environmentally friendly practices in order to enhance overall sustainability in both groups of countries. In other words, promoting fisheries production through sustainable and regenerative approaches is essential.

Moreover, where EU13 developing countries lack the capacity to implement fully sustainable and renewable systems, they may adopt moderate environmental taxation policies in the fisheries sector. Such policies would impose relatively low taxes on fisheries activities that contribute to marine waste. The revenue generated from these taxes should be reinvested into long-term infrastructure development to support sustainable fisheries.

Finally, the scope of this analysis could be broadened by incorporating additional determinants such as tourism, globalization, and financial development, all of which influence the sustainability of the EU fisheries sector and marine environment. Furthermore, future research could extend this study to individual countries such as China, or broader regions such as ASEAN, BRICS, the G7, or the Next Eleven economies, using more advanced econometric techniques. Like all empirical studies, this modeling approach has limitations. Due to data constraints, the analysis could not fully capture additional unexplored benefits of fisheries governance that may require further investigation. Moreover, given the scope of the present study, the model could be extended to other country contexts to achieve a more comprehensive global understanding. Although modern econometric techniques were applied, the study still faces limitations. For example, the results are based solely on econometric analysis and are restricted to a limited set of variables covering the period from 1996 to 2022. Future research could explore the effects of public-private investment in fisheries and ocean energy on fisheries production using updated datasets and additional indicators. Second, the current framework is limited to EU countries, which could be improved by incorporating other high-emission economies. Third, future studies could examine structural changes over time by analyzing the economic complexity of fisheries governance across different periods.

References

1. Drakopulos L (2022) Privatizing the fisheries observer industry: neoliberal science and policy in the U.S. West Coast fisheries. Geoforum 131:116–125. https://doi.org/10.1016/j.geoforum.2022.03.004
2. Van Der Zwet A, Connolly J, Huggins C, McAngus C (2023) Network resilience and EU fisheries policy engagement in third countries: lessons for post-Brexit governance. Br J Polit Int Rel 25(1):121–138. https://doi.org/10.1177/13691481211067146
3. Connolly J, Zwet AVD, Huggins C, McAngus C (2022) The governance capacities of Brexit from a Scottish perspective: the case of fisheries policy. Public Policy Admin 37(3):342–362. https://doi.org/10.1177/0952076720936328
4. Galappaththi EK, Susarla VB, Loutet SJT, Ichien ST, Hyman AA, Ford JD (2022) Climate change adaptation in fisheries. Fish Fish 23(1):4–21. https://doi.org/10.1111/faf.12595
5. Chudik A, Pesaran H, (2015) Large panel data models with cross-sectional dependence: a survey. [Online]. Available: https://academic.oup.com/edited-volume/41981/chapter/355335206?searchresult=1&utm_source=TrendMD&utm_medium=cpc&utm_campaign=Oxford_Academic_Books_TrendMD_0. Accessed May 20, 2026
6. Flaten O, Lien G, Koesling M, Løes A-K (2010) Norwegian farmers ceasing certified organic production: characteristics and reasons. J Environ Manag 91(12):2717–2726. https://doi.org/10.1016/j.jenvman.2010.07.026
7. Breitung J, Pesaran MH (2008) Unit roots and Cointegration in panels. In: Mátyás L, Sevestre P (eds) *The econometrics of panel data (*in advanced studies in theoretical and applied econometrics, vol. 46*)*, vol 46. Springer Berlin Heidelberg, Berlin, Heidelberg, pp 279–322. https://doi.org/10.1007/978-3-540-75892-1_9
8. Andrews DWK (2005) Cross-section regression with common shocks. Econometrica 73(5):1551–1585. https://doi.org/10.1111/j.1468-0262.2005.00629.x
9. Cerrato M, Sarantis N (2002) The cross sectional dependence puzzle. London Guildhall University. [Online]. Available: http://fmwww.bc.edu/RePEc/res2002/Cerrato.pdf. Accessed May 20, 2026
10. Balsalobre-Lorente D, Shahbaz M, Roubaud D, Farhani S (2018) How economic growth, renewable electricity and natural resources contribute to CO2 emissions? Energy Policy 113:356–367
11. Bekun FV, Alola AA, Sarkodie SA (2019) Toward a sustainable environment: nexus between CO2 emissions, resource rent, renewable and nonrenewable energy in 16-EU countries. Sci Total Environ 657:1023–1029. https://doi.org/10.1016/j.scitotenv.2018.12.104

12. Isham J (2005) The varieties of resource experience: natural resource export structures and the political economy of economic growth. World Bank Econ Rev 19(2):141–174. https://doi.org/10.1093/wber/lhi010
13. Mehlum H, Moene K, Torvik R (2006) Institutions and the resource curse. Econ J 116(508):1–20. https://doi.org/10.1111/j.1468-0297.2006.01045.x
14. Leite MC, Weidmann J, (1999) Does mother nature corrupt? Natural resources, corruption, and economic growth. International Monetary Fund. [Online]. Available: https://books.google.com/books?hl=en&lr=&id=UFYYEAAAQBAJ&oi=fnd&pg=PA3&dq=Leite,C.,+Weidman n,+J. +(1999). +Does+Mother+Nature+Corrupt:+Natural+Resources,+Corruption,+and+Eco nomic+Growth+IMF+Working+Papers. +International+Monetary+Fund.&ots=o29DeYoDjX &sig=TVLC-xIqLLL9VCD6nJ30s1TCTho. Accessed May 20, 2026
15. Machado JAF, Santos Silva JMC (2019) Quantiles via moments. J Econ 213(1):145–173. https://doi.org/10.1016/j.jeconom.2019.04.009
16. Koenker R (2004) Quantile regression for longitudinal data. J Multivar Anal 91(1):74–89. https://doi.org/10.1016/j.jmva.2004.05.006
17. Luo Y, Wu J, Xu Y (2022) Can self-governance tackle the water commons?—causal evidence of the effect of rural water pollution treatment on farmers' health in China. Ecol Econ 198:107471. https://doi.org/10.1016/j.ecolecon.2022.107471
18. Kelkar N, Arthur RI (2022) A review of governance and tenure in inland capture fisheries and aquaculture systems of India. [Online]. Available: https://books.google.com/books?hl=en&lr=&id=wL1qEAAAQBAJ&oi=fnd&pg=PP1&dq=Kelkar,+N.,+%26+Arthur,+R. +I. +(2022). +A+review+of+governance+and+tenure+in+inland+capture+fisheries+and+aquaculture+sy stems+of+India.+https://doi.org/10.4060/cb8615en&ots=43o6zSOa5g&sig=lUJVzKdVu4w fxr1-PtME1opbuaU. Accessed May 20, 2026
19. Sultana A et al (2022) A review of fish diversity, decline drivers, and management of the Tanguar Haor ecosystem: a globally recognized Ramsar site in Bangladesh. Heliyon 8(11):e11875. https://doi.org/10.1016/j.heliyon.2022.e11875
20. Okafor-Yarwood I, Kadagi NI, Belhabib D, Allison EH (2022) Survival of the richest, not the fittest: how attempts to improve governance impact African small-scale marine fisheries. Mar Policy 135:104847. https://doi.org/10.1016/j.marpol.2021.104847
21. Bender M, Bustamante R, Leonard K (2022) Living in relationship with the ocean to transform governance in the UN Ocean decade. PLoS Biol 20(10):e3001828. https://doi.org/10.1371/journal.pbio.3001828
22. Hou N, Zhu Q, Yang J, Zhang D, Liu W, Chang H (2021) The impact of environmental governance on the development of fishery economy—the intermediary role of technological innovation. Sustainability 13(20):11378. https://doi.org/10.3390/su132011378
23. Collins AM (2022) Empowerment, rights, and global food governance: gender in the UN committee for world food security. Globalizations 19(2):220–237. https://doi.org/10.1080/14747731.2021.1877006
24. Leeuwis C, Boogaard BK, Atta-Krah K (2021) How food systems change (or not): governance implications for system transformation processes. Food Sec 13(4):761–780. https://doi.org/10.1007/s12571-021-01178-4
25. Ahwireng AK (2022) Small pelagic fish for food: governance and performance of small pelagic fish value chains for food security and nutrition in Ghana. Anderson Kwasi Ahwireng
26. Haysom G (2021) Integrating food sensitive planning and Urban design into urban governance actions. Urban Forum 32(3):289–310. https://doi.org/10.1007/s12132-021-09417-9
27. Chen J, Glibert PM, Cai W-J, Huang D (2023) Editorial: eutrophication, algal bloom, hypoxia and ocean acidification in large river estuaries, volume II. Front Mar Sci 10:1225903. https://doi.org/10.3389/fmars.2023.1225903
28. Spijkers J, Mackay M, Turner J, McNeill A, Travaille K, Wilcox C (2023) Diversity of global fisheries governance: types and contexts. Fish Fish 24(1):111–125. https://doi.org/10.1111/faf.12713

29. Jarvis RM, Young T (2023) Pressing questions for science, policy, and governance in the high seas. Environ Sci Pol 139:177–184. https://doi.org/10.1016/j.envsci.2022.11.001
30. Guggisberg S, Jaeckel A, Stephens T (2022) Transparency in fisheries governance: achievements to date and challenges ahead. Mar Policy 136:104639. https://doi.org/10.1016/j.marpol.2021.104639
31. Jentoft S, Chuenpagdee R (2022) Interactive learning and governance transformation for securing blue justice for small-scale fisheries. Admin Soc 54(7):1255–1282. https://doi.org/10.1177/00953997211073947
32. Dudayev R, Hakim LL, Rufiati I (2023) Participatory fisheries governance in Indonesia: are octopus fisheries leading the way? Mar Policy 147:105338. https://doi.org/10.1016/j.marpol.2022.105338
33. Acuña-Alonso C, Varandas S, Álvarez X, Martinho A (2023) Analysis of the evolution of a fisheries management plan based on environmental governance: living laboratory in the Olo River, Portugal. Fish Res 260:106595. https://doi.org/10.1016/j.fishres.2022.106595
34. Aguión A et al (2022) Establishing a governance threshold in small-scale fisheries to achieve sustainability. Ambio 51(3):652–665. https://doi.org/10.1007/s13280-021-01606-x
35. Linke S, Gillette MB, Jentoft S (2022) Institutionalizing injustice? Aligning governance orders in Swedish small-scale fisheries. In: Jentoft S, Chuenpagdee R, Said AB, Isaacs M (eds) *Blue justice* (in MARE publication series, vol. 26), vol 26. Springer International Publishing, Cham, pp 529–547. https://doi.org/10.1007/978-3-030-89624-9_28
36. Ynacay-Nye A, Hisano S, Suryawan AS (2023) Alternatives to sustainable seafood certifications: transitions of small-scale fisheries governance in northeastern Japan. J Rural Stud 97:269–280. https://doi.org/10.1016/j.jrurstud.2022.12.002
37. Day S (2026) Pacific Salmon fisheries management: an (unusual) example of polycentric governance involving indigenous participation at multiple scales. The Center for Growth and Opportunity. [Online]. Available: https://www.thecgo.org/books/the-environmental-optimism-of-elinor-ostrom/chapter-3-chapter-3-pacific-salmon-fisheries-management-an-unusual-example-of-polycentric-governance-involving-indigenous-participation-at-multiple-scales/. Accessed May 20, 2026
38. Sorensen JA, Kincl L, Weil R, Dzugan J, Christel D (2022) Fisheries governance and associated health implications: current perspectives from US commercial fishermen. Mar Policy 141:105119
39. Anderson LG (ed) (2020) Fisheries economics: collected essays, 1st edn. Routledge. https://doi.org/10.4324/9780429288500
40. González-Gaya B, García-Bueno N, Buelow E, Marin A, Rico A (2022) Effects of aquaculture waste feeds and antibiotics on marine benthic ecosystems in the Mediterranean Sea. Sci Total Environ 806:151190. https://doi.org/10.1016/j.scitotenv.2021.151190
41. Kim JH, Vaughan A, Kincl L (2023) Characterization of musculoskeletal injury risk in Dungeness crab fishing. J Agromedicine 28(2):309–320. https://doi.org/10.1080/1059924X.2022.2068715
42. Charter M, Carruthers R (2022) Products from waste fishing nets: accessories, clothing, footwear, home ware and recreation. [Online]. Available: https://research.uca.ac.uk/5734/1/Products-from-waste-fishing-gear-March-2022.pdf. Accessed May 20, 2026
43. Mozumder M et al (2022) Sustainable utilization of fishery waste in Bangladesh—a qualitative study for a circular bioeconomy initiative. Fishes 7(2):84. https://doi.org/10.3390/fishes7020084
44. Rahmizal M (2022) Analysis of Indonesia marine fisheries with economic growth, population and effort effectiveness. Eur J Formal Sci Eng 5(1):53–61. https://doi.org/10.26417/ejef.v1i1.p17-22
45. Setiyowati H, Nugroho M, Halik A (2022) Developing a blue economy in Depok West Java, Indonesia: opportunities and challenges of neon tetra fish cultivation. Sustainability 14(20):13028. https://doi.org/10.3390/su142013028

46. Coppola D, Lauritano C, Palma Esposito F, Riccio G, Rizzo C, De Pascale D (2021) Fish waste: from problem to valuable resource. Mar Drugs 19(2):116. https://doi.org/10.3390/md19020116
47. Abo-Taleb HA et al (2021) Effect of a new feed daphnia magna (Straus, 1820), as a fish meal substitute on growth, feed utilization, histological status, and economic revenue of grey mullet, Mugil cephalus (Linnaeus 1758). Sustainability 13(13):7093. https://doi.org/10.3390/su13137093
48. Shamsuddin M et al (2022) Effects of stocking larger-sized fish on water quality, growth performance, and the economic yield of Nile tilapia (Oreochromis niloticus L.) in floating cages. Agriculture 12(7):942. https://doi.org/10.3390/agriculture12070942
49. Palomino E, Pardue J (2021) Alutiiq fish skin traditions: connecting communities in the COVID-19 era. Heritage 4(4):4249–4263. https://doi.org/10.3390/heritage4040234
50. McKinley E (2022) A blue economy for whom?: linking marine social sciences with blue economy discourse. In: Blue economy. Routledge, pp 13–25. [Online]. Available: https://www.taylorfrancis.com/chapters/edit/10.4324/9781003280248-4/blue-economy-emma-mckinley. Accessed May 20, 2026
51. Nunes AJP, Masagounder K (2022) Optimal levels of fish meal and methionine in diets for juvenile Litopenaeus vannamei to support maximum growth performance with economic efficiency. Animals 13(1):20. https://doi.org/10.3390/ani13010020
52. Starkweather KE, Keith MH, Zohora FT, Alam N (2023) Economic impacts and nutritional outcomes of the 2017 floods in B angladeshi S hodagor fishing families. Am J Hum Biol 35(1):e23826. https://doi.org/10.1002/ajhb.23826
53. Karimi MS, Khezri M, Khan YA, Razzaghi S (2022) Exploring the influence of economic freedom index on fishing grounds footprint in environmental Kuznets curve framework through spatial econometrics technique: evidence from Asia-Pacific countries. Environ Sci Pollut Res 29(4):6251–6266. https://doi.org/10.1007/s11356-021-16110-8
54. Alharthi M, Hanif I (2020) Impact of blue economy factors on economic growth in the SAARC countries. Maritime Bus Rev 5(3):253–269
55. Wang Q, Wang S (2022) Carbon emission and economic output of China's marine fishery–a decoupling efforts analysis. Mar Policy 135:104831. https://doi.org/10.1016/j.marpol.2021.104831
56. Gou L, Yang S, Ren L (2022) Coupling analysis of the quality and quantity of marine economic growth from the perspective of high-quality development. J Coast Res 38(1):218–228
57. Peng D, Yang Q, Yang H-J, Liu H, Zhu Y, Mu Y (2020) Analysis on the relationship between fisheries economic growth and marine environmental pollution in China's coastal regions. Sci Total Environ 713:136641. https://doi.org/10.1016/j.scitotenv.2020.136641
58. Ayilu RK, Fabinyi M, Barclay K (2022) Small-scale fisheries in the blue economy: review of scholarly papers and multilateral documents. Ocean Coastal Manag 216:105982. https://doi.org/10.1016/j.ocecoaman.2021.105982
59. Suh D, Pomeroy R (2020) Projected economic impact of climate change on marine capture fisheries in the Philippines. Front Mar Sci 7:232. https://doi.org/10.3389/fmars.2020.00232
60. Nunan F (2020) The political economy of fisheries co-management: challenging the potential for success on Lake Victoria. Glob Environ Chang 63:102101
61. Anwar A, Siddique M, Dogan E, Sharif A (2021) The moderating role of renewable and non-renewable energy in environment-income nexus for ASEAN countries: evidence from method of moments quantile regression. Renew Energy 164:956–967
62. Ehigiamusoe KU, Lean HH, Smyth R (2020) The moderating role of energy consumption in the carbon emissions-income nexus in middle-income countries. Appl Energy 261:114215. https://doi.org/10.1016/j.apenergy.2019.114215
63. Dogan E, Seker F (2016) Determinants of CO2 emissions in the European Union: the role of renewable and non-renewable energy. Renew Energy 94:429–439. https://doi.org/10.1016/j.renene.2016.03.078

64. Apergis N, Payne JE (2010) The emissions, energy consumption, and growth nexus: evidence from the commonwealth of independent states. Energy Policy 38(1):650–655
65. European Commission (2022) Joint Research Centre., Clean energy technology observatory, ocean energy in the European Union: status report on technology development, trends, value chains and markets: 2022. LU: Publications Office. [Online]. Available: https://data.europa.eu/doi/10.2760/162254. Accessed Dec. 30, 2025
66. World Bank. Fish to 2030: prospects for fisheries and aquaculture
67. European Commission. The European Green Deal–European Commission. [Online]. Available: https://commission.europa.eu/strategy-and-policy/priorities-2019-2024/european-green-deal_en. Accessed Jun. 08, 2025
68. Alsaleh M, Wang X (2025) The integration of ecosystem-based management in EU fisheries governance: challenges and opportunities. Econ Gov 26(4):527–558. https://doi.org/10.1007/s10101-025-00338-y
69. Alsaleh M, Yuan Y, Longqi S, Abdul-Rahim AS (2024) Retracted article: do worldwide governance drivers affect the blue sustainability practices? An empirical study of the fisheries sector. J Coast Conserv 28(1):7. https://doi.org/10.1007/s11852-023-01001-6
70. Pedroni P (2004) Panel cointegration: asymptotic and finite sample properties of pooled time series tests with an application to the PPP hypothesis. Econ Theory 20(03). https://doi.org/10.1017/S0266466604203073
71. Kao C, Chiang M-H (2000) On the estimation and inference of a cointegrated regression in panel data. In: Advances in econometrics, vol 15. Emerald (MCB UP), Bingley, pp 179–222. https://doi.org/10.1016/S0731-9053(00)15007-8
72. Sarkodie SA, Strezov V (2019) Effect of foreign direct investments, economic development and energy consumption on greenhouse gas emissions in developing countries. Sci Total Environ 646:862–871. https://doi.org/10.1016/j.scitotenv.2018.07.365
73. Koenker R, Bassett G (1978) Regression quantiles. Econometrica 46(1):33. https://doi.org/10.2307/1913643
74. Binder M, Coad A (2011) From average Joe's happiness to miserable Jane and Cheerful John: using quantile regressions to analyze the full subjective Well-being distribution. J Econ Behav Organ 79(3):275–290. https://doi.org/10.1016/j.jebo.2011.02.005
75. Canay IA (2011) A simple approach to quantile regression for panel data. Econ J 14(3):368–386
76. Phillips PCB, Sul D (2003) Dynamic panel estimation and homogeneity testing under cross section dependence. Econ J 6(1):217–259. https://doi.org/10.1111/1368-423X.00108
77. Pesaran MH (2021) General diagnostic tests for cross-sectional dependence in panels. Empir Econ 60(1):13–50. https://doi.org/10.1007/s00181-020-01875-7
78. Levin A, Lin C-F, James Chu C-S (2002) Unit root tests in panel data: asymptotic and finite-sample properties. J Econ 108(1):1–24. https://doi.org/10.1016/S0304-4076(01)00098-7
79. Im KS, Pesaran MH, Shin Y (2003) Testing for unit roots in heterogeneous panels. J Econ 115(1):53–74. https://doi.org/10.1016/S0304-4076(03)00092-7
80. Westerlund J (2007) Testing for error correction in panel data*. Oxf Bull Econ Stat 69(6):709–748. https://doi.org/10.1111/j.1468-0084.2007.00477.x
81. Kremers JJM, Ericsson NR, Dolado JJ (1992) The power of cointegration tests. Oxf Bull Econ Stat 54(3):325–348. https://doi.org/10.1111/j.1468-0084.1992.tb00005.x
82. Galappaththi M, Armitage D, Collins AM (2022) Women's experiences in influencing and shaping small-scale fisheries governance. Fish Fish 23(5):1099–1120. https://doi.org/10.1111/faf.12672
83. Kim S, Kim YT, Oh LS, Kim HJ, Lee J (2022) Marine waste upcycling—recovery of nylon monomers from fishing net waste using seashell waste-derived catalysts in a CO_2 -mediated thermocatalytic process. J Mater Chem A 10(37):20024–20034. https://doi.org/10.1039/D2TA02060B
84. Jones EC, Ong C, Haynes J (2022) Disaster-related food security and past general governance strategies in a worldwide sample. Weather Clim Soc 14(1):3–18. https://doi.org/10.1175/WCAS-D-20-0138.1

Index

M. Alsaleh, *Fisheries Sustainability in the EU*, Urban Sustainability,
https://doi.org/10.1007/978-981-92-4496-6

M

O

P

R

S

T

W

Kolibri 360 GmbH
Ferdinand-Jühlke-Straße 7
99095 Erfurt, Deutschland
produktsicherheit@kolibri360.de